The Foreigner Group

THE FOREIGNER GROUP

BY

CAROLUS LÖFROOS

ATH
ANTELOPE HILL PUBLISHING

First printing 2022.

Cover art by Friedrich v. Drake.
Edited by Sebastian Durant.
Formatted by Taylor Young.

Antelope Hill Publishing
www.antelopehillpublishing.com

Paperback ISBN-13: 978-1-956887-49-5
EPUB ISBN-13: 978-1-956887-50-1

CONTENTS

FOREWORD

Most people believe memories of certain strong and powerful moments are never forgotten. I used to think so myself, especially as I'm gifted with an unusually good memory as it is. Since then, though, I have been through and seen and experienced many things which I would have thought impossible to forget, yet I have forgotten about much of it all the same.

In order to really be unable to forget something, it must be something more than just a simple memory. It has to be connected to something: the more common, the better—a taste, a sound, or a scent. Something which alone can bring you back in time to a completely different place, allowing you to meet, feel, and perceive something in reality long gone like it's right there again.

Something as simple as the scent of freshly cut grass, for example, will always bring me back to my summers as a young boy, as will the sound of a Nintendo NES and an old television set just the same. The smell of freshly cut pine and its sweet resin will always bring me right back to when I worked in the forest with my father as a young boy. The smell of salty sea water and the bloom of algae takes me right back to the year I trained to be a Jaeger and the beautiful summers of the Finnish archipelago.

Then there is the smell of cold winter, something I know I had other memories attached to before. I don't remember any of them anymore. Instead, that smell now always brings me to the winter in Ukraine, February of 2015. It doesn't matter where I am or what I do, whether it's a smoke break at work or just walking the dogs during a cold winter morning. As soon as the air hits my nostrils and I smell the winter cold, my mind brings me right back every time. It's not without a certain feeling of unease, but overwhelmingly with an almost tearful joy. A nostalgic reunion, like meeting an old, old, and beloved friend—reliving a dream I have already once dreamed. I was the one who ate the ripest of fruit and drank the finest wine, and now I starve and thirst for them forever. Inside I cry out of fear I may never taste them again, but all the same I find myself indescribably elated knowing that I at least get to smell it once every year. I am glad I at least know what they actually taste like. As smooth as water, war has run down my throat like perfect single malt whiskey, and my tongue has tasted hot fierceness of armed combat. And I get to taste and relive it, at least a little, once each year.

Each time I feel that scent, part of me is back over there again. I stand there, feet firmly planted against the fertile but now frozen soil below. Around me lay the large, wide-open fields full of hundreds, perhaps thousands of ravens, all of them croaking loudly how nothing beyond their lands are illuminated by the star of prosperity. I am there again, as the hereditary enemy's heavy artillery thunders by the horizon and then casts its fire down on us, making the stiff earth shake so violently that the graves are filled back in before they are even dug. Once more I can hear the raging cracks of rifle bullets striking through the air around me, and the violent sound of diesel engines and tank tracks squeaking around the next corner. The simple scent of winter fills my mind with all other kinds of smells like that of gunpowder, iron-rich blood, burned cordite, fuel, and burning houses. I can hear the sound of machine guns cycling in automatic fire, roaring engines, howling shells, and thundering explosions echoing back from inside my eardrums already rubbed by the noise. My neck stiffens and the sinewy muscles in my fingers tighten, instinctively searching for a rifle to carry.

I am at peace. I am awake, but I dream of the small village of Shyrokyne, where for a few days I was the best man I will ever be, fighting and wildly throwing myself against the rage of death, side by side with others who were exactly the same. We, simple good men bringing out the monsters inside ourselves to greet the other monsters coming against us. We who went forward with nothing to win or gain, but did so anyway simply because it was what was expected of us. Because it was what we expected from ourselves.

There has been a lot said and written about us, from all different kinds of places. About the Western volunteers who fought for Ukraine during those days. Vladimir Putin once called us a "NATO legion." At one point, the US politician Max Rose managed to claim that we were seventeen thousand strong, citing the CIA—all of us, of course, right-wing extremists and Nazi terrorists. Who we were and what role we played have been immortalized in writing many times before, but never really by anyone who was actually there.

I was there, and even though it's been several years since, I remember very well that we were not that many. We were just a few men, a motley crowd of warriors with different reasons to join the Foreigner Group. I began writing our story down some time ago as a sort of therapy. A way to try and free up my mind by getting all memories down on paper, leaving space for other, new ones instead. After a while, and being told this is a good story, I decided to try and make it for others to enjoy, as well. To enjoy just like I did.

Carolus Löfroos
Norrfjärden, November 12th, 2021

POST-SCRIPTUM

A lot has happened since this manuscript was finished. The initial attempt to publish this story was met by fierce attacks from leftist political activists, often claiming that the story's mere connection to the Azov Regiment made the book a dangerous document full of lies and Nazi propaganda. These attacks, like they often do these days, proved successful, regardless of how empty the claims were.

Then, only a mere month after the initial publishing of this book was canceled, the Russians finally (and as had long been expected) began their large-scale invasion of Ukraine. They did so as to, using Russian president Vladimir Putin's own words, "de-Nazify Ukraine." While I myself was surprised by the very poor choice of timing, planning, and execution of the attack, the West as a whole seemed more in shock by the invasion itself. As if they had intentionally been looking the other way for eight long years and refused to listen to those of us who all this time tried to warn them the attack would happen. As if they had completely missed the first Russian invasion of Ukraine, which began way back in 2014, and everything which had happened since.

It's a fact that the long war in Donbas will forever be overshadowed by the Russian invasion of Ukraine in 2022. I do, however, believe that nobody can understand the latter without knowing the first. It is my hope that this story not only proves an interesting read on its own, but also that it may give some much needed understanding about not only the wars—and Ukraine and Russia themselves- but also the beginnings of the notorious Azov Regiment, which became legendary during their stubborn defense of Mariupol and the Azovstal steelworks. I will forever stand humbled, having had the great honor to fight alongside these men, and I would like to dedicate this book to the defenders of Mariupol. You will never be forgotten.

Glory to Ukraine. Glory to you, the heroes.

Also, a special thanks to everyone with whom I've personally served in the war, and to Robert, who kept saying I should write these stories down to paper.

Carolus Löfroos
Jämtland, August 6th, 2022

MAP OF THE MARIUPOL AREA

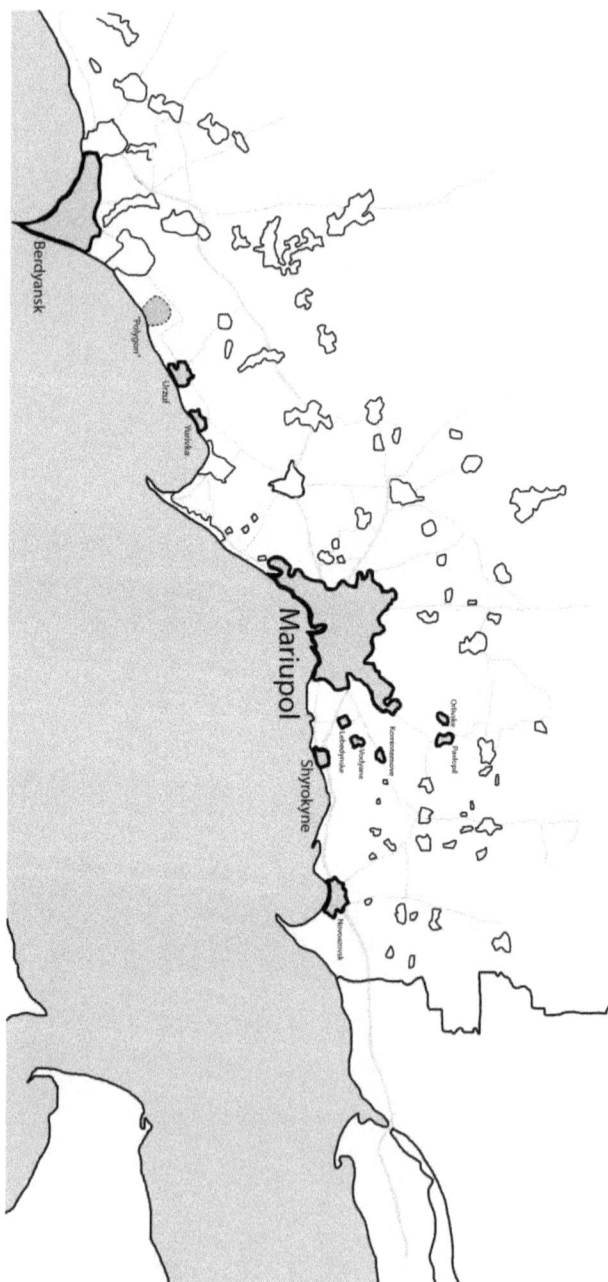

CHAPTER 1: THE BUILDUP TO WAR

February 2015

The sun had not yet risen over the horizon, but the sky was already gradually starting to brighten. The fog was dense over the field that lay between the heights behind us and the ancient, picturesque little village in front.

"Shyrokyne," I mumbled to myself. "What use do we have of these ruins, anyway?" I was then reminded that Chris was still there, somewhere, inside.

There was no snow in sight, but the tall grass and ground were nevertheless dressed in all white from the previous cold night's frost. The breezes, smelling of fresh sea air, were interrupted at times by the smell of smoke from the smoldering houses which had burned down during yesterday's deadly fighting. Bear jumped down off the eight-wheeled BTR-3 armored car, and a collective sigh of relief was drawn, knowing that he would not continue forward riding on its roof.

The soldiers gathered behind the T-64BV tank and the BMP-2 in a shallow ground depression in front of the entrance to the village.

"What is the plan, anyway?" I asked Långström. He looked at me glumly, then toward the village, and sighed.

"I guess it seems like we'll just attack along that road. Fuck do I know, really?"

"That's the plan? Just a frontal attack? It will never work. Couldn't they come up with something better?"

Långström just shrugged. "It was you guys who wanted to do this. What do I care? You were the ones who wanted to go back in and look for Chris. I said it was pointless. There are too many Russians, regardless of whether we have a good plan or not."

A Ukrainian assigned to the second column came walking through the fog from the other side of the road. In his hands he carried a large glass jar. The faint light from the sky shone through the jar, which glimmered in a red hue.

"He found strawberry juice in the abandoned houses?" Cuix asked with a laugh.

The starving and thirsty group cheered loudly. Fabien went first and had a hearty drink, perhaps with the hope that it could have been wine. The Greek put his AK-74 down, cleared his throat, and took a somber tone.

"My friends, gentlemen…and women. Don't want to exclude any of you if you identify as such." He smiled jokingly before continuing: "I just want to say that it has been an honor to serve with all of you. I'm not joking now. I am serious—I mean it. I've really had a great time getting to know all of you. Seriously. Honestly."

Richter stood quietly, peering out at the village without saying anything. He observed the surroundings in complete silence without even a worried expression on his face. Not even the smell of fire and smoke made his nostrils move.

Tjeck was shaking. Not from the cold, but from fear.

"How are you doing?" Fabien asked in his thick French accent. "Are you okay? You can do this, right? It's okay. You come with us only if you know you can do this."

Tjeck smiled nervously and struggled to utter his words. "I think artillery is the hardest for me. That sound it makes—it is difficult for me. Enemy bullets will be easier, I think." He stammered as he spoke, looking away. His facial muscles twitched involuntarily as he tried to squeeze out a confident smile.

I looked down the road, toward the village. The Russian force had to be twice as big as ours, and they had several armored vehicles, including tanks. We wouldn't stand a chance. Everything from the past year had boiled down to this point, to this very moment. We had been given an impossible task, and the choice was to either back away from it and retreat, or move forward, anyway, and fail, likely dying in the process. We could not win, but we couldn't back down, either. We would press forward and give everything we had left. We stood no chance, but in deciding to proceed regardless, we were as free as any men could ever be.

"The fog is starting to dissipate," I said to Långström. "This clusterfuck needs to hurry the fuck up. If we are going to do this, we must do it now, otherwise they'll bring their artillery on top of us."

Långström looked at me and laughed. "Yeah. You take the lead position on the right side of the BMP."

"In the very front?" I asked. "What in all the—"

"Fuck it, does it matter?" my squad leader interjected. "I'll be right behind you. Then Bear and the Greek behind me with Tjeck, Fabs, Cuix, and Richter behind the BMP. Metro's squad are taking the left side. Remember to keep an eye on the windows."

Metro was a new name without a face to me, but I didn't think much of it. I had already become comfortable with dying, but the thought of being the absolute first one to die made me uneasy.

Långström didn't respond to my dissatisfaction and continued: "Be sure to gun down any Russian bastard you see before he manages to send an RPG into our BMP. We have to keep it alive as long as possible."

2

Our BMP-2 infantry fighting vehicle started its engine. Metro's squad was ready on the left side and we lined up on the right.

"Glory to the heroes!" cried the Greek sarcastically.

The engine exhaust heated the air next to me, making it very warm and pleasant. I dreamed of just sleeping in a warm bed again, about taking a shower in really hot water—or even better, sitting in a hot sauna. Långström tapped me on the back of my helmet, and I turned around to see what he wanted.

"We are going to die! Do you understand that?" he shouted over the engine noise.

"Yes," I replied bitterly, "that's pretty evident."

Långström had finally returned to his normal combat ecstasy. He was sharpened and back on the chopping block again.

"We're all going to die! It's so fucking great!"

The coarse diesel engine of the fighting vehicle roared away, spewing thick white smoke just like the exhaled air from our own heated lungs. The steel tracks started squeaking, the chassis jerked backward, and our BMP began rolling forward. Forward toward the village, toward the enemy, with a squad on foot on either side. All toward a certain death. All determined to face it as men rather than retreat as cowards.

One year earlier, February 2014

The violent, black-clad Ukrainian Berkut riot police had just opened fire on the large crowd of protesters occupying Maidan Nezalezhnosti, or Independence Square, in central Kyiv, the capital of Ukraine. The number of dead and injured was unclear, but the bloody asphalt hinted that it would be many. The protests had been taking place since November 30th, 2013. They had been triggered by President Yanukovych's sudden reversal—he had decided to scrap his plan to seek closer ties with the West and would instead seek closer economic ties with Russia. In the eastern and southern oblasts, where much of Yanukovych's base was located, this decision was not disappointing: it was met with cheers. In contrast, Ukrainians living in western oblasts and urban areas felt cheated and greatly disappointed. Protests started in many cities, with the biggest demonstrations taking place in Kyiv. The harder these protests were repressed, the more they grew. Those that began as mere hundreds quickly became thousands, and soon grew to tens of thousands.

There had always been violence, but it definitely kicked into a new gear when the president's guard switched from water cannons and rubber bullets to opening fire with Kalashnikovs and sniper rifles. I sat in my apartment with an old childhood friend of mine, Anders, and followed the developments via YouTube, in the company of cold beers and smoky Laphroaig.

"We really should go there," I said.

Ordinary media discussed the crisis often, but their coverage rarely provided any serious understanding of what was going on. In addition to YouTube, only Motgift, a Swedish alternative media source, provided viewers with up-to-date information. Their Swedish correspondents, on location in and around the protests, provided daily updates. Their photographs and descriptions of people struggling to rid themselves of the Russian yoke moved me deeply, especially because of my own Finnish background.

The protests continued to increase, getting ever larger with each escalation of violence. Though many would die before it was over, Yanukovych soon fled the country. Everything seemed to end as quickly and suddenly as it had started. I felt a little bit sad over having lost the opportunity to embark on a great adventure. A great calm laid itself above the horizon, and the story seemed to be at an end. The calm, however, would soon prove false. The story hadn't even begun yet.

While all this was going on, I was participating in the annual winter exercise with the Home Guard, the main territorial defense units of the Swedish Armed Forces. I could feel my phone vibrating in my pocket while I lay in my foxhole dug out of packed snow. There was little time to answer the call, being busy fighting off an attacking opposing enemy force. The large blank fire adaptor attached to my Ksp 58B machine gun gave off beautiful flames, flashing into the thick, wintery woodlands surrounding me.

The enemy was easily pushed back, as is the norm for such an exercise. There are clear good guys and clear bad guys, and the officers in charge decide who wins the day. One of these officers, referred to as the "blue-yellow," did, however, declare one man in our squad as "severely wounded" during our glorious defense. The day had warmed up enough that the snow was both deep and wet. I left my machine gun in my foxhole for a weaker soldier to look after while I assisted in evacuating our casualty. While we struggled to drag the adult man through deep snow, I cursed the inopportune moment for someone to get "hit" in such a location in such conditions.

"Wouldn't it be better if people died at more convenient times instead?" I asked the others.

After an effort involving copious amounts of swearing, we finally mounted the casualty onto a sleigh which would take him the rest of the way. I wiped the sweat from my forehead and buttoned up my snow blouse and uniform as I returned to my defensive position. The sun had shone brightly during the fighting, but by the time the evacuation was complete and I was back in my machine gun nest, it had already begun to set. The exercise was taking a brief pause, so I had the opportunity to check my phone and return that earlier call. It rang only a few times before the other line connected. Before I could ask, I heard a voice saying "Jansson is dead."

I was stunned for a short moment before asking the caller to repeat, unsure that I had heard the message correctly. The voice just repeated the same thing. I sat myself down in the snow under a spruce, completely oblivious to my surroundings. You are never really fully prepared to hear about the death of a friend or relative, but in this particular instance, it became even stranger, even ironic.

Just days after Ukraine's now ex-president's escape to Russia, images of Crimea began showing up in news outlets. The large peninsula, which extended from southern Ukraine out into the Black Sea and divided it from the Sea of Azov, went from a place known only to connoisseurs of history to headline news.

Crimea's geography and milder weather had made it a popular vacation spot since the Soviet era—it also made it an ideal location for a naval base, and the Russian Federation's Black Sea Fleet was based out of the Crimean port city of Sevastopol. Much like the United States' operations from Guantanamo Bay in Cuba, the naval base in Sevastopol had been leased by Ukraine to Russia since the countries split apart in the early 1990s.

The media was reporting that "mysterious soldiers" had quickly begun taking control of the peninsula. The soldiers did not carry any nationality designations or insignia and were thus treated as unknowns. Anyone with a small amount of knowledge and understanding could, of course, see that these were Russian soldiers. They moved around in armored cars like the GAZ Tigr and were all wearing standard-issue Russian uniforms with the new digital Flora camouflage. Their small arms were also Russian, most notably their Pecheneg machine guns. Russia was lying outright, claiming that these soldiers were not theirs, but local self-defense forces, spontaneously organized by the population out of concern over the precarious political situation in the country. Despite this clearly being a lie, ignorant Western media hesitated to point it out. In not doing so, they not only permitted the confusion to continue, but intensified it.

Most pictures of the "little green men," as the international media soon began to nervously call them, showed them calmy patrolling and guarding key areas. In the few cases where the green men did not wear uniforms, but civilian clothing, their equipment was even more revealing. They were described as "local rebels" and were dressed accordingly in Adidas trousers. There was not, however, enough three-striped athletic gear in the world to cover up their 43mm GM-94 pump action grenade launchers, which they wielded openly in the street during tactical movements. This weapon was not something you saw in the daily news, and the few times it did show up, it was almost exclusively in the hands of Russian special forces units. It was obvious that Russia was the main player in these events, but no one dared to say it out loud as long as official Russian sources denied it.

Vice News had previously followed and reported on the events of the Euromaidan—as the protests had come to be called—in a very professional and straightforward manner, and their reporter Simon Ostrovsky now continued these reports directly from Crimea. It became increasingly clear how Russia, taking advantage of the power vacuum in Kyiv, seized the opportunity to conquer Crimea. The most painful thing to watch by far, however, was Western media and politicians. In a mixture of incompetence and meekness, they collectively refused to even acknowledge what was going on, leaving an open goal for Russia to continue scoring into day in and day out. Ukrainian soldiers stationed in Crimea were confused and without leadership, left completely alone. The highest-ranking Ukrainian officers in Crimea had, probably for the purest of Judas silver, betrayed them and defected to the Russians. Without any orders to follow, Ukrainian soldiers and sailors were forced to give up their weapons under threat of violence.

It took only a few weeks for the Russians to hold a referendum. The polling stations were all controlled by itinerant armed men. These were no longer Russian soldiers, but more like armed *gopniks*—paid bandits who could perform tasks that Russian soldiers would not. Other types of bandits, fifth-columnists, and Russian

5

sympathizers from a number of countries—including Finland, Belgium, and Bulgaria—slid around to observe the spectacle. In their respective home countries, they were, of course, well-known as tinfoil hat-wearing lowlifes and provocateurs obviously bought by the Russian state. Through Russian media, especially internationally-oriented ones like Russia Today, this pathetic lot was dressed up and made to look like a group of international independent election observers. For those who wanted to believe it, or simply didn't know better, they certainly appeared as such. Between free grog, biscuits with Russian caviar, and "services" from expensive prostitutes, they told the cameras that the referendum—where almost 97 percent of voters voted "Yes"—was one of the fairest democratic elections they had ever seen. Just like that, Crimea—together with the strategically important naval port of Sevastopol (which was all that really mattered to Russia)—was torn from Ukraine and incorporated back into Russia without anyone really having time to grasp what happened.

The weak pope in Rome read a prayer in which he asked God to let peace rule over Ukraine, after which he let two children each release a white dove from his window in the Vatican. The birds were immediately attacked and killed by a crow and a seagull—or the two-headed eagle, if anyone dares to believe in symbolism. The pope himself patted the children on the head and led them back into his room, a perfect foreshadowing for how the outside world would respond to what was now about to happen. He had made a mediocre symbolic effort which failed in front of his very eyes. For the pope's conscience, however, it was good enough and he withdrew from the public view, lazy and content.

As Crimea was changing flags, I was visiting the Åland Islands, located between Sweden and Finland. My best friend from my Finnish military service, Jansson, had unexpectedly and suddenly passed away and was about to be lowered into the ground. I had written a short obituary for him containing the following:

> I always felt great security and honor in knowing that if war was to come, I would fight shoulder to shoulder and eye for an eye by your side. And now, suddenly, you, the one among us who led with the highest courage, the most honest mind, and the kindest heart, are gone. The one among us who was most immortal no longer exists.

After the song and hymn, the coffin was carried out to the cemetery to be lowered into the ground. Jansson had been the sniper team commander in our Coastal Jaeger company, and even after we left the service, his interest in that particular subject had remained strong. He had talked for a long time about how he would one day buy a TRG-42 in .338 Lapua Magnum just like the one he had as a sniper in the army. Together with the roses of the other mourners, I threw down a Lapua Magnum cartridge as the coffin was lowered into the ground.

After greeting and thanking his relatives and friends, I went back to the grave once the mourners had left. I just sat next to my friend and opened the small bottle of Lagavulin I had inside my jacket pocket, which we shared together.

We had always agreed that we belonged to a minority that perhaps had a slightly greater interest in war than the average person, but without becoming completely obsessed with it. Even after we went civilian, we had strived to greet each other from time to time to drink Pilsner and talk about how much we looked forward to one day fighting a war against the Russians together. We would go over how we would beat them and plunder everything they held dear. We would, most of all, take back what was rightfully ours—Karelia, Vyborg, Petsamo, and all those other places which had long since been stolen. We joked and joyously dreamed of reconquest. Perhaps we dreamed of even more, maybe even Greater Finland.

It was March, and most of the snow had already melted away. The still somewhat cold breeze was offset by both the sun's rays and the good whiskey. I told Jansson that I thought it was a shame that he did something as stupid as dying at a time like this, when our boyhood dreams seemed to be approaching by leaps and bounds. Jansson did not respond. The entire time that I sat there with him, I had a constant feeling that he was ashamed. I took another swig of our good single malt and poured the last one out for him before getting up. I told him I missed him now and always would. I snapped my heels in front of the grave, gave him a stern salute, and turned away, leaving my friend for the last time.

I made my way back to the Finnish mainland. At my uncle's house in the countryside, which happened to be my mother's old parental home, I helped with spring cleaning by hauling out garbage and old furniture and throwing it into a fire outside. I then took the tram from Karis toward Hanko as Jansson and I had done so many times before, getting off at the Dragsvik stop. I passed by the Uusimaa Brigade's garrison area, walked toward and through the nearby training area and the Galoppskogen and Baggby shooting ranges. With the exception of some XA-180 Pasi armored cars and other military vehicles scattered about, the forest was empty.

I passed by old foxholes where we had learned basic infantry combat skills, as well as the small clearing where seven years earlier Lieutenant Westerholm, under vague rape threats, forced us to partake in strange CS gas rituals. I passed the small bushes where Jansson threw me off his back, just before he collapsed himself, after having carried me while running in full CBRN equipment under the hot summer sun. It was a great trip down memory lane.

I sat down next to one of the foxholes to rest my feet and picked up a few 7.62x39 rounds lying in the sand. I moved them around in the palm of my hand, felt the aged brass against my skin. There were a lot of memories collected in this otherwise inconsequential and bland little forest. Everything we had done here, everything we learned. Why did we learn it? Was it all for nothing, or would all this knowledge eventually be used? And if so, how? And when?

Meanwhile, in Ukraine, unrest was beginning to flare up more and more, especially in the eastern parts of the country. It was said that the regions with Russian majorities wanted to break away from Ukraine and become part of Russia, much like Crimea had done. Protests turned into riots and organized street brawls. It began to look as if the whole of eastern Ukraine, complete with its important industrial base, would be conquered in the same way as Crimea, for Russia's greater geopolitical pleasure. The scenario was the same, but the conditions were very different. Crimea was a surprise, where no one had time to react, but this time, the element of surprise was gone. Large numbers of Russian "little green men" would be unable to play a role, since there was no longer any easy or covert way to bring them into the Ukrainian mainland.

The well-oiled planning and preparation that characterized the Crimean operation had been replaced by chaos, and the green men had been replaced by two main groups. The first was a large but completely disorganized civilian mob fueled by the propaganda spread by Russian-language media. Its most powerful weapon was the lie that "fascists," as the Ukrainians in the west were now called, wanted to ethnically cleanse Russians from Ukraine. The mob soon took on a life of its own and proved extremely difficult to control. The second group was composed of cells of criminal bandits led by Russian intelligence services and special forces operators. They carried out strategic operations in the wake of the chaos created by the wild mob, taking over strategic locations such as police stations and government buildings. Even though the Russians no longer had the element of surprise, the Ukrainian military was unable to restore order. The effects of twenty years of decay, often caused deliberately by a corrupt and Russian-influenced political leadership, were palpable. Even the simplest security challenges proved insurmountable for the poorly trained soldiers and their completely incompetent leadership and organization.

At home in Sweden, summer was closing in, and it was once again time for the second annual Home Guard exercise, the largest of the year. This year, it would take place at the shooting ranges outside Härnösand. The focus this year was on having soldiers qualify on the Ak 4B rifle, so life was simple, albeit somewhat boring. As one might have expected, the situation in Ukraine was a recurring theme of discussion. Just about everyone understood that Russia was about to try to seize the eastern parts of Ukraine, if not the entire country. The tone was nervous, but also cheerful. We joked about the possibility of going to war if the situation escalated. Would it be in Sweden, or maybe Finland? Maybe even Ukraine proper—who could know for certain?

In Ukraine, more and more volunteer units were beginning to take shape. As the regular military—and by extension, the state itself—failed in their core tasks, free and independent men created their own militias to fill the vacuum. While Swedish alternative media like Motgift had begun to reduce their Ukraine coverage, a group of Swedish volunteers remained in Ukraine. From this group, a small number of activists had enlisted in the so-called volunteer battalions. In addition to the Right Sector (a broad coalition of right-wing groups that had been an important player during the Euromaidan), many other independent groups were beginning to form their own battalions. Swedish volunteers had particularly close ties to two of these: Aidar and Azov. Both groups were very small, and while the total number of volunteers is difficult to estimate, it was clear that none of these so-called battalions were battalion-sized. These humble beginnings would not last long, however, and their numbers would quickly increase in the coming months.

Aidar and Azov, along with many other independent battalions, quickly began operating in the Wild West that was eastern Ukraine. They provided general support to the military but were primarily deployed to the more dangerous sectors. Where there was a serious risk of violence, the military would simply fail to follow orders, mostly because they didn't trust their chain of command. The volunteer battalions, on the other hand, were driven by a self-confidence that stemmed from their fervent Ukrainian nationalism. This confidence allowed them to not just put up a hard fight, but, most importantly, it allowed them to make difficult decisions that the bloated and poorly led military could not. While heavy weaponry remained with the Ukrainian military, the volunteer battalions became their reaction force, momentum and, in practice, their real leaders and the spearhead at the front.

As the summer reached its peak, I was once again attending another Home Guard exercise. The situation in Ukraine had escalated in recent months, along with the tone among the soldiers. Many believed it was incomprehensible that no one was helping Ukraine put an end to the situation, essentially giving Russia free reign. The EU countries had issued strong declarations of solidarity, but nothing more. In heated discussions, increasing numbers of Swedish military personnel begun comparing Ukraine's situation with that of Finland in 1939—few people, aside from Swedes, had volunteered to help Finland during their 105-day Winter War against the Soviet Union. While the conflict in Ukraine was not yet a real war—just an insurgency, rather—the language was clear. Many would stand with Ukraine if the situation deteriorated further. This sentiment was no longer just hinted at, it was outright stated. "Together with the EU and NATO, we must back Ukraine," was a common opinion, and many were willing to back this up with action, should a real war ever materialize.

By late summer, the Ukrainian military and the volunteer battalions were slowly beginning to regain control of the situation. While the Russians had hoped to seize all the important cities of eastern Ukraine, in most urban centers, the pro-Russian mob had been met by an equally strong or stronger pro-Ukrainian equivalent. In others, pro-Russian elements had only barely succeeded in taking power, only to be quickly driven out when the volunteer battalions arrived. Ukraine had soon recaptured almost all their territory in the east. The pro-Russian mob and the bandit force, now calling themselves "separatists," had only managed to gain a real foothold in two cities—Luhansk and Donetsk. The war seemed to be over before it even began. Then, once again, something unexpected happened that changed everything.

Reports began trickling in of heavy shelling of Ukrainian positions from the Russian side of the border. In use were all types of heavy artillery, including the extremely destructive MLRS rocket artillery. The bombardment forced Ukrainian troops out of border regions and left the borders completely undefended. Then, seemingly out of the ether, the "separatists" that had once been equipped with small arms were now rolling around in tanks and other types of heavy equipment. In reality, these "separatists" were regular Russian units—a desperate attempt to take back the initiative in a quickly deteriorating situation. Now, with this action, the Russian-backed insurgency had escalated into what was essentially a "real war" between Ukraine and Russia. In the ensuing chaos, Ukrainian units were quickly forced back, taking huge losses in a very short period of time. Russian artillery bombardments were in some areas so intense that they reportedly wiped out entire companies and battalions—hundreds of men—in minutes. In rapid bursts of flame, both tanks and men were transformed into distorted piles of metal, flesh, and ash. With each passing hour, the inexorable slaughter continued.

In August, the Russians encircled the Ukrainians in the small town of Ilovaisk, located between the separatist stronghold of Donetsk and the Russian border. The Ukrainian force was virtually annihilated. Nearly two-thirds of the one thousand men surrounded there were killed or injured. Large amounts of Ukrainian military equipment were either destroyed or abandoned. The Azov Battalion, which by now had grown in size to resemble a real battalion, managed to break out of this pocket before it was too late. But for the Aidar Battalion, things did not go so well. I could find photographs taken by their enemies showing Aidar soldiers being ambushed during their retreat. Around their burning trucks on the hot summer asphalt lay burned and mutilated naked bodies. The "separatists," who were not Russian-speaking Ukrainians, but mercenaries from other countries (particularly Russia), proudly posed next to their contorted remains. They took their time desecrating the corpses, cutting off their ears and disfiguring their faces, clearly taking great pleasure in it.

Both Jansson and I had tattooed the Finnish Coastal Jaeger insignia, the golden head of a white-tailed sea eagle, on our upper arms. We did this partly to commemorate this part of our lives and partly out of pride from having overcome the rigors of military life. Primarily, it served as a reminder to never be a coward when it really matters. The Coastal Jaegers had trained us to always go first and foremost, regardless of the risk to our own lives. As non-commissioned officers, it was particularly important to be the ones who lead, the ones who step up to a challenge when no one else dares. During wartime, the Coastal Jaegers were supposed to step up whenever the otherwise competent Finnish navy could no longer cope. During our last and final exercise in Finland, we were ordered to attack part of the Armored Brigade. It is well-known that light infantry, as we were, has difficulty defending against heavy infantry and armor. Having light infantry attack armor seemed crazy indeed, especially considering that we barely carried enough APILAS anti-tank weapons to destroy their tanks and fighting vehicles. The point of the exercise was not to achieve victory, however. Rather, it was to imbue the men with an aggressive mindset and to put them at ease with the idea of unquestioningly taking part in a fight they know cannot be won.

Seeing brave but unprepared Ukrainian soldiers slaughtered while resisting a vastly superior force was the final straw for me. Everyone talked loudly about how they would do something, how they, too, would help the Ukrainians, if only the circumstances were right. Everyone talked about how EU countries, NATO member states, and the United States should do something. But now, as burned and desecrated corpses lay smoldering on the asphalt outside Ilovaisk, they only continued to offer excuses. "Someone else needs to do something," was a common refrain. I looked in the mirror and did not see myself anymore. I only saw a member of the gray, cowardly masses offering excuses, and I refused to continue being a part of it. I knew that if I gave in to cowardice in this moment and did not go join that war, I would be just like the others and never amount to anything. I would forever be a fake person, a fake soldier, and a fake Coastal Jaeger.

During the Euromaidan, I entertained the idea of a quick in-and-out, twenty-minute adventure. It never really caught my attention, however, and was over before I could seriously consider the notion. During the initial stages of the conflict in eastern Ukraine, the motivation was the same: a desire for adventure, but the Ukrainians seemingly took control of the situation rather quickly. There was no point in throwing away a fairly good life just to play a part in the final chorus. The temptation was always there, but the motivation to embark on an adventure had never been sufficient.

Ilovaisk was the turning point for me. Thousands of brave but ill-equipped Ukrainian soldiers broke before a now overpowering enemy. Figures were circulating that claimed that more than half of the regular Ukrainian military was destroyed. They were on their own, despite all the supportive talk from high-ranking politicians, the pope, and my friends in the Home Guard. They only offered empty words, and I couldn't stand for it.

It was like a fire raging within me. I didn't understand where it came from, but I had no choice but to follow it. I knew that I belonged in the war much more than I belonged in Sweden, spending my days waiting for the next meaningless Home Guard exercise. For the first time in my life, I knew I was going to go to war, and I knew it would be soon.

CHAPTER 2: VOLUNTEERING

I began inquiring about which Ukrainian units I could apply for. I initially looked at the regular military but gave up on them fairly quickly. The terrible impression I had of their performance so far in the war was only boosted by unclear messaging and their lack of effective communication with potential volunteers. Anyone could see that their enormous losses since the Russian incursion could not be attributed simply to the Russians' superior equipment and numbers. It was obvious that their poor performance was at least partially explained by their lack of organization and training, as well as sheer incompetence.

I then turned my attention to the independent volunteer battalions. Most were based on or were directly linked to Right Sector. Here, too, communication was difficult, and I had trouble finding clear information about volunteering. Instead of trying to get answers from Ukrainians, I turned to Westerners that I knew had become heavily involved in Ukraine. By pure luck, I had come across an anonymous Swedish man on social media that was currently serving in the Aidar Battalion. He provided guidance on how I could volunteer for the unit—it was surprisingly easy. You had to travel to Kyiv at your own expense, look for people associated with the battalion, and inform them of your intention to join. The battalion would then provide you with arms and send you into the conflict zone to rendezvous with your assigned unit. Aidar would allow just about anyone to join, as long as he provided his own basic equipment and was physically able.

The last part was good to know, but also worrying. I never asked him directly, but it was obvious that the Swede I was speaking to was younger than me. Between the lines, it was evident that he had some military background and experience, but it was limited. I wanted to serve with professionals, not amateurs.

The Azov Battalion had long been open to accepting foreign nationals, but had emphasized the need for these to be professionals. This was well-known in

Sweden, as several columns had been written about the battalion, and among the Swedish "right-wing extremists" that had enlisted in its ranks. It was also easy to find in Azov's own official channels, which were readily accessible online.

Descriptions of Azov volunteers in Swedish media were numerous and problematic. Sweden was headed toward the 2014 parliamentary elections, and political observers predicted that the nationalist, right-wing Sweden Democrats would be even more successful than they had been in the 2010 election. The 2010 election had shocked the media and political establishment to its core, and they were determined to do everything possible to prevent it from happening again. To frighten the public, corrupt Swedish media wrote extensively about the "threat" from the far right.

Making things more difficult for the Swedish Azov volunteers was the ongoing phenomenon of the Islamic State, which in 2014 was living through its golden age. Muslims from all over the world had traveled to Syria and Iraq to volunteer, including many immigrants living in Sweden. The official number of so-called "Islamic State travelers" reached three hundred in 2013, the last year official statistics were kept. In all likelihood, the responsible government agencies stopped publishing their data once they noticed that it had damaged public confidence in established parties and greatly benefited the immigration-critical Sweden Democrats. To shift public attention away from the Islamic State and immigration, authorities and their media puppets began focusing on Azov and its many Swedish volunteers. The idea was to paint nationalists, as well as the Sweden Democrats, as dangerous extremists akin to the Muslim immigrants that had fought for the Islamic State. Hardly a single story was written about the Islamic State or Islamist terrorism without Ukraine and the danger posed by Azov's Swedish volunteers being mentioned.

I could maybe get away with joining the regular Ukrainian military, but if I chose to join Azov, my life would never be the same again. I would be lumped in with the media narrative of "Nazis" and right-wing extremists and essentially become a "white jihadist," as the volunteers were sometimes called. It was likely that I would never be able to get a normal job again, and I would even be exposing myself to possible criminal charges. Was I willing to face terrorism charges in order to help Ukraine and participate in the war as a volunteer?

I reasoned that the main threat was going to be Russian bullets and artillery, certainly more than being called a "Nazi" in the papers. Azov was not just the easiest unit to apply for, it also gave me the impression that it was a disciplined military force, at least compared to other volunteer units. They had made a name for themselves in the battle for Mariupol, a large city in southern Ukraine. The Ukrainian military had made little progress in retaking the city until Azov arrived and, serving as the volunteer spearhead, helped achieve victory. They also had managed to escape the devastating encirclement in Ilovaisk, potentially showing that Azov had a strategically skilled leadership. Additionally, there were many foreign volunteers in the battalion. Certainly not as many as the media had claimed in order to amplify the imaginary right-wing threat, but it had to be more than a handful. This meant that other Western volunteers had managed to serve in the unit without speaking Ukrainian or Russian. I anticipated that I could do the same.

With these calculations, I had made up my mind. Azov was the most suitable unit to apply for, and so that's where I would start. Now it was just a matter of finding out if they would have me.

The Swede who had become the media's poster boy for foreign volunteers in Azov was named Mike. According to the newspapers, he had been a long-time professional soldier in the Swedish Armed Forces, where he had served as a sniper for some time. He had been involved with Azov since the beginning of the war and seemed to have some sort of lower leadership role within the battalion. I had occasionally chatted with him to seek first-hand information about events in Ukraine. Now I asked how I could volunteer for his unit.

He answered quickly, and after a few short follow-up questions, I felt that I knew everything I needed to know. Azov sought both combatants and instructors (to train their Ukrainian recruits), and the terms were similar to Aidar's—the battalion would provide food and a rifle; everything else would be my responsibility. I was told to contact the acting coordinator of foreign volunteers, a veteran of the French Foreign Legion that had previously fought with the Croatians in the Yugoslav Wars. Gaston, as he was called, needed some kind of military resume to determine if I had the required skills and experience. I quickly typed one out and emailed it to him:

General:
Age: 25 years. Military rank: Sergeant. Physical condition: Capable, healthy in general.
Military background:
Finland: *Conscription service with the Uusimaa Brigade in 2007, followed by non-commissioned officer training, and then Coastal Jaeger squad leader with a focus in anti-tank combat. Also served as an instructor. One year in total.*
Sweden: *Guardsman rifleman with the Medelpad Battalion, later machine gunner and squad leader. Also served as an instructor. Six years in total.*
Other: *Engaged in hunting and recreational shooting for many years.*

And that was it. I looked at what I had managed to put together and felt that it might not look like much to the world, but I hoped for the best.

Anders, my friend and fellow Maidan enthusiast, one day went out to a shooting range. It was essentially just a large gravel pit, far outside civilization, where we could usually conduct our business relatively undisturbed. We had talked about Ukraine several times before, and he had been somewhat cautiously interested in it. However, our personal lives had developed very differently, and he now had his first child on the way. He was no longer interested, though he had a difficult time in saying so.

Aside from zeroing the old Weaver scope on my German .300 Winchester Magnum bolt-action rifle to prepare for moose season, I had no particular reason for being at the range. Anders was trying out some new handloads, but at six hundred meters instead of my eighty. As he climbed to the edge of the pit to shoot down toward the targets, I stayed near them. There were large piles of timber stacked along the roads passing through the pit. Occasionally, a car or some old lady walking her dog would come this way, and the rifleman would be unable to see them. Therefore, it was necessary to maintain radio contact between the ground spotter and the rifleman situated above, to warn him about movement around the target area.

The range was clear, and I gave Anders the go-ahead to start firing. He started with his Remington 700 in .308 Winchester and then stepped it up to the TRG-42 in .338 Lapua Magnum. The sound of the bullets cracked like whips as they passed above me. I tried to take in as much of it as possible from my place down in the pit. I turned in different directions to see if it was possible to hear differences in a bullet's direction. I moved from open terrain to between the timber lengths to see if the surroundings had any effect on the bullets' sounds. I cupped my hands over my ears to imitate the effect of wearing a helmet. It was indeed very easy to hear differences in a bullet's angle and point of origin, as well as the difference between calibers. The .308 wasn't nearly as loud as the heavier Lapua rounds.

When I got home from the range, I found that I had received an email response. My nervousness subsided pretty quickly as I saw that Gaston had found my military background perfectly adequate. I was approved for service and combat. All I needed to do was travel to Kyiv with as much of my equipment as possible. From there, I was to contact him again for further instructions. Suddenly, I was in the clear. I was in. All I needed to arrange now were the logistics and travel practicalities, as well as, it would turn out, spiritual matters.

My salary as a volunteer would be small or non-existent. I had some savings but not enough to last very long. To create a financial nest egg for myself, I began selling off things that I had accumulated over the years. The possession that I got the most money for, but also felt the worst to let go of, was my revolver. It was an antique and permit-free, yet fully functional .44 caliber Remington New Model Army, often incorrectly called a Model 1858. With mixed parts and worn markings, it was not exactly in the finest condition, serving as more of a beater or a "fun gun." After a few shameful offers, I finally received a reasonable price proposal, and in just a simple handshake I added another twelve thousand Swedish krona to my little war budget.

Through Mike, I had been told that the battalion could provide me with a combat vest and bulletproof plates if I didn't already have some. It would also be possible to buy other equipment on location in Ukraine if I did not wish to haul everything all the way from Sweden. Above all, I learned that Ukraine was damn hot in the summer, and I made sure to take this into account when ordering clothes and other equipment online. Ukrainian soldiers did not have a strict uniform system, and their clothes were often military surplus from other countries. Looking at pictures of Ukraine's terrain, I judged that the most appropriate camouflage pattern would be similar to MultiCam, which was, unsurprisingly, already a common choice in Azov. The British frequently used MultiCam, and since they would often sell off surplus in near-new condition, I managed to get a good deal on a uniform set from the Finnish surplus store, Varusteleka.

It was a little embarrassing when, in early September, I showed up to the annual moose hunt dressed as some kind of military LARPer. It was, however, as good of a field test for my equipment as I would be able to get. I wished to spend my time in the forests of Jämtland evaluating my shoes or clothing and correcting any fit or chafing issues. It was certainly preferable to anticipate and correct problems here than having to do it on the Ukrainian steppe. Moose hunting (as we do it, anyway) is not entirely different from war. It's a lot of early mornings and waiting around, often in cold winds without very much happening. Then, in sudden bursts, we would hear the sounds of gunshots, which were sometimes followed by bloodshed. The time between the sudden bursts of action gave me plenty of time to think.

I mostly thought about death. I constantly found myself thinking about every possible way I could die, how it would happen, and how it would feel. I could be stricken by a bullet, for example. I could spend my final moments watching myself bleed to death through a large groin wound with no help in sight. Perhaps I would get my leg blown off by a mine? The pain would come first, then panic. I would start losing body temperature as the blood left my body, until I was finally embraced by the darkness. I could get hit by bullets in all kinds of places. Maybe in the stomach, or some other important internal organ? I pressed my fingers into my liver and wondered how painful it would be to have it pierced by a piece of hot copper-jacketed bullet.

There were other ways to die than bullets. Artillery was a common battlefield weapon, and, during a botched light mortar exercise during my conscription, I had heard and seen how hot shrapnel could tear violently through otherwise tough tree trunks. The Russians were also using thermobaric weapons—basically flamethrowers. I wondered what it would be like to burn to death, how terrifying it would be to go out that way. Perhaps I would be the lucky guy who simply got hit in the head and was done with it all in less than a moment, but if so, then what? What would come after the darkness?

In vain, I tried to understand things completely unimaginable. In my head, I went through dozens, maybe even hundreds, of different scenarios related to my possible death, endlessly trying to come to terms with each one of them. If I died in Ukraine, I wouldn't even have the chance to see the new *Mad Max* movie, set to come out in 2015. I had actually been looking forward to that.

My reasoning was that these mental exercises would make me more prepared for the coming journey, not just in the specific events I imagined, but in any scenario where I would have to face the risk of death. I knew that when discipline is lost and panic takes over, all is lost—doesn't matter if it's an individual soldier or an entire army. Medieval armies sustained the majority of their casualties not during a battle's organized combat, but after, as soldiers panicked and fled, only to be run down by enemy cavalry. Similarly, line infantry suffered the most—not from the hail of enemy bullets, but when their ranks broke down and the soldiers dispersed. Modern-era infantry were most vulnerable when they were pinned down by enemy fire, unable to muster the courage to respond to it. Even though it was a terrible feeling, I would prefer to deal with the panic and anxiety sooner rather than later. If I had that part resolved, I thought I would have a better chance of coping through future situations. It's still difficult to describe, but I saw it as a personal "Golgotha walk" that I needed to complete to become an effective and disciplined soldier.

After my decision to volunteer, I tried to meet with as many friends and family as I could in Sweden. It was important to me, even though very few of them knew about my plans. Next, I intended to return to Finland to do the same thing there. The plan was to take an early morning bus from my small northern hometown of Sundsvall to Stockholm, where a ferry would take me to Helsinki. There, I had arranged to meet some old friends from the army and quit my job.

The evening before my departure, I gave some final thoughts to my current military service in Sweden. I had left all my military-issued equipment at home, including my Ak 4B battle rifle. Should my enlistment in Azov become public knowledge, this would cause a complete shitstorm. By leaving the crown's equipment and giving a spare key to a friend, I had hoped to avoid having the military police break down the door of my apartment. Nevertheless, I still felt that I owed my platoon commander some kind of warning. We were, after all, good friends and had known each other for many years. I would feel terrible if my actions got him in trouble. Therefore, I decided to contact him that evening (instead of from Finland, as I had originally planned) to give him time to prepare in case trouble erupted. After informing him of my intentions, I explained how, if I were to be found out, authorities could easily enter my apartment and pick up the rifle, combat equipment, and all the related gear. My commander seemed very confused, but to my great relief, told me that he respected and understood my decision.

After a close to sleepless night, I was awakened not by my alarm clock, but by the phone. It was my platoon commander. He was calling from his car, already heading over to my place. After being awake all night, he had made the decision to inform his superiors about what I was about to do and was now coming over to relieve me of my rifle.

It took me a moment before I understood exactly what he was saying and briefly entertained the notion of asking this rat to kindly turn into oncoming traffic. Thankfully, though, I quickly regained my composure. It was I who had chosen to tell him, and thus it was I who had put him in this situation and should understand the actions he had taken. In a difficult position between one of his soldiers and his commanding officers, I should have known that his loyalty would remain with his chain of command, no matter what direction his moral compass actually pointed toward.

My commander appeared, and we exchanged a few words. It looked like he had slept even worse than I had. He apologized several times for his actions but said that he had no choice. I could see in him that there was some kind of internal conflict; he had reluctantly chosen to follow protocol even though he was unsure if it was morally right. I handed the gun over to him and received a "good luck" and a hug back.

Based on this very bad decision, my initial plan had to be changed fundamentally. My platoon commander had informed company command, who told battalion command, who passed it up to the commander of Västernorrland, I19 Regiment—the internal security and military intelligence unit. In a single phone call, the number of people who knew about my plan had increased from a trusted handful of friends to hundreds of potential hostiles. My stealth advantage was gone in a moment, and there was nothing I could do to get it back.

As I was on the bus, I called my company commander, Karl-Alfons. He was widely regarded as an asshole and a phony who only looked out for himself. He had a reputation for leaving in the middle of exercises for personal business and pleasure, on one occasion resulting in his soldiers going hungry in the field. He was also known as a person who had no moral qualms about making some low-ranking soldier a scapegoat whenever he made some colossal mistake. With the exception of a few other career-hungry personalities, he was almost unanimously the most hated person in the company. Very few of the hundreds of men and women in the company would ever dare say that out loud, however. Nevertheless, I still thought it would be harmless to try and explain my decision to him. At the very least, I would get a hint of what the higher command was thinking and what I should expect.

What followed was a series of political monologues, where he mechanically let out one pre-programmed sentence after another. Even though the conversation took place over the phone, I could practically taste the lies and pathetic rationalizations on the tip of my tongue, the mechanical rattle of a lazily written script. It did, however, tell me that the Home Guard and their command in Härnösand would try to silence this. They would try to spread a rumor that I got a job in Finland and was therefore dismissed from the Home Guard. This was something that they were supposedly doing for my sake, out of the kindness of their own hearts. They thought that I should be thankful for it and ignore the fact that it was simply a move to avoid what they saw as bad publicity. I was not supposed to understand it was all just another lie.

The ferry arrived in Helsinki around midday, and I proceeded with my plan, checking off a few final important errands and then meeting up with old friends. Once these errands were done, I headed over to the central train station in the Brunnsgatan district. Down the street, opposite the station's large stone entrance, lay an Irish pub, the Dubliner. I had on many occasions spent time there while waiting for friends or train departures, and if the Dubliner was a good enough place for a beer before the train to Karis, then it would also be a good enough place for a beer before what could be my last trip out of Finland.

After securing a table conveniently located between the bar and the smoking room, I ordered a pint of Guinness—one of my many bad decisions, it would turn out. My previously well-controlled thoughts and reflections about death now ran wild. In addition to the fact that these thoughts were omnipresent at this point, the fear that I had previously tried to come to terms with began to hit me harder and harder, radiating throughout my entire body. In my mind, I was still able to stop the panic from showing, but in reality, it had already begun manifesting itself in a physical manner.

In situations of extreme stress, the body shuts down or inhibits certain functions in order to be able to focus all available energy on the most important ones. In combat, whether it is war or just an ordinary bar fight, the body will limit a person's ability to use small muscles and will prioritize access and control to the big ones. The neck, eyes, and fingers get numb and become stiff while the thighs, hips, and arm muscles are given priority access. I had a hard time entering my card's pin number as my fingers and hands shook uncontrollably.

As soon as the pint was full to the point of spilling over, it was displayed on the counter in front of me, and I quickly realized that I would never get it to my table using a single hand like a normal human being. I adjusted my strategy, and attempted to do it using both hands, concentrating every ounce of willpower. It was nevertheless an unprofitable venture. My "Golgotha walk" had now taken over, and there was nothing I could do to get my pint back to my table unscathed. The only thing I could do was down it at the bar, order a new one, and hope that the alcohol would sufficiently relax my mind and muscles before the second pint was filled. The strategy worked, at least to an extent. As I drank more pints and new ones were filled, there was certainly less spillage, though holding my beer still required a two-handed grip and complete focus. I was ashamed as I sat in my corner, drinking like a retarded child with two hands around my glass. Shortly, a couple of old friends met me at the table.

"You really look like shit," one of them said in a tone that was unusually serious for him.

"I guess I look the way I feel then," I replied, finishing off my current pint.

A little later, I was on my way to board the ship when I passed what seemed like a familiar Viking Line employee. Behind the thick black beard, I thought I recognized the young man. I had not seen him in several years, but I was sure it was one of my mates from the Coastal Jaeger company. We looked at each other for a moment, each man trying to place the other.

"Löffe?" the man asked.

"Koskinen?" I responded.

"What's all that shit you're carrying? That's got to be, what, thirty, forty kilos? Are you going to war or something?" he thundered loudly in his deep voice, laughing. My laugh in response was a lot more convoluted and fake.

We did not have time to say very much aside from how nice it was to be able greet and see the other, after which I continued on. Once on board the ship, I looked for a secluded place, unrolled my sleeping mat out on the floor, and laid down to try to sleep by my backpack. In a number of hours, I would be in Stockholm and then, soon after, in Ukraine.

CHAPTER 3: THE ROAD TO UKRAINE

Getting to sleep was impossible, but I managed to fall into some sort of half-awake state, not entirely there, but still aware of my surroundings. I wavered between the dream world and reality for a while until I felt that I had sufficiently recharged my mental batteries.

The trip to Stockholm would take fifteen hours. Going to the bar to pass the time was an option I eliminated relatively quickly. It was a little too early in the day, and above all, my war funds were already quite thin. I sat up on the light and compact, albeit still comfortable and warm, sleeping pad that my friend Anders had donated to me before leaving, along with a few other things. The clothes and other equipment were feeling heavier than I expected. Together with my two slightly smaller extra pouches that I had attached to my belt, the weight was about forty kilos. I loosened the straps and opened the lid of my LK-70 backpack and dug into it. I had packed only the bare essentials, with the exception of a few meaningful extras, like *The Tales of Ensign Stål*—a collection of poems by J. L. Runeberg about the Finnish War 1808–1809. It was in paperback form and very worn, as I had read it many times since I had gotten it from my grandfather on my tenth birthday.

As I had always done since I was a child, I started the book with my favorite poem: the one about Sven Duva, the large but annoyingly stupid soldier who, in a comical way, consistently fails to accomplish his missions. In the climax, however, he saves his platoon (and in doing so, the greater Finnish army) as he single-handedly stops the Russian horde at Virta Bridge.

In the poem, a single platoon of Finnish infantry is tasked with defending the bridge. A larger force of Russians appears on a hill on the other side of the bridge and with two quick volleys kill several of Duva's comrades. When the Russians then begin their assault down the hill toward the strategic bridge, the Finns lose heart and flee from the overwhelming enemy. All except for the complete idiot

Duva, of course, who as usual did the exact opposite. As the commander ordered a retreat, Duva mistakes the "retreat" for "bayonet," and charges the Russians alone, halting them in the middle of the narrow bridge. In a simple mistake, this broad-shouldered giant mongoose ceased to be the comic figure that couldn't fit in at home or in the army. As he stepped up onto Virta Bridge, he ceased to be a simple fool and was transformed into the legendary, axe-wielding Viking at Stamford Bridge. The man who always pulled his plow backwards with his shirt buttons done all wrong suddenly became the Spartans at Thermopylae. Duva stopped the horde but died in the process, shot through his best quality—not his head, but his heart.

Thoughts of death still wreaked havoc on my soul, and even though the beautiful poems calmed my mind for a short while, I soon couldn't help but obsess over death again. The little peace the poems gave me was fleeting, and I felt like they were whispering in my ear that I had forgotten something.

In the ideation of myself bleeding to death alone in the grass of Ukraine, it finally came to me. Even if everything faded to black for me, the rest of the world would continue as usual. I was almost entirely insignificant, but those around me would remain. I had been so preoccupied with thoughts of death and dying that I hadn't even written a last will.

I pulled my pencil and notepad out of my backpack and stared at the blank sheet. I had never written a will before. I did not even know what one looked like. Would it be a kind of letter or a legal form listing my things and the names of whom would receive them? I touched the pen to the paper only to pull it back again. I thought for a moment and tried again, but the pen kept coming back without leaving a single word.

I put down the paper and leaned against the backpack's hard but comfortable metal frame. I attempted to find some kind of peace. I was only twenty-five years old. Should I die, I would have no significant wealth to leave behind, and without children or siblings, I also had few people to leave what little I had.

After a moment of thought, I decided that I didn't care how the will would end up looking. The important thing was for it to be simple and not create any confusion. I started writing and committed myself to keeping it short. Firstly, I wanted it to be clear that I was deeply remorseful for any grief my death would cause. Then, I followed up with a simple list of where I wanted my few assets to end up.

I had an old friend that enjoyed the woodlands and had always wanted to buy his hunting permit, but never seemed to have the money for it, always putting it off for "next summer." I therefore decided to leave him my guns, in part to help him out financially, but above all to encourage him not to wait until "next summer" indefinitely. The only stipulation I had was that he could never sell my grandmother's brother's gun, which I considered a family heirloom. With the other rifles, he was allowed to do as he pleased.

Otherwise, I stated that most of my other assets, especially my apartment, should be sold. Most of the money would go to one of my cousins, for him to manage until his son was of age. The rest would be distributed among the children of other cousins and friends, with the amount based inversely on how big of silver spoons the children had in their mouths when they were born.

The evening approached, and the ferry began navigating through the Stockholm archipelago. I packed up what I had in front of me, rolled up and fastened the sleeping pad, and threw all my shit back onto my back. During the day, I had gone out on deck number eight to smoke and had settled in right next to one of the exits. In the evening chill, however, a cigarette was best smoked in the smoking room next to the gateway on deck number five. I squeezed myself and my rather large backpack through the normal-sized door of the smoking room. The few benches inside were all occupied, mostly by experienced Finnish alcoholics and their terrifyingly ugly women, so seating was out of the question. Taking everything off my back for a simple cigarette was also not ideal. There was space at one of the small, round tables with ashtrays on it, so I backed up against it. Tiptoeing, I rested my backpack on it to relieve me of its weight. I lit a John Silver cigarette and looked out the window. The islands passed—larger, more plentiful, and with bigger concentrations of settlement the further we went. Soon we passed into Stockholm's city limits. I had always loved coastlines and the sea, and this archipelago had a particular charm that could not be found anywhere else—not in cities, countryside, or mountains. The beautiful landscape gave me a moment of peace, and I watched it with pleasure as I lit my next cigarette.

The ferry soon started docking in the harbor. I snuffed out my third cigarette and lifted my backpack off the table. The weight immediately began to cut off the circulation to my shoulders. I did a throwing movement with my upper body and tightened the shoulder straps, then did the same with the waist belt to put more weight on my hips. I could feel the blood finally flowing back into my arms again. I turned around and met the pair of eyes I had felt observing me for a while. I was wearing black boots, jeans, and a flannel shirt. By itself, this was perhaps not particularly remarkable, but combined with the overpacked LK-70 on my back, the green gas mask bag for my camera, and the old army medic bag on my hips, I stood out from the normal environment—enough for a Finnish alcoholic to have locked his target radar on me.

"Are you traveling to Ukraine for war stuff or something?" the drunkard slurred indistinctly but loudly.

I understood that a quick-witted joke would not sound as convincing as simply telling the truth, and I knew that experienced alcoholics could smell a liar as well as they could smell an opened bottle of Jaloviina.

"Yes, it's definitely the plan," I replied, and was greeted by a smiling set of dirty teeth.

"That's good! That's very good!" replied the drunk as he clapped his hands in the air. "Those Russian bastards! One should just shoot and kill them, and if they don't die, then you should shoot them again!"

I nodded and quickly squeezed out the door as the alcohol-driven monologue about Russians, murder, and the Winter War became a self-playing piano in the smoking room. As the door slammed shut behind me, I also overheard people discussing how Swedes were a bunch of cowards and pussies. The soundproofing prevented me from hearing the context and wisdom of the message, but I felt like I understood its core as I hurried across the bridge to the Swedish mainland.

CHAPTER 4: KYIV

I checked the messages on my phone to see a new notification. It was Kristian, saying he had already arrived and was waiting at the benches near the entrance.

We did not know each other very well, but we had mutual friends. He had been, like me, someone who merely entertained the thought of going to Ukraine, but he was also the only one among his peers who did not back out of it, having become more serious about going the more serious the situation in Ukraine became. Our way of looking at the entire conflict was reminiscent of each other, and when I had told him that I was going for real, he was quick to say that he would also come along.

Kristian was a bit younger than I was and therefore had a shorter military experience, but it was still sufficient. I also got a good impression of him as both knowledgeable and driven, which made him good soldier material. However, I had some concerns about the youthful, sometimes naive manner in which he often avoided taking things seriously. I had many times subtly given him a chance to back out, sometimes even asking him bluntly if he had really thought everything through. I had reminded him many times that we might actually die, that everything could very well end. I had told him, though never in detail, how this was something which had haunted me for a long time. When I asked him if he had also thought about it, his answers were generally evasive and nervous. As uneasy as it made me, he was still determined to follow.

Despite the fact that both our backpacks were over the permitted maximum weight, the staff let them slip away on the luggage belt, though not without cautious looks. We both boarded the plane that would take us to Oslo for a change and then on to Kyiv.

The sun had set while we were in the air, but soon the capital of Ukraine started showing up outside the window below us. In the compact darkness, a pattern of buildings and electric lights began to pile up on the horizon. To me, a simple Northerner, Stockholm and Helsinki were huge in their size already. Seeing the huge city of Kyiv with its almost three million inhabitants from above was a perspective that, for me, became almost inconceivable. I had never seen anything like it, and the extent of the city was difficult to grasp. As we descended toward the airport of Boryspil, just outside the city, I wondered if what I was experiencing was the same feeling the Nordic traders and Vikings had when they first approached this place in their ships. Had Kyiv given them the same magnificent impression then as the great city gave me now, a thousand years later?

After going through passport control, we tried to withdraw money in the local currency: the hryvnia. We tried several different ATMs, but none of them allowed withdrawals. Kristian, who was a more experienced world traveler than me, had prepared a little better and brought US dollars with him. With these, we managed to pay a taxi driver to drive us from the airport to the address where I had booked a hostel for the night.

Kristian started to seem nervous. The taxi driver looked like a typical Eastern European bandit, and the multi-lane road from the airport to the actual city was long and devoid of life. On the sides of the road, there was not much more than forest lit by the streetlights. The rugged old Soviet car smelled of alcohol and cigarettes.

"What if he drives us to some fucking dump and just kills us?" Kristian attempted to joke, but there was obviously more to it than just humor.

We finally began entering the city proper, passing lots of large billboards along the roads through the huge Soviet-era apartment complexes. The traffic around us was heavy with simple old cars and trucks, all of them looking like they would soon collapse in different ways. Kristian's gaze flickered around. While the terrible anxiety I had been carrying for weeks had ceased the very moment I stepped on the plane, it seemed as if his had only just begun as we departed.

The taxi soon stopped at the address I had given. The location was correct, but the dark alley looked more like a scene from a horror movie than the hostel I had seen in the pictures. The driver assured us it was the right spot, and the street and the metal sign on the stone wall next to it confirmed it, as well. We stepped out and explored the area. There were no signs about bed and breakfast, nor any other such amenities, and the worn steel gates in the courtyard were all locked. We quickly had to come up with another plan. I knew that Independence Square was not very far from where we were and that it should be downhill. Soon, we came across some young people crossing the street in front of us. I shouted at them, asking which direction Independence Square was from where we stood. They quietly discussed among themselves how to handle my shouting at them, seemingly uneasy with some unknown person confronting them in English. Soon, the designated alpha of the small herd, the tallest young man of the group, took charge and pointed down to the street they had come from.

"Is this way. Not very far."

We thanked them and continued in the designated direction. I soon began to recognize the surroundings from the videos I had seen from the riots last winter. The square opened up in front of us, and the large, gold-plated Independence

Monument, on its tall white pillar, rose high above all the large buildings around. There were lots of candles and flowers along the low walls and the fountain. The cobblestones, which were used as weapons during the chaos, were still missing in some areas. In the middle of the square, there were long placards with pictures of all the "Heavenly Hundred," as they were known—the protestors who fell to the bullets and other deadly violence of the riot police at the same spot less than a year before.

We were still in need of a room for the night, and after scouring a few hotels around the square, we finally found a relatively cheap room with two beds at Hotel Ukraine. From the window, we had a perfect view of the entire square, and from the thick, antiquated television, we could enjoy the incomprehensible local entertainment. I called my recruiter Gaston to explain that Kristian and I had both arrived—it was our first and only direct contact apart from a few emails. On the other side of the phone, I was greeted by a friendly voice which sounded pleased with the message. We were then instructed to rest up properly and were given two new local phone numbers which we were to dial the next morning.

The TV was still incomprehensible. The only channel that played Western film was dubbed. I sat down on the bed and folded my trusty laptop. It was almost ten years old, and had been on life support half that time already. But it was sufficient for Internet browsers and, on a good day, to watch movies—at least if you knew the complicated booting procedure. As nice as those features were, I primarily brought it to back up my Canon camera's memory, as I was hoping to take a lot of pictures during this adventure.

As I watched the latest news, a recently published video appeared in my feed: another Swede who had apparently just come to join Azov gave a short interview with them. The man said that he was just over thirty years old and introduced himself as "Leo." He spoke English in the video—while not bad, his English was still mediocre for a Swede. His gaze looked nervous, and he gave a very confused impression at best. He did not look like a military man in the face, but rather a tramp, especially as a small and weakly filled tattoo loomed under one of his eyes. He was saying some incomprehensible nonsense about how he had come to fight for Ukraine because Putin was a communist—hardly the analysis of someone with gifted political insight. I was hoping it was all a bad joke.

The next morning, we got up and had breakfast at the hotel before I called the numbers Gaston had given me. The first one got us nowhere, and the latter brought us to a phone operated by a woman who obviously didn't speak a word of English. As it turned out, however, Azov had a small headquarters building about a hundred meters to the other side of Independence Square, so we went straight there instead.

It was housed in a yellow, couple-story stone building. High up hung a huge flag on the wall. It depicted the black silhouettes of a group of armed men surrounded by the text "Chorny Korpus," or the "Black Corps"—sometimes translated to the "Men in Black," a parody of the Russian "little green men." It was Azov's first unit name and insignia. Under the large flag hung two slightly smaller Ukrainian flags on each side of the large steel gate, both with Azov's latest insignia in the middle. Against a yellow background, the black sun lay in white with a black wolf hook in the middle. These two then rode on top of a blue wave of water. Above went a black bow with Cyrillic text in white: "АЗОВ"—or Azov, in Latin letters. The battalion's insignia, an aesthetic perfection where the colors matched the shapes perfectly, was already infamous due to the clear right-wing radical vibes. But in my eyes, the anxious Western media reporting on the symbolism only gave it an extra sexy shimmer.

Kossatskiy, the Azov HQ building in Kyiv, adorned in Chorny Korpus flags.

I rang the intercom next to the big blue steel door.

"Yes?" answered a voice from the other side.

"Do you speak English?" I asked.

After a moment of silence, the voice answered: "What do you want?"

"We're volunteers from Sweden," I replied. "We want to join Azov."

"You want to volunteer?" replied the voice in surprise.

"Yes. Can we come inside?"

"Eh…wait one minute."

We waited. I don't recall how long, but it was enough for me and Kristian to start talking about it: had the voice gone radio silent and ignored us in hopes we'd leave? What would we do then? We decided we had time to stay a while longer, and soon enough the gate opened. In front of us was a dirty and dimly lit small hallway, which led up to a larger entrance hall via a small stone staircase. The guard who met us held a baton and wore a black uniform. He was young, slim, and slightly taller than me at my own one hundred eighty centimeters. His face was slightly angular and his nose markedly large and crooked. Despite the buzzed cut, one could suspect his hair was brown, and even though his blue eyes were wide open, it was clear that they would normally be quite narrow. To the left of the entrance itself was a simple desk where the guard was otherwise stationed.

"You can sit here," he said, pointing to a semi-broken sofa and pair of armchairs next to a table on the other side of the room. "A commander will come soon. Please, sit down and wait."

Kristian and I put our backpacks on the floor and sat down as instructed in the sofa, which was much more comfortable than its decrepit appearance would indicate. The commander lingered, and the young guard was clearly having difficulty sitting still by the desk on the other side of the room. He soon came over to us to speak in his broken English. It was not particularly bad grammatically, but his sparse vocabulary limited him considerably. Regardless, it was readily apparent that he was excited to talk to us, despite giving the impression of having a shy personality. I assumed that the rare chance to speak English with someone who knew it outweighed his nervousness.

The Ukrainian fighters usually took on nicknames, or *nom de guerres*—war names. This tradition went back a long way but was especially linked to the Ukrainian Insurgent Army during World War II. Nicknames had been used by their fighters so that they could not be identified as easily by their enemies. The same concept still applied today, both from necessity and tradition. The young guard's name was Roman, and he had not yet really figured out his nickname.

"I like bears, so I think I want to call myself 'Bear.' But I have not decided yet," he said.

I looked at the tall, thin, young Ukrainian and thought it ironic, as he in no way resembled the animal in question. He was approximately eighteen years old and extremely curious and inquisitive. After asking us who we were and what military background we had, I asked him about his. Like many Ukrainians, he had none at all, other than playing airsoft. He said that in doing so, he understood a little about combat and tactics, but he was not so naive as to think it was the same thing as real combat.

When we asked how he spoke such good English, he replied, after shyly apologizing again for how bad he thought it was, that he had studied at the university before the Maidan protests. When the revolution had begun, he was drawn into it. He himself was born in an independent Ukraine but knew through his family the story of the Russian and Soviet times. He did not want to see his Ukraine brought back there. He had soon joined the protests and had followed them all the way, even when the police had begun gunning people down. After this, he returned to his studies, but now, after Russia attacked his country, he had joined Azov. "Bear" had not been here long, just a few days. His temporary job at the desk was only part of his recruitment period, and he was eagerly waiting to be sent to the front line with the other future soldiers.

Finally, a commander descended the stairs toward us. He did not speak English himself, but Bear was eager to interpret. Kristian and I were given a couple of forms to fill in. It was simple A4 paper where we would write down a mix of personal and other information. "What are your political views?" was one of the questions. Kristian tapped the pen on the table and looked at me.

"What the hell am I supposed to write here?"

"Just say you're a Swedish nationalist or something," I said. "Just don't say you're a commie or some other stupid shit, and it should be fine."

The commander was highly suspicious, reading us with his eyes. The real plan had been that Mike, whom I had also talked to, would have been here by now to verify us as legitimate, but he wasn't around and wouldn't answer his phone. Kristian was worried, but as we had filled in our forms, I unpacked my rucksack, taking out some neatly packed clothes. They were all Finnish and Swedish military surplus: shirts, long johns, and the like. It was not much, but it was what I had at home, and I knew that a warm sweater would do much more here for a soldier than it would unused at my place. As I explained this, handing the clothes over together with the forms, the commander's gaze changed immediately.

"Give these to some of the guys heading to the front," I told him. A slight but friendly smile showed up as he nodded in response before leaving.

The friendly gesture was one step on the way; however, we still were not allowed to leave the building before our identities had been verified. Several hours had passed, and Bear's incessant babbling, however nice at the beginning, increasingly become an annoyance. Luckily for us, the doorbell rang again, and Bear hurried away to answer it. The walkie talkie made its distinct cracking noise, and he called the commander. After some deliberation between the two, Bear opened the door and two men in their late thirties or early forties stepped inside. One was light in complexion, with narrow glasses over his blue eyes. As he lifted off his black cap, his slightly dark blonde hair was revealed. It was thin, his hairline high. His chin was rounded off by a short beard—about two weeks' worth of growth. He wore a shirt over a dark leather jacket and large trousers with leg pockets. A laptop bag and a camera hung over his shoulder and neck. He had "journalist" written all over him. The other man was significantly darker, both in skin and hair color, with glasses, brown eyes, and black stubble. On his head, he wore a black cap, a glossy down jacket, and well-fitting dark blue jeans. He was tall, strongly built with broad shoulders, but still slim, boasting a much better posture than the journalist.

The dark one acted as an interpreter between the journalist and the commander for a moment, before the two newcomers were referred to the couch just like we had been. The dark one swapped his regular spectacles for reading glasses as he soon got his own note to fill in. The journalist smiled and asked us if we spoke English, to which both me and Kristian responded yes. The journalist introduced himself as Sandor, a Hungarian war correspondent who followed the war as a freelance journalist.

"Off the record," he said, to clarify that he did not intend to write any of this down.

He asked us what we were doing there, since we didn't look like the others he had seen—we actually looked like soldiers. We briefly explained that we had both completed military service long before this. Sandor repeated that he could see it, but thought it hardly explained our good English. He asked if we had studied at university or college. We told him no, none of us had any higher levels of education outside of the military.

The journalist looked confused. "I don't think I've come around any Ukrainian who speaks as good of English as you two, and especially none with military experience, as well."

"Eh, we're not Ukrainian," I said.

"You are not Ukrainian?" Sandor asked, increasingly confused.

"No, we only arrived from Sweden yesterday."

Sandor's eyes widened. He slammed his arms out in front of us from his armchair on the other side of the worn table, where the dark one was still struggling with all the tricky questions on the form.

"You're from Sweden?" he shouted. "What are you even doing here? How old even are you?"

I replied that it's exactly as we talked about earlier, when he mistook us for Ukrainians. Russia had begun to take bites out of Europe, and other countries were next in line. We saw the war as a duty—foreigners or not.

"You have to get out of here!" he shouted. "It's hell out there. You idiots simply don't understand!"

The dark man left his form and told Sandor he was going out for a smoke, then wanted to find somewhere to eat. Sandor, still up in flames about everything, said he would accompany his interpreter, who apparently also intended to join Azov as a volunteer. Before he left, however, he made sure to quickly write down his phone number in his notebook, tear the page out, and give it to me.

"Give me a call, okay? I want to meet you again before you go to the front line, you crazy fucking Swedes."

They left, and we stayed behind again. Kristian's laughter grew increasingly nervous as we talked about our most recent meeting. I glanced at the wall to the right of the couch. It was like a small makeshift shrine with photographs of the Azov fighters who had fallen so far. There were not that many—only about ten or so, but I recognized some of the faces from the news on the war and remembered reading stories about some of them online during the summer and autumn. Beneath them, a few candles burned. The floor was littered with a few flowers, too. Both older and younger men were pictured. Mostly young, some just eighteen years old. I imagined that my own photograph might hang there one day. The anxiety that had been eating away at me was almost gone. I could see my face among the dead on the wall beside me, and I was neutral toward it. For all that mattered, I was hanging there already.

I received a text message on the phone. It was Mike. The plan had been that he would meet up in the morning, and now it was already evening.

"I'll be there in five. Tor will come, too," the message said. About twenty minutes later, the intercom rang again. I heard the door open and got up to go see. Two men entered into the dimly lit hallway. I only saw the face of one of them. It wasn't Mike, and I had no idea what this "Tor" guy looked like.

The man, who looked to be about forty, was slim but fit and dressed casually, yet still like a bit of a snob: neon orange, white, and light blue sneakers together with some kind of well-ironed trousers and suit jacket. His face was tanned and looked worn with a stubble that alternated between dark and gray. His hair was dark and short, a few centimeters long or so. His blue eyes reflected the light of the dull ceiling lamp above, and even though I couldn't really see them, I felt that he was looking at me.

"Tjena!" I heard suddenly, delivered in what sounded like a clear Södermalm dialect—an area in Stockholm known for its politically correct and terrible upper-class, inner-city yuppies. The man quickly moved up the stairs and stretched out his hand.

"Tjena. Tjena! Tor! Tor. I am Tor. Hello!"

Tor and Mike had returned to Kyiv a few days before Kristian and I arrived. A cease-fire agreement between Russia, the West, and Ukraine had just been hammered out in Minsk, the capital of Belarus, and they had been given leave from the fighting. Mike had made himself very famous by marketing himself on social media, primarily on Twitter, but also through appearing in countless interviews. In the West, the reactions were mixed; Swedish media, in particular, simply loved to hate him—the Nazi warrior. In Ukraine, on the other hand, perhaps not surprisingly, he found massive popularity. So much so that he and Tor didn't have to take a train back from the front line; instead, they got a lift in the local oligarch Serhiy Taruta's luxurious private jet. It did not end there, either, as they—for the moment—lived in a luxury villa outside the city center, generously lent to them by a wealthy Ukrainian sports star.

"I was going to fart in bed this morning but ended up shitting myself," Tor laughed, telling how his stomach was in total revolt due to weeks of stress and the prolonged lack of normal food.

"It just squirted! Everything flowing out of me! Shit everywhere!" he continued. "It's like being a small child again!"

Mike said that it would probably be a while before our papers went through the bureaucratic system and we were approved for the front. He suggested that we switch from Hotel Ukraine to a cheaper alternative and started looking for one on his phone.

"If you don't want to live here, of course, at Kossatskiy," as Azov's local HQ was called.

"No, fuck that, man. Find them a hostel," said Tor. "It's revolting here."

We had been around and looked at the place. There were bunk beds available which were not entirely awful, but the standard of everything was very Spartan, to say the least. Kristian was particularly hesitant after visiting the toilet facilities, which may not have been entirely up to five-star standards. We agreed that a cheap hostel would be a fair thing, in case we had to spend several days waiting in Kyiv.

"This is a good place!" said Mike. "It's called ZigZag. Not far from here, just past Arena, next to the university. It's within walking distance to the Maidan."

We decided to take a taxi there, saving us the inconvenience of walking with our heavy rucksacks. Tor fussed with the driver, who wanted forty hryvnia instead of the usual thirty-eight. I tried to say that the difference was comparable to a penny and not worth arguing about. Tor, on the other hand was very determined, not for the sake of the money, but for the principle that no one would scam him, no matter how big or small. After some deliberation, we finally got a taxi for thirty-eight hryvnia and left for ZigZag. Kristian and I booked a room with four beds that we got all to ourselves. There were no windows, and it felt a bit like a scrub, but there were toilets and showers right outside, and everything was clean and tidy. While we were arranging our accommodation, Mike took the opportunity to ask the young girl at the reception if she recognized him. She didn't, but after explaining he was the famous Swedish sniper from Azov, her face lit up, and she asked if she could take a selfie with him. It was obvious that the image of Mike as someone who enjoyed all the attention he got was not just something the media made up—he really did love being the center of attention.

Kristian and I met the journalist, Sandor, and his dark friend the next day. The dark one was a Georgian named Gia. He had also just arrived in Ukraine from Holland, where he had lived for a long time with his wife and children. As a child, he had tried in 1991 to take part in the short war in northern Georgia against the separatists in South Ossetia after his father had gotten involved. He later tried to return and take part in the Russo-Georgian war in the same region back in 2008, but never had time to take part in any battles. The Georgian army had collapsed in less than two weeks against Russian supremacy. Now, like many other Georgians, he had gone to Ukraine seeking a chance for revenge.

Gia's English was deficient, and he did not seem particularly bright. He did, however, make up for this with a funny, confident, and charming personality. Sandor was a bit more of a thinker. Kristian and I spent a lot of our time waiting for a go-ahead with these two, and in the evening, Mike and Tor also used to show up to make company. After having dinner at a restaurant underground beneath the Maidan, Sandor would promptly invite us all to a shot of horseradish vodka.

"That's the best thing Ukraine has to offer," he said, leaving me thinking only about how terrible the rest of the country must be. After tormenting myself with that abomination of a drink, I needed a snus to relax myself.

"Is that snus?" Sandor asked, now in complete symbiosis with the horseradish vodka. "I've heard of that! It's Swedish tobacco you put in your mouth, right?"

I replied that he had been correctly informed and handed him the small box. It was Probe Whiskey snus, loosely scattered inside the box rather than in portioned-out bags. It was my personal favorite for many years and the strongest standard snus that existed at the time.

"Would you like to try?" I asked, but before I had time to explain to him how to shape it into a small ball, let alone do it for him, the childishly eager drunk had dug his Hungarian fingers straight into the can and run a not inconsiderable amount of the contents into his entire mouth. With a troubled grin and teary eyes, he fought bravely for a few seconds. Soon, however, with snus everywhere and vomit dripping from his mouth, he found himself defeated by the tobacco and retreated hastily toward the restrooms.

He returned soon after, whiter than usual, with a sweat-stained forehead, saying, "never again."

I offered Gia a try. He analyzed the box for a while and looked at me.

"Make one for me as you make for yourself," he said, "but small."

I did as he asked and baked a small one suitable for a child or beginner and gave it to Gia, explaining how he should bring it up under one side of his upper lip. Maybe he wasn't as stupid as I had initially thought. Even though his attempt fared better than the Hungarian, even the great Georgian had to give up after a few minutes of complaining of burning sensations and dizziness.

Kristian was increasingly nervous. He held up appearances, but beneath the surface, the same anxiety I had gone through before was boiling within him. Tor saw it, too, and when I went away for a smoke, he followed me to ask how my friend was doing.

"You guys don't really have to do this, you know," he said. "None of us really have to do this. You can go home if you want, and that's all right. Nobody here will look down on you guys if you want to go home."

He told me about how they'd ended up barraged by the Russian artillery just prior to the Treaty of Minsk and the cease-fire. Azov had moved toward Novoazovsk, a small town in the south very near to the Russian border. Suddenly, the vanguard there had come into contact with regular Russian army tanks and mechanized infantry rolling across the border, headed west. Azov, like the other volunteer battalions, lacked heavy weapons and were forced to withdraw. They regrouped in defensive positions at the small village of Shyrokyne between Novoazovsk and the important Ukrainian port city of Mariupol. There, all hell had broken loose. Suddenly, the artillery had rained down on them—not sporadically and randomly, but massively and precisely. There had been heavy mortars and MLRS rocket artillery. Many had panicked and broken down from industrial-scale violence. Their squad paramedic, an Irishman called Papi, had been sitting in a car which crashed during the retreat. He was a mercenary with many years of experience, but amid the car crash and the artillery, even he simply fled, despite a wounded friend remaining in the car. Papi just left, even leaving much of his stuff behind.

"If you don't really want this, you do not need to do this," he repeated.

I responded that I was well aware of the risks, but that he might want to talk to Kristian about it.

The next night, we kept it a bit more chilled, just me and Kristian. I ordered a beer for me and a Long Island iced tea for him, the most disgusting grog I knew. Now face to face, I explained everything to him: how bad I had felt before and how bad I could see he was now. I told him, just like Tor, that it was okay. No one would look down on him or think any lesser of him if he left at this point. He had nothing to prove. My situation was different: I had been exposed, and everyone already knew I was down here. But for him? No one knew. He could return to his life as if nothing had happened, as if he had been away for a few days on any weekend trip.

Kristian booked a flight ticket home. The aircraft would leave early the next morning. The alarm sounded and he got his things in order. We said goodbye and good luck to each other. I laid back down and continued to sleep, greatly pleased in knowing he was going back home to his normal life instead of staying against his inner wishes.

CHAPTER 5: STRANGER AMONG STRANGERS

I spent the next day—my first one alone—listening to music during my walks through central Kyiv. From the hostel, it wasn't far to the large Taras Shevchenko Park, which was directly adjacent to Kyiv's large university also named after Shevchenko. The building was large and powerful in its Greek style, with tall pillars at the entrance. It was also colorful, contrary to what might have been expected, painted entirely in a strong wine-red color. The pillars and the sharp edges shimmered in the sunlight as everyday people passed in the beautiful park outside. Even though it was September, it was still warm outside. At home up north, I would probably have been forced to bring out at least an autumn jacket by this time, but here in Ukraine, summer was still hanging in there.

As I headed in the other direction toward Independence Square, I just followed the main road toward Arena as the first landmark. It was a large, rounded building that was encased in its own part of the surrounding roads. On the inside of the large, circular, arena-like structure was an open, likewise circular plane. From there, the center was surrounded by expensive and luxurious taverns, nightclubs, and the occasional strip club.

One had to be able to navigate the underground to pass the Arena. Large stone steps at the end of the sidewalks led down below beneath the many roads. The underworld was as big as the world above, and it was easy to get lost when there were no real landmarks to follow. In practice, it was a single huge shopping center, and if you made a wrong turn among the myriad of small shops, you ran the risk of getting completely lost very quickly. At the point where you then finally found an ascent to the fresh air above, it was far from certain that you would recognize your surroundings once you hit street level. Several times I had to try to turn around to retrace my steps before I finally got up on the right side of the road, from where I could continue my way toward Independence Square.

Soon, I came to the large Khreshchatyk street that went through this part of the center. It was several lanes wide, and on weekends, it was often closed to car traffic, which created an unusually free and vibrant city center. On both sides of the wide road, it was surrounded by large buildings, many built with strong stone in a mighty Stalinist style. Entertainers and musicians performed at regular intervals, ranging from simple troubadours to talented classical musicians with key harp-like instruments. It was difficult to imagine this country was currently at war.

I had arranged a meeting with Sandor and Gia again, at the same underground restaurant where I had previously introduced them to the wild nature of snus. Gia had timely forgotten that he was married and tried his luck fishing at the bar. Sandor had long since stopped trying to police him morally and was standing alone smoking when I showed up.

"Where is Kristian?" he asked.

"He had to go home," I said. "This thing didn't really suit him."

Sandor lit up with a big smile and clasped his hands together as if in a prayer. "Oh! Yes! That's such a goddamn good thing to hear! I told him to go back so many times, but I never thought he would!"

"Huh?" I replied. "You talked to him about this, as well?"

"You can't be pissed with him," Sandor said calmingly. "He simply did not want to do this. He was only staying because of his bad conscience."

Sandor leaned forward, put his hand on my shoulder, and took a deep breath from his cigarette, smoke pouring out his nostrils while continuing: "He felt like such a traitor, leaving you alone here. I told him he was too young, and…"

It dawned on me that we had both thought and acted on the same goal, but completely independently of each other. I therefore cut him off mid-sentence and explained that I also had told him to head back home. The Hungarian stepped back, surprised.

"Really? So, then you are not pissed off?" Sandor asked.

"No, not a bit. Why would I be?"

"He was so worried that you would see him as a traitor. That was the only thing keeping him here."

We laughed at the irony of it all for a while, the sheer mess of misinterpretations before we both, relieved at having resolution, smoked our cigarettes and stepped into the bar. Gia was still busy trying to charm a woman some distance away.

"Damn," Sandor sighed, "I think that's the third night. You know, I only met that guy the same day I met you guys. He says he is here to fight, but I'm getting bad vibes from him."

We each ordered a beer and sat down at a table.

"So," Sandor began diplomatically, "when are you going home, then?"

I was a tad unsure if he was serious or not. After silently trying to resolve this for a moment, I explained my situation in a little more detail. How, unlike Kristian, I was not here without people already knowing, and more importantly, that the authorities did, too.

"You do know these guys are Nazis, right? You can't fight with them. Why don't you join the army instead?"

"I've looked into the army, but it's no use. Besides, they're useless. Azov seem to be good, and I'm joining them because of that. Besides, whatever they're called, I don't care. I'm already branded like cattle, and I'll never get away from that. Even if I went home now, my life is already fucked. They will always call me a Nazi, as well," I explained, "and if I go home now, everything has been for nothing."

Sandor listened with a worried expression to my description. "But you're not a Nazi?" he asked. "You don't seem like one, anyway."

"Who and how I am doesn't matter," I said. "The only thing that matters is what stamp I have, the brand, and I already got that."

Sandor tapped his finger on the table and scratched his forehead, trying to find a solution to a situation while struggling to even understand it.

"All that's irrelevant, though," I continued. "I came here because I wanted to be here, and even if I could go home, I still wouldn't do it. I came here to fight, and I'm staying to do it."

Sandor looked at me resignedly, looked down into his glass before finishing it. He put the empty glass down on the table and looked at me again.

"Yeah, I get that, and I'm not against it. They're fucking Russians. And you're a crazy fucking Swede," he said. He stood up and continued, "I'm going to get more beer. Do you want a horseradish vodka, too?"

I said no but found myself getting one, anyway. Shortly afterward, Gia joined us back at the table.

"It's not going well," he said in his weak English. "The women are so beautiful, so beautiful. At first it goes well, but after a while, everyone gets quiet and then they go."

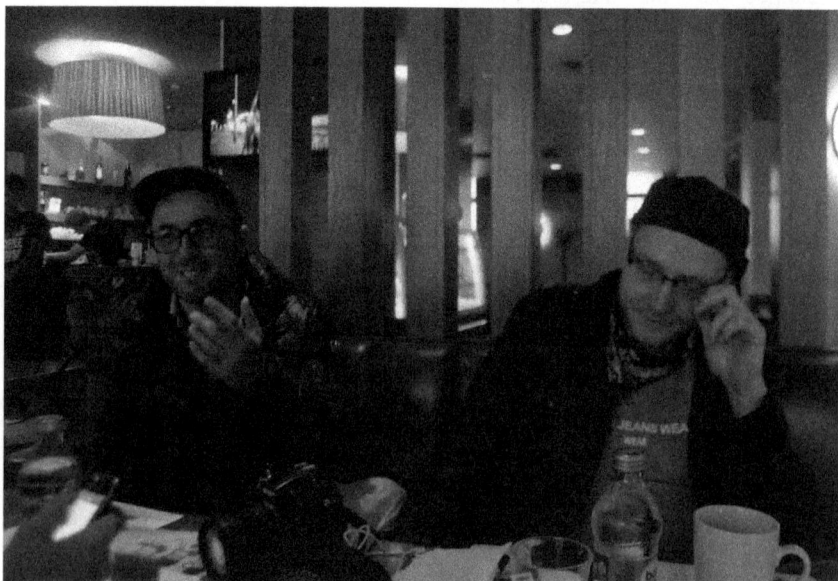

Gia and Sandor at the bar.

He grabbed his beer and drank. Sandor sighed deeply, but suddenly he looked at Gia with a different look, as if he had come up with something.

"You're a huge fucking idiot, Gia, you know that?" he said.

Gia put down his beer and looked surprised.

"My wife is home. I love her, and she knows nothing. And I love her. Why am I an idiot?"

Sandor sighed again. "You're a huge idiot, Gia, because you still have the ring on your finger."

When most of the taverns had closed, we parted and I headed back to the hostel. It was dark but still warm outside. Just as I was about to enter the code for the gate, I noticed a bright light in a basement staircase down to the left of the entrance. The green sign above the stairs read "Gorky Bar," with a picture of the old Russian communist author of the same name next to it, holding a beer. I went downstairs to see if it was really open in the hope that I might be able to get one last beer before bedtime.

The door was Indeed open, and I stepped inside to see half a dozen small tables and chairs with a small bar counter in the middle. Most of the interior was dark wood, and behind the counter, music was playing through a small TV connected to a laptop. The overall impression I got from many other small details indicated that this was a student bar, which fit well with the fact that it was close to the university. The students, however, seemed to have left. Besides myself, the only visitor was a middle-aged woman sitting alone at a table in the corner, sleeping with a beer still in her hand. The only others were the bartenders. One was a slender, dark-haired man around his mid-to-late thirties. The other, a younger woman with light red hair. The woman greeted me in Russian as I stepped through the door.

I sat down at the bar, looked at them both, and asked if they spoke English. The male bartender looked at the young woman to refer the question in that direction, in that way answering for the both of them. The reddish woman looked at me with her big green eyes and replied a little cautiously that she indeed did speak a little bit, but not very well.

"Enough for me to order a beer, anyway, I hope?"

She laughed and replied, "Yes, yes, I can do that," before she informed me that they would close soon, so it had to be the last.

"Fine by me," I said, explaining that I also needed to wake up at some point the next day.

In the absence of other guests to serve, we had plenty of time to try to communicate with each other, where the woman also acted as an interpreter between me and the male bartender. The woman asked me if I was here as a tourist, or if I was here on another business.

"I have volunteered for Azov," I replied.

"Azov—it's the army?" asked the woman. The man also opened his eyes widely at the sound of the notorious battalion.

"Yes. I'm here to join the war."

"You're going to war?"

The woman's eyes widened even more, and the man turned to her to confirm if he had indeed heard correctly. I tried to explain, now in the simplest English possible, how I ended up here, and the woman listened carefully. From time to time, she took short breaks where she'd translate what I'd said to her male colleague. The man placed three small glasses on the counter.

I tried to say that I hadn't ordered anything more. The answer came in Russian. The young woman quickly tried to translate to me as the man opened a bottle of vodka from under the counter and started pouring.

"Dima says we should drink now, all three."

I tried to stop him and said that I had already been out all evening. To turn on the vodka now would be an unwise decision. Dima, as he was apparently called, firmly waved away any and all of my attempts at restraint. He took a glass. The woman and I understood that it wasn't worth resisting, thinking a single little one couldn't be too bad. Dima raised his, and we followed. He gave a short speech and emptied it, after which the woman and I did the same. After coughing a little bit, I asked the woman what he said. She struggled to try to explain and soon gave up, despite my insistence she continue trying. "It was very beautiful" was the best description I got.

Dima, who during the ultimately fruitless translation attempt had been away and thrown the alcoholic woman onto skid row outside, had now locked the door. It was four in the morning, and the bar had closed. The young woman who introduced herself as Zenya presented an ashtray on the bar counter and lit a cigarette while Dima poured three new glasses of vodka. My continued protests were something he no longer even bothered to wave away. Instead, he just replied, "I don't speak," and kept pouring.

As the alcohol kept flowing, the language barriers came down. Zenya explained that she was Russian and had not lived here for very long. She had no real opinion about the war itself—who was right or wrong. Previously, she had held an image of the Ukrainians as a bunch of provocateurs and assholes who created problems where none existed, but since she moved here, her opinions had changed. She had instead begun to see the conflict from their perspective, as well. Now she just saw it as unnecessary, created out of nothing: "I just pray it will end now, before it gets worse."

Dima was from Donetsk, the now de facto capital of the Russian-backed separatists, where he had a wife and children. Zenya had a hard time translating everything. I understood that a lot must have disappeared in translation, but at least I understood that he now worked here to be able to support his family on the other side of the front line. He didn't seem to care about borders, flags, and national anthems. He was just a simple father who wanted a roof over his head and food on the table for his small family. However, he respected those who went to war because they thought it was right, and he wanted to show this. He not only constantly poured the vodka, but also my beer, which never went dry. By the time I was finally forced to thank him for the hospitality and hesitantly head toward my bed, the sun had already begun to rise.

After waking up the next day, I went back to the bar. Zenya and Dima were already behind the counter.

"Hello again! You still live?" Zenya asked with a big smile as I slowly moved forward to sit down at the otherwise empty bar counter.

"Yes, but only so much," I said and turned to Dima, staggering on my very poor Russian: "Never again vodka, please."

Dmitriy smiled as he polished the beer glasses, and Zenya leaned over the bar toward me.

"But maybe another beer? It's past twelve."

"No, thank you," I said. "Not yet, but a cup of coffee would be nice, if you have."

Zenya withdrew and looked ashamed. "No, unfortunately, we have no coffee here."

I pointed to the kitchen. "I see the coffee maker over there. Of course you have coffee."

Zenya's eyes fell down at the floor. "Yes, yes, we have coffee, but it's very bad." She hurried to the other side of the bar and dug out a small card which she handed to me. "Here, this is a good coffee place. You see the address here. It's just down the street a little."

I pushed the card back to her.

"If the coffee is this bad, I really have to try it," I said. "Pour me a cup now so I can see if it's as bad as you say."

She took a step back, turned around, and put the card back in the pile where she had taken it. Then she went out into the kitchen and poured the coffee into a simple little white cup on a plate and fetched it for me. It was plain black coffee, straight from the glass bowl. No freshly done "Americano," as was otherwise served in restaurants and cafes in the city.

I took the cup and tasted it. The taste was distinctly dry, with hints of old bark—quite rough, and definitely very cheap. Both the aroma and the taste were very familiar to me.

"Not very good, huh?" she asked.

"This tastes exactly like Finnish coffee!" I replied, without even trying to hide how excited I was by it. I explained how people in Finland drink more coffee than any other people, but at the same time drink perhaps the worst coffee in the world. I finished my cup and asked for another one.

The Finnish coffee was a boost of energy like nothing else, and after finishing the second cup, I asked what I owed. Zenya said it was free, as they did not normally serve such brew, anyway. I checked my wallet and found a fifty hryvnia note, far more than what a cup of coffee cost at a cafe.

"Free," I said. "For a cup of authentic Finnish coffee in Ukraine, I will happily pay extra." I then left the note under the cup and bid goodbye to the more than ever confused young Russian girl.

I met up with Tor and Mike outside the university. As far as combat equipment came, I didn't have much with me. Tor also needed to supplement his current setup. With Mike as our guide, we went to a major airsoft and militaria store in central Kyiv, Militarist. The supply was sporadic, and it was obvious that business was going well. Equipment and clothes were sold faster than they managed to fill the shelves.

I found a very simple plate carrier, a simplistic vest mainly designed to put bulletproof panels, or "plates," into—one over the chest, and one on the back. It lacked any form of its own shrapnel protection but was light and airy. I had understood that the plates were something I would be assigned at the front and asked if they would fit in the carrier. Mike concluded that they would probably be fine and I gave it a further review. The carrier really was in its simplest form—a chest and back piece with MOLLE-type straps over both, and simple but sturdy clip-on straps at the waist sides. Its simplicity meant fewer things to break, and what few straps that held everything together seemed sturdy. Tor bought a similar one, as he also wanted something simple and small to replace his current heavy Kevlar combat vest.

"When we ran away from the artillery, I was close to calling it quits, giving up several times," he said. "The goddamn vest was just too heavy. Had it not been for the fact that it was difficult get out of, I would have thrown it away during our run."

I had told everyone that because I had previously mainly been a machine gunner and enjoyed it, I would like the same role in Ukraine. Mike had given an okay to this request, and I therefore arranged my setup with pockets suitable for machine gun belts instead of rifle magazines. The mixed and meager offerings meant I had to improvise a little bit, but after a while, I managed to find what I assumed I might need. Mike asked me if I knew how the PKM, the standard Soviet-era machine gun, worked. My experience with it was limited, but I had trained with it in Finland on the rare occasions we used it instead of the homegrown KvKK M62 guns.

"It was simple enough to learn. I don't believe there was much to forget about its operation," I said.

"Do you know any other weapons from around here?" he asked.

"Kalashnikovs, of course. We also trained with the NSV, the heavy machine gun. This was only briefly, though, and I don't really remember much about it."

"It doesn't matter. I'll talk to some people you can refresh your knowledge on it with. Then I'll fix it so you get to train on the other heavy weapons, as well. For example, those 20mm automatic cannons in the APCs."

I had a bit of a hard time placing what he was talking about. The 20mm was a typical Western caliber and weapon—not as common in the Soviet arsenals.

"You mean the KPV? The 14.5mm heavy machine guns in the BTR IFVs?"

"Yes, exactly that! We'll fix it so you can learn them."

Mike picked a small notebook out of his back pocket and wrote something down—I assumed a reminder about everything we just discussed. I still thought it was a bit strange that he would mix such weapons, for someone who had been here for as long as he had, and in such a position. I thought it should be common knowledge that Soviet armored vehicles had 14.5mm machine guns in their turrets, not 20mm. Or was it just me who allocated brain matter capacity for such? I discarded my questions on the subject and enjoyed myself with the thought that I would soon get to play with really big guns. Or at least I would if I actually got sent to the front at some point.

Mike and Tor accompanied me to the hostel where they also decided to check in to my room. The other beds were vacant, and it was otherwise a fairly comfortable existence, other than that it was properly enclosed with no windows. After leaving mine and Tor's new equipment, we then went out to find something to eat. After a walk toward the Maidan, we turned into a tavern and took a table. The waitresses were slow, even by Ukrainian standards, and we had plenty of time to talk while our stomachs churned.

The group I was to belong to was newly created. Earlier, everyone had belonged to something called the "Båtsman Group." It was in the same class as a company by size, and the name came from its Belarusian boss. As a child, he had always liked the dog in "Seacrow Island," a Swedish 1960s TV show for children, and therefore he himself now went by the very same name: Båtsman, the Swedish name for the naval rank of Boatswain. Mike had recently had the chance to break out of this unit and begun to create his own company, specially designed for the foreign volunteers. Even though they were still very few, there was a hope among some that the number would grow.

There were two more Swedes who had been down there a long time already, "Sonic" and "Mikola." Sonic was in his early thirties and had a background as a conscript reconnaissance soldier. It was, however, highly uncertain whether he would be back in service, as he was now hospitalized in Kyiv. He was the man who had been injured in the car crash during their retreat through the artillery barrage near Shyrokyne, the one whom the Irishman Papi had left behind. The thick-skulled Swede, however, had taken care of himself, crawling out of the crashed car to continue on foot, all the way back to friendly lines. I was reminded of how I recently had seen articles in the Swedish evening news media about how a Swede had been injured during fighting in the east. According to the headlines, he would have shouted "Slava Ukraini"—Glory to Ukraine—as he was carried into the ambulance.

However, this was a misunderstanding—either unconsciously, or created as propaganda. As Sonic arrived on his own after his long and lonely march, during which he had filmed his bloody face with the mobile phone to try to make a last "I love you" message to his young son, he had indeed shouted the words as described. This was, however, to identify himself as a friendly, a cry so that his own friends would not open fire on him, being the last soldier coming back from Shyrokyne. Later, the doctors would say that Sonic had walked all the way, not only alone, but with a fractured spine. It was highly unclear how restored he would be after such hardships, if he would ever walk again, let alone fight.

The second Swede, "Mikola," which he chose to call himself, was also in Kyiv fixed on leave just like Mike and Tor. Mikola had not done any military service but had spent a short time in the French Foreign Legion. Together with Mike, he was the Swede who had been in Azov the longest, before the unit was even founded. He had come because he was interested in the revolution. When the war began, he, as he said it, more or less just slipped into it. Unlike Mike and Tor, Mikola had not been riding any private jets from Mariupol to Kyiv. He had instead taken the train from the front together with the one they called "the Greek."

41

"The Greek is a bit special..." said Mike before laughing.

How so, I wondered.

"He's a former legionnaire and all that, but he's also very...Greek," said Tor.

I didn't quite know what to make of the information and asked that them both to explain in more detail.

"You'll see, to put it simply, when you meet him—" said Mike.

"Or hear," Tor filled in.

"He's just a very...different character," said Tor, looking at Mike laughing. "Yeah, yeah...you've never met anyone else like him before, I'm sure."

"He's just not like everyone else, but he's a very fine soldier," said Tor. "Remember that, and don't let his personality take over your first impression."

I didn't understand much of what it was about, other than the fact that the Greek was a difficult person to define.

There was a Frenchman left in the east who was also injured during the retreat from Shyrokyne. He was called "Barzuk," or the Badger, and had fortunately not been as badly blasted as Sonic. It was mostly minor shrapnel wounds, but some of it concentrated in his face around his eyes. Therefore, he was also at the hospital, where the doctors were trying to save his sight.

"He was fat as hell before he came here," said Mike, "fatter than I was! Båtsman put him as being a chef!"

"Did he have any military experience?" I asked, to which both replied no.

"I think he was a lawyer or something," Mike laughed while Tor continued the description.

"He was struggling, sure. However, he always did as he was told without protest. He has lost a lot of weight since then, and he has been doing really good."

Mike agreed and said that he had the potential to become a good soldier after all.

"I just hope he doesn't lose his eye. That would just suck," said Tor.

"Then there are a couple of Belarusians who came quite recently," said Mike. "We don't know them very well. We only call them Bulbash Bolshoi and Bulbash Malinki so far."

"Bulbash—ain't that like, potatoes?" I asked. "Yes, one is long and thin, and the other small and wide—that makes them Big Potato/Small Potato."

Mike himself I already knew a lot about from the media. He had, after his conscription as a sniper, served as a professional soldier for a number of years. Then he had gone over to the Home Guard, the same battalion that I was in since we both originated from the small town of Sundsvall. I had never encountered him before, though, as he was just kicked out of the armed forces about the same time as I got in. In connection with being charged with assault, his involvement in the Swedish Resistance Movement leaked out. SMR was not only considered a neo-Nazi movement, but rather *the* neo-Nazi movement in Sweden at the time, and as such, his career in uniform had been definitely ended—at least in Sweden. Many were those who had a lot to say about him, but there was rarely any substance to it other than him being a Nazi, which is why I didn't put much weight into it.

"Those young guys, the Bulbash boys, I mean," Tor filled in. "I'm honestly a little worried if they even know what's front and back of a Kalashnikov."

He said that this, the volunteer battalion's poor standards, frightened him. During his first mission, a soldier had been shot in the throat from friendly fire right behind him, leaving him with blood splattered all over.

Tor was significantly older than I first guessed. With his forty-eight years, he was the oldest man in the group. He was originally from Kiruna, in the extreme north of Sweden, but had grown up working class in Stockholm's suburbs (hence his dialect), where he had engrossed himself in skinhead subculture. He brought this sting with him as he returned to his child home during military service as a Lapland Ranger at I22 regiment. There he had, among other things, caused trouble when kicking the door in on the communist party newspaper "Norrskensflamman" offices during a leave.

After his youth and military service, he had changed his life significantly. After trying life as a simple worker and family man, he had divorced and focused on entrepreneurship. He went from working class and skinhead to presenting himself as an educated libertarian. While I had been dicking around as a nineteen-year-old, throwing smoke grenades at the Sundsvall Street Festival many years earlier, he had been on the stage promoting his IT companies. He later settled in Tallinn, Estonia to expand his business and even venture into the construction industry. It was there in Estonia, in the direct vicinity of Russia, that he saw what happened in Ukraine as a direct threat. By now, his businesses were mostly running independently. His children had already reached adulthood. Therefore, without any other serious commitments, he had left everything he had and traveled to Ukraine in late summer to pick up a weapon of war for the first time since 1985.

The food had served us well, and the cheap beer continued to do so, as well. We sat in a place called Porter, a chain of rock bars that littered Kyiv. The one near the Maidan, next to the metro station on Kyiv's central street, Khreshchatyk, was something special. The general British working class style and punk-like feeling, along with the furnishings, were supplemented by what looked like a museum. Not only on the underground culture in general, as is often the case, but with artifacts from the Maidan as well as the ongoing war. From the ceiling hung flags: Ukrainian, Georgian, and Chechen. There were also well-known military units like Azov's and Right Sector's—the Ukrainian nationalists' black and red-colored flag. Even more obscure groups were represented, such as Misanthropic Division's black flag with white Kalashnikovs and a Totenkopf.

The walls were full of photographs, signs, and knick-knacks from the Maidan and the war—clothes, helmets, riot shields. In one place, spent casings from a 30mm grenade machine gun had been glued together to form a wolf hook and the text "A3OB." Opposite on the other side of the aisle, together with tubes from empty RPG22s and other similar things, hung a 120mm heavy mortar shell from a string connected to the tail. I saw that the impact fuse was missing, so I stuck my finger in the body itself to feel—mostly as a joke for myself—if it was live. To my surprise, my fingers met the explosives still inside. The fucking thing was live! I loved this bar!

The restaurant was full of people in motion, and on the small stage, a live band played classic rock and blues music mixed with what seemed to be patriotic songs, both in English and Ukrainian. The man who sang and played guitar wandered around the stage like a crooked giant. It was immediately apparent from his posture and abnormally shaped face that the entertainer was suffering from gigantism. Despite his grotesquely big fingers, he managed in some strange way to strike the chords right. His voice was phenomenal as a result of the disease, and so was his singing. Whether it was a Ukrainian nationalist song or Dire Straits, he performed the song with a darker and deeper voice than any other white man could ever muster.

Other Azov members dropped by, and our company grew bigger and bigger. Most Ukrainians wore Azov's original all-black uniform. It was generally no longer in frontline use, and had instead been altered to the role of an on-leave uniform. One of the men had come from Russia as a volunteer, like many others like him. Putin's grip on Russia was strong, and it was difficult to oppose him without major risks—including direct danger to one's life. Therefore, many Russian nationalists, for various reasons, had seen an armed struggle for Ukraine as their main, perhaps only, way of seriously doing anything against a regime they felt oppressed by—even though it also meant that they would be forced to raise weapons against their own countrymen.

I talked a little with one who called himself Homer. He was in his mid-forties, sinewy and slender in body. His face was broad with an unusually thick base for his nose, which then narrowed between the squinting, deep-seated little blue eyes. His skin was very light, and his hair was light to the point of being almost white. I asked Homer in English what he was doing or working with in Russia before coming to Ukraine. My question in English confused him. "What did you do for a living?" he interpreted literally, as in what he did "for life," for the sake of enjoying life.

"In my hometown, Moskva, I stab niggers and throw them in canal." He looked down at the glass in front of him and was silent for a moment. "Now, instead, I have come here, and shot my friend. It feels…different…"

I was reminded of Tor's story about how the man behind him was shot in the neck during a friendly fire incident and now linked it together with what I just heard.

Another of the men—Sober, as he was called—sat across from me. He was broad and tall. His beard was black, and both his eyes and smile showed confidence. I had met him once before, together with Mike in a nearby park a few days earlier. When Sober was going to buy a cup of coffee, he had been quick to offer both Mike and me, whom he did not know at all, each a cup from his own pocket, as well. Together with his all-black Azov uniform, he carried a small waist bag. To me, this thing was something that my mother wore when I was kid, and then continued to wear long after it was no longer fashionable. With the exception of immigrants in Sweden slinging drugs in recent years, Ukraine was the first time I saw people carrying it unironically since the 1990s. After a while I could not help but to ask about it.

"You have decent pockets in both your jacket and trousers. What do you really need a waist bag for?"

Sober looked back at me, confused at first.

"Waist...bag? Ah! This, what for? I show you."

He unbuttoned the belt straps and cleared the many empty beer glasses away to make room. He lifted the bag onto the table and pulled the zipper back, exposing the contents. I leaned forward to get a good look: an old style of cell phone, a keychain, some hryvnia notes in smaller denominations, and three hand grenades.

I looked up at Sober, where he allowed himself a slight grin beneath the thick black beard. The natural consequential question, after seeing that the contents included hand grenades—grenades *plural*—was, of course, why he was carrying multiple hand grenades.

"Is simple!" he replied with a lively tone of voice. "If they try take me alive, I say fuck you, pull pin, go blyat."

With that, I understood that kidnappings were something which occurred around here, and found myself pleased with that answer. I could imagine worse excuses to wear an otherwise pretty gay-looking fanny pack, and with that, my curiosity had shifted focus. I had never seen a real Soviet hand grenade other than in pictures before, so I asked Sober if I could take a closer look. With a simple combined gesture of hands and face declaring the request approved, he picked up one of them hand handed it over to me.

All three were RGD-5s, and I immediately recognized them as such. The RGD-5 together with the F-1 were the two most common Soviet and Russian hand grenade models. The F-1 was somewhat similar in shape and size to a really thick spruce cone. It was a defensive hand grenade at half a kilo in weight, holding a slightly smaller explosive charge in favor of a thick and heavy metal shell which created a lot of shrapnel spreading over long distances. The idea being that, while on the defense, you do not have to throw long distances and therefore can afford to focus on power and a larger blast radius instead.

The RGD-5 I held in my hand was its smaller, offensive cousin. The outside of the grenade body was round and smooth, only somewhat larger than an ordinary chicken egg, and it discarded the thick and heavy cast iron shell for a simple, thin metal casing, bringing the weight down to half that of the F-1. Instead, it compensated with almost twice as large an explosive charge. As an offensive grenade, it is more important that it could be thrown a far distance accurately. It could also be argued that if it ended up sending shrapnel far away, it could pose as much or even more danger to the man throwing it than the intended target. Due to its larger explosive charge, the RGD-5 had an even greater effect at a radius of a few meters than the defensive grenade, but unlike the F-1, it was virtually harmless outside it.

For Sober's purpose, the RGD 5 was an excellent choice, especially as he carried three. In case someone would pull him into a car, he would easily blow up both the car and everyone in it without any chance of survivors, but with little risk for anyone outside.

I examined the grenade carefully. The pin was of an extremely simple design and was seated firmly as it was split in two metal wire pieces, each having been bent outwards to lock on the other side of the fuse head. It would take serious force to pull it, but it would also be easy to bend the two pieces of pin together, making it an effortless pull. The detonator, pin, and spoon came as a single unit and was easily unscrewed from the body of the grenade. As I screwed the grenade back together again, I noticed how the guests at the tables around ours quickly but discreetly began to move away. I was wondering if it was closing time already; it wasn't even two o'clock in the morning. Though I soon realized it might also have had something to do with me playing with a live hand grenade in one hand, drinking beer from the other.

The next morning, the local press reported the previous night's brawls. Among other things, there was some story about how someone had knocked out the front teeth of a policeman using a hand grenade for brass knuckles. There was nothing to indicate that this was done by Sober, though. After all, he was far from the only man in the Ukrainian capital carrying hand grenades in his pockets.

CHAPTER 6: WAITING

Tor and I were at the hostel, struggling to attach all the magazine pockets and the like on our brand-new, still-stiff MOLLE carriers when Mike entered the small four-bed room. He had the keys for group's car, a small Chevrolet-manufactured 4x4 Niva, and had been out running a few errands around the city.

"They are a such a bunch of despicable little monkeys!" he shouted indignantly as he sat down on his bed and opened up his laptop. "One of these fucking idiots almost crashed into me when trying to pass by on the freeway, being just inches away."

Mike continued on to tell us how he caught up with the Ukrainian driver and cut in front of him to stop the car. He afterwards got out of his car, proceeded toward the bad driver in the middle of the street, and tapped his Makarov against the driver's window. As the Ukrainian nervously cranked the window down, Mike, together with his pistol, gave him a few lectures on traffic rules and general behavior.

Incidents like this one occurred quite often when Mike was out by himself. It was indeed true that the traffic sometimes seemed to lack common rules, but unfortunately, neither Tor nor I had been lucky enough to witness a confrontation like the one just described. The story itself was good and entertaining on its own, though, which is why we laughed our hearts out, anyway.

Tor and I returned to our business while Mike continued his work on the computer. He was sketching a diagram of his future foreign volunteer company.

"Do you remember what the symbols for company and platoon levels looked like?" he asked me.

"No idea, man, I just carry the machine gun," I replied. "Those kinds of things never got stuck in my head."

Mike quit struggling with the boring parts of it and proceeded to sketch a unit patch instead: a special insignia the members of the company would carry to distinguish themselves from the rest of Azov.

"Screw this, I'm getting hungry," said Tor, tired of struggling with his plate carrier and pockets. His guts had finally begun to come back to function again, and his appetite had increased. "Let's go find something to eat."

Although all the food, even in fine restaurants, was still cheap, the long stay in Kyiv really started to eat through my wallet. With the current armistice, I understood very well that I was not going to see any action immediately, maybe not even at all, but I couldn't afford to be a tourist here for any length of time. I was living well above my budget and had done so for a long time. I needed to get down to the eastern parts of the country and the front as quickly as possible. Mike had the car and had said several times that we would go "the next day," but with every new morning, a new unforeseen obstacle or important errand delayed it until the next one.

Mike accompanied us to a new restaurant of his choice. We ordered some steaks and waited for the food while Mike was on the phone.

"Abdullah is coming over here," he said, putting down his phone as the waitress brought the food in.

Shortly afterward, just as we had finished ordering our third beer, Abdullah showed up. "Abdullah" did not really look like his *nom de guerre* would suggest. He was young and slender in body, wearing some simple gray training pants and hoodie a couple sizes too large. His face almost seemed a little childish, especially with his blonde hair, which appeared when he pulled the hoodie back. The hair was cut in a pot shape like that of a child. The name, the clothes, the general look—all three made me wonder if it was a strange joke I was exposed to.

Abdullah spoke good English and put out his hand to greet us. I looked around a little suspiciously but followed the gesture and put forward my own. The handshake definitely broke from the previous pattern. It was firm, and the skin on his hands was hard. My own hand, despite Abdullah's childish appearance, was smaller than his.

He had previously belonged to the same Båtsman group as the others but was now in the process of switching to Mike's newly formed company. There he would do general management, and above all, act as an interpreter. As he sat down with us, I had the time to look a little bit closer at his hands as he laid them on the table. They were very rough, but not like a farmer's or a factory worker's. His fingers and knuckles were all gnarled, with thick calluses and scars. Abdullah was obviously someone who liked to fight, did it often, and had been doing it for a long time. Some small wounds and bruises which had not yet healed also meant that his latest fighting was a rather recent event.

The restaurant was pretty dead, so Mike suggested we go to the nightclub.

"Finally! Finally, a place where you can watch some girls!" shouted Tor in a tone of ecstasy.

After getting out of the taxi we took from the street, we stood in the queue outside the Coyote Ugly, which Mike and Tor in tandem had selected as appropriate. At the entrance, the bouncers took one look at Abdullah and refused to grant him entry, recognizing trouble when they saw it. With Abdullah not being allowed in, only us foreigners passed through the metal detectors. The inside was flashing with lights, and high bass vibrations passed through our lungs and chest. Beautiful Ukrainian women danced on the tables of the bar. Tor was overjoyed, smiling like a child in Willy Wonka's chocolate factory. I was mostly disappointed that there did not seem to be normal beer for normal people on the menu, but only gay drinks.

We sat down at the bar. Mike ordered a whole bottle of vodka, a bucket of Red Bull and three glasses. The drink was okay, but after getting my grog kicked all over me for the second time by the dancing girls at the bar table, I was honestly quite fed up with everything. I considered arranging a taxi for me and my Red Bull and vodka-smelling pants to head back to the hostel. Or even more preferably, the quiet and peaceful little student bar next to the gate entrance.

Then the mood at the nightclub suddenly changed. The music was lowered, and one of the girls picked up a microphone. She shouted something in Ukrainian and ended with "Slava Ukraini, heroyim slava!"—glory to Ukraine, glory to the heroes. Everyone repeated after her, several times. Ukrainian flags began to be waved as patriotic songs was sung in harmony by all the staff and clients. The song echoed, and the patriotism was palpable. I saw a young man near me singing loudly with his hand tied to his heart. When the song was over, I took the opportunity to approach him for a chat.

I asked him what his thoughts on the war were and got a clear answer. With fire in his eyes, the young man talked about how Russia was indeed both a threat and the aggressor, but Ukraine would win. He would never stop fighting, and Ukraine would leave this war victorious.

"A few of us here are going to the front line soon!" I shouted at him to overpower the loud, resurrected music. "Come with us, let's win this war together!"

The young man's eyes stopped glowing. He explained that he was currently studying at the university and did not have the opportunity to enlist and fight in the war.

"Fuck, man! The university will still be there, but only if you win the war!" I yelled at him.

The young man looked down at the floor and explained that he had tests and exams he promptly had to study for; there was simply no way for him to go to war. He also had family to think about. I looked at the miserable coward in front of me with contempt. All the songs, slogans, and flag-waving were nothing but a charade for him, and all the others around like him. He, a Ukrainian, was no better than the Swedes I had left at home. I hated them all.

The next day began and proceeded as the one before, waiting for a front that never seemed to get any closer. Mike, Tor, and I were near Independence Square. Discussions were going on about where we should go for today's evening meal as Mike's phone rang. After a short conversation, he put the phone in his pocket and turned to me.

"You should probably hurry up now. Your bus to the front will be departing very shortly from Kossatskiy," Mike told me. Kossatskiy was Azov's Kyiv base.

I jumped. "When exactly?" I asked.

Mike laughed. "I don't really know. It may be in half an hour, or five minutes," he replied. "Everyone else is there waiting right now, so you should hurry."

"But...fuck! I still have all my shit at the hostel," I moaned in a slowly escalating feeling of panic.

"You still have time!" said Tor, pointing to the vacant taxis lined up across the street. "Take a cab there, and we'll walk to Kossatskiy while you pack your things. When you get back, we'll meet up outside."

Mike laughed as I dodged incoming traffic while running across the street toward the taxis. "This is just like any other army! First you wait for a long time, then you have a giant stress overload!" he shouted after me as I pulled open the passenger door on the first car in line.

While the taxi was waiting outside the hostel, I ran up the stairs. After quickly packing up all my equipment, I then hurried back to the taxi and then toward the Maidan, checking the time every thirty seconds or so. I had been waiting for days which felt like months to get to the front. It was as if my chance was about to slip through my fingers. Who knew how long I would have to wait before the next bus left?

I paid the driver, got out of the cab, and hauled myself and my overloaded and poorly packed rucksack up the street toward Kossatskiy in a frenzied pace. I soon saw the street outside the HQ was full of people, affording me to slow down a bit and relax. No frontline tourist tour bus seemed to have departed yet.

I relocated Mike and Tor among the crowd and threw my ruck on the side of the road. Hundreds of Ukrainians around us covered the narrow street. The scene was similar to the prelude to a big football match, but militarized. Most of them were young, obviously new recruits. Many wore uniforms, but no uniform pattern existed. There was German Flecktarn and British DPM jackets, both woodland and desert mixed with Ukrainian pattern pants, to name only a few. Whatever was available was used. Many wore balaclavas so as not to be recognized by potential enemies, especially the men who originated from the now enemy-controlled territories, since threats against family members were common. One man wore a hockey mask, like Jason in the *Friday the 13th* horror movie franchise. In addition to the soldiers, there were also many girlfriends, wives, and family among the masses, ready to send off their fathers and sons.

"Hey, there's Mikola!" said Mike. "Come along here. I'll introduce you and he'll be your guide," he said and began to zig-zag through the crowd.

I grabbed my backpack and pressed on to follow. There was a glimpse of a man wearing Azov's black uniform further away in the crowd, his head higher up than most others.

Hell, I thought, *is that Dolph Lundgren over there?*

The sun had begun to set, and his short, well-combed hair shone golden yellow in the streetlight. His blue eyes were focused on a smartphone which, despite its size, disappeared into his huge hands. Mike shouted at him, and Mikola looked up toward him. In the light, his strong and clean-shaven jaw was contrasted against the background. While he was looking for whoever called his attention, his profile in the light looked exactly like the German ideal man from the Waffen-SS and the Wehrmacht's old recruitment posters.

Mike introduced me. Mikola smiled and held out his hand.

"Hello! Welcome to Ukraine!" he said and reached out, my hand disappearing into his.

I had rarely seen a bigger unit of a man. I stayed with Mikola while Mike ran on to talk to others in the crowd. We talked further, to start getting to know each other. Mikola's voice was dark, dull, and monotonous, and his tone seldom changed. When he laughed, the sound was both unusually loud and almost mechanical in its execution.

Behind Mikola, another man approached. Unlike most others, he was civilly dressed in jeans and a simple dark outdoor jacket. He was only slightly shorter than me, but in contrast to the giant I spoke to, he appeared to be considerably smaller. His head at first looked completely shaved, but as he approached, a short but clear black stump around the sides revealed that he was also bald. The black and marked eyebrows hid equally dark eyes beneath them. His chin and upper lip were hidden by a thick, black, full beard.

"What the fuck is this!" shouted the man as he threw off his large black duffel bag on the ground. "Look at these people! They have no idea what they're doing, have they?" he continued in what was clearly more of a monologue of his own than an attempt to start a conversation. The speech was conducted very quickly in a strong accent while he gestured around wildly with his hands.

"H-h-horrendous is what is!" he shouted, after stammering on the H—not so much because he seemed to have a real speech impediment, but because he seemed to have difficulty in shaping his tongue correctly.

Mikola pointed at me and had me introduced as the new Swede. The man with the dark beard looked at me, put out his hand to shake mine, and continued to speak, now at a slower yet still brisk pace.

"A-ha. They said someone new was coming. They call me Greek."

I had already heard a lot about him through Tor's and Mike's descriptions. Now that he was standing in front of me, I felt a little underwhelmed. Maybe it was because of the contrast to Mikola that made him look even smaller, or just because he looked so simple and bland in general.

Mikola asked what my military background was, and I told him briefly while he and the Greek listened to every detail.

"Coastal Jaegers?" the Greek asked. "They're amphibious, huh? Like marines"?

"Yes," I replied, "something like that. They're light, seaborne infantry."

The Greek said that he had also been amphibious light infantry in Greece at the beginning of his career.

"Were you there as a professional soldier, or just a conscript?" he asked.

"No, just regular conscription service, for a year."

The Greek looked a little bit disappointed but then stopped and said, more as a statement than a question: "Well, you have a pretty good conscript army in Finland, anyway, right? Not like us niggers down by the Mediterranean Sea. Our conscripts are a bunch of fucking monkeys. Stealing shit, and...fuck! Shit. They're niggers. I should know."

I had some trouble understanding everything he was saying. The Greek spoke great English and his vocabulary was huge. However, this combined with his accent and sometimes strange pronunciations meant that full concentration was required to keep up, especially as he got aggravated and sped up the pace of talking.

"Opa!" said the Greek, prolonging the iconic Greek expression for emphasis to make us aware of the surroundings in which he looked around. "Gentlemen, the horde is moving."

As he had said, the crowd was indeed moving—upwards, along the street from Independence Square in the direction of the beautiful Sofia Cathedral, likely the location from where the buses would depart. We quickly hauled on our equipment and began to follow the "horde," as the Greek had correctly described it.

Suddenly, a loud bang was heard behind us—some kind of explosion. The sound waves bounced between the solid house walls and hit everyone's ears multiple times. Mikola backed away, and the Greek was about to throw himself on the ground. It turned out to be one of the many football hooligans in the league who had thrown a banger a bit behind us. I laughed a little at the Greek, who looked around nervously before I quickly withdrew my grin. I realized that both he and Mikola, unlike me, were veterans. It was not long since the two had been under Russian artillery in the east and experienced things I had only read about and imagined. To instinctively take cover at the sound of a sudden explosion was completely normal, and the only idiot around was me, who had, full of naive, boyish feelings, just been laughing about it.

As suspected, a number of buses were lined up by the road next to the cathedral, and the hordes gathered around them. My new acquaintances and I threw our bags in the trunk and looked for seats inside one of the buses. The buses were big and roomy, but there was no chance everyone would fit, especially not with their own private seat. Outside, Ukrainian recruits tenderly said goodbye to their girlfriends. The man with the hockey mask waved a torch burning in bright red around. Someone carried a large stereo loudly playing true Norwegian black metal.

Soon, the buses were full to the breaking point and we started rolling. Mikola said that now there are only sixteen hours left.

"Sixteen hours?" I asked in surprise.

"Yes, or eighteen maybe. Depends on several factors."

I had never taken the time to check the distances properly.

"Ukraine is a big country," he replied just before he sat down in his seat.

I leaned back in mine. The stereo playing black metal had followed us on board, and Burzum's debut album echoed inside. In the aisle sat and lay the soldiers who had not been about to find proper seating. It was hot in the bus, and sweat ran down my forehead. I was finally on my way to the front line.

CHAPTER 7: URZUF

Bathory's *Nordland I & II*, and the debut album with the goat on the cover. Mayhem, Watain, and Nifelheim. Darkthrone. Thyrfing and Finntroll. *Battles in the North* by Immortal. The music which thundered through the cramped bus could have been played directly from my own teenage collection of burned CDs. It was at first a nostalgic element, but as the clock kept ticking past midnight and never seemed to cease, it turned into an iridescent moment. I did not know what to expect the next day and very much preferred to sleep. Hours passed and, with the exception of piss breaks where the buses stopped at the roadside or some gas station, it was an express trip toward Europe's eastern front. Sometime during the night, the music and the noise of the hooligan recruits gradually began to subside. Now it was instead the roads, which got worse and worse the further east we rolled, preventing me from sleeping. The bus shook and bounced around. Sometimes I thought we had run off the road, but no, it was only the driver who tried to turn past a large hole in the asphalt but instead drove down into another.

The sun had begun to rise as the buses turned into a gas station. It looked like it would be a sunny and clear day with no clouds. The gas station was open, and there was an opportunity to head inside to look at its fine wares. Being extremely thirsty after the night, water took priority. Looking through the fridge with refreshments, I picked out a bottle of water and took a chance on what appeared to be some kind of odd energy drink. I also found a couple of processed and wrapped croissants. I had not eaten anything since breakfast the day before. Though I couldn't understand any of the words on the plastic packaging, the pictures led me to assume that they would contain some form of chocolate—a fast source of energy.

The half-liter bottle of sweet energy drink only made me thirstier, and I proceeded to down most of the water afterward, as well. I then proceeded to viscerally smash the croissants—which really did have a filling of sweet chocolate mousse—into my facial cavity. With a stilled hunger and with blood sugar levels on the way up, I saw how the bus turned off the main road to the right. The open fields on the sides of the road began to be replaced by populated area. I leaned out of my seat and turned back toward Mikola and the Greek, who were sitting a few rows behind me.

Mikola also leaned out toward the hallway and looked at me: "Welcome to Urzuf!"

Urzuf was a small village located right by the coast along the Sea of Azov, midway between the two major cities of Mariupol in the east and Berdyansk to the west. There were few large apartment buildings here. The houses were small and simple single-family cottages or villas as a rule. The only larger buildings that existed were the church and a number of brick Soviet-era government buildings. Since the village was located by the beach, the people living there subsisted on summer tourism. Thus, there were many restaurants and small shops around, compared to what the small population otherwise may have used. Despite the fact that the war had been going on since May, Urzuf had been open to all during the summer. The fighting had taken place just an hour away, but despite the machine gun fire and shelling, the tourists had been in place as usual. All the veterans I had met so far had fond stories about how wonderful the free time during their leave in the little tourist paradise had been during the summer.

The picture they'd painted for me, which took place only a few weeks ago was, however, very different from the one I saw outside the bus window now. What not even the war had managed, the approaching autumn had now done, just as it had done all the years before. The holidaymakers and tourists who were not disturbed by the rumble of the artillery in the distance had set off as the first cold autumn wind sailed in from the sea across the shoreline. The amusement park with its Ferris wheel, a common sight for larger Soviet cities but unusual in small villages, stood desolate as the sea winds blew through its spokes. The streets were empty and the taverns had taken down their signs. Urzuf's hibernation was fast approaching.

From the bus window, I saw a long white stone wall on my left side. It was beautifully painted in what reminded me of Swedish folk art, full of green twigs and colorful flowers. The bus continued and stopped at a large steel gate. This one was just as beautifully adorned with more flowers around a Ukrainian flag. The flag itself contained an extra detailed interpretation of Azov's insignia in the middle. On each side of this shield stood two painted Ukrainian Cossacks with drawn sabers. The bus stopped, and the gate began to roll to the side to open up the entrance to the base.

"You know this is a Greek village, right?" the Greek asked.

Mikola interrupted him and said that no one cares what or where Urzuf comes from.

"Urzuf is Greek!" the Greek answered him firmly. "We Greeks colonized and owned this whole area once upon a time!" He looked at me and asked if I noticed the sign as we turned in from the main road.

"No," I said, "I must have missed it."

"Well, it had a Greek flag on it, anyway. This land used to be Greek!" He clapped his hands as if to show how all this really belonged to him, that his steps went over the land of his ancestors. "We owned all this. And rightly so! Once we've sucked enough money out all of you 'real' Europeans, we'll take it back. I can promise you that! Just give us more money, and then you'll see!"

A slightly sly grin broke through the Greek's serious facade, and I began to understand what Mike and Tor had been talking about. He was quite the joker. As we passed through, I was, however, too busy looking around at the inside of the walls to listen to any further romantical talk about a Greater Greece. A large paved road led in from the beautifully decorated main entrance. To the left was mostly open lawn with a few trees here and there, while the wall continued along the entrance on the right. All three of us followed the road with the horde. On the left side, parked by the road, stood a picturesque collection of assorted vehicles. What stood out the most was an orange-painted T-72 tank chassis. It lacked a turret and instead sported a tall superstructure with large windows. The Greek noticed that my eyes were fixed on the tank-like vehicle as we passed. He put one hand on my shoulder while leaning forward next to me and pointed at it.

"You see that? Fuck, they ripped off a good chunk of that stupid thing!" he said at a fast pace. "Biletsky, the battalion commander, he resides in some kind of fantasy that Azov will get tanks from the government! And then some scammer tricked them into buying that piece of shit to train tank drivers! Ha!"

The Greek continued to talk about how much money it had cost, laughing about how the cash might have been better spent on things like uniforms or proper body armor for everyone.

"You get it? He thinks the government will give Azov—the far-right Nazi terrorist battalion—fucking tanks? What a fucking dreamer, huh? Now they're sitting here with this pile of useless Soviet scrap metal, and the fucking thing doesn't even run! I've never seen it drive! Ha!"

I did my best listening to the monologue, but the Greek's fast-paced speech made it difficult to keep up. We passed the apparently useless training tank and continued past the other vehicles. There were ordinary cars of various kinds and a meager mix of army trucks, all in varying levels of decay with punctures, shattered windows, bullet and shrapnel holes through the chassis. One of the Ural truck's canvas roofs over the flatbed had begun to cave in. Next to these followed a number of refurbished garbage trucks that had formerly been civilian vehicles. The rear parts had been reconfigured, either for troop transport or with armament in the form of anti-aircraft guns. One of the garbage trucks for troop transport was painted all black with double white stripes over the front cab and the side doors, the standard marking to identify Ukrainian vehicles and armor. The sides and tires both had armor plates welded onto them. There were also additional large armor plates on hinges that could be lifted up to cover the front and sides of the cab, then at the expense of much impaired visibility. The glass windshield had some bullet holes in it, suggesting the extra armor plating wasn't always used.

"What the hell is that thing? Mad Max Ukraine?" I asked the Greek.

"Ah, you mean Cookie?" he replied after following my gaze to the heavy black vehicle. "It's an old garbage truck. The Ukrainians armored it up themselves. I doubt it's really bulletproof, but it looks the part, anyway."

Garbage truck-based self-propelled anti-aircraft gun.

"They call it a cookie?" I replied, and the Greek confirmed that I had heard correctly. "That's dumb. It should be called Mad Max," I said.

The Greek looked at the black truck, pondered in deep thought for a short moment before deciding to agree: "You're right. 'Cookie'—what is that? It's a stupid name. 'Mad Max' is a lot more fitting."

The only thing that resembled an actual military vehicle was a BTR-60PB, an eight-wheel armored infantry fighting vehicle. It was way obsolete, though, and had not been considered modern since my father did his military service in the early sixties. Behind it hid a smaller and even older armored car, a six-wheeled BTR-152 from the early 1950s. Unlike the BTR-60, this antique didn't even have a proper roof to protect the soldiers riding inside from shrapnel.

"That one was donated to us from a private museum," said Mikola.

Behind some trees, another armored vehicle appeared. It was an MT-LB, an all-metal tracked artillery tractor and troop transport. However, it was not the standard variant with a simple 7.62mm turret opposite side of the driver. This was the 6MA variant with a centrally located turret similar to the one on the BTR-60PB, armed with both 14.5mm and 7.62mm machine guns. The entire chassis had a large bolted steel grille around it, designed to provide extra protection against RPGs and similar shaped charge-styled anti-tank weapons.

While observing the surroundings, I followed the horde blindly toward the beach.

"Hey!" I heard Mikola shout. "We're going this way."

On the right side, next to some yellow brick buildings, there was an opening in the wall through which Mikola and the Greek were on their way through. I turned around and hurried toward their position. When we came out through the wall on the other side, we took a left past a dirty gray-white brick building next to the wall.

"This is the diner," Mikola explained. "You're not picky about food, are you?" he asked.

I explained that I probably ate most things.

"We'll see, we'll see," he laughed in his usual monotone way.

"The food is shit. Quite honestly. I'm being serious. It's even worse than Greek army rations, and they pay, like, one Euro per sample, or something," the Greek added. He then became silent for a short moment (though longer than usual for him), pausing for some reflection, before continuing: "No, wait. It has to be even less. They would never spend that much cash on our useless conscripts."

The size of the building gave a small indication of the size of the base itself. It looked like it could contain a kitchen and enough seating to serve food for a slightly smaller infantry battalion, up to five or six hundred men or so. Any more than that, and the small diner would have a hard time serving lunch to the last man before dinner began.

Meals were served daily at certain times of the day in three batches. As a rule, it was also open otherwise for those who wanted to simply drop by for a cup of tea or similar. It looked simplistic, but a dining hall is a dining hall, and a good surprise as it was. However, the food can still be bad, I thought, as long as it actually exists.

After passing the diner, we came out to a number of yellow two-story barracks. The buildings and the area in general were called "Trupnik," and had served as hostels before the war. The elongated buildings held to a simplistic standard roughly equal to the diner, and the rooms varied in size, which was usual for hostels. It had everything from small two-bed rooms to larger rooms for ten or so. The Greek and Mikola had a room on the upper floor of one of the barracks.

Mad Max.

"He can take Papi's bed," the Greek told Mikola. "That bum won't be coming back, anyway."

Mikola showed me to the room on the lower floor, explaining that I would be sharing it with Tor and Barzuk, the Frenchman injured by shrapnel. It was unknown to anyone if Barzuk was back from the hospital, but a simple knock on the door would provide a good start to finding out. Mikola banged on the thin metal imitation of a solid wooden door, which soon opened. Barzuk was indeed back, and Mikola introduced us.

There are, for starters, basically two kinds of Frenchmen. One is light-colored and difficult to distinguish from Northern Europeans. The other is a darker Mediterranean variant, which could easily be confused with Italians and Greeks, as well as people from the Middle East. Barzuk belonged to the latter part of the population. He was in his mid-to-late twenties, approaching thirty. His short hair that bordered on getting long was black. It married together in front of his ears with a half-long and thick full-coverage, equally black beard. The beard framed his face without interruption, except for a short discontinuation between it and the otherwise equally thick mustache. His eyes were brown with large wrinkles underneath. The white of his left eye was still red with blood from the injuries he received from the Russian artillery. The rest of the left side of his face was also scarred, with large stitches and blue-green skin where the shrapnel hit and then had to be pulled out of his skull by a doctor.

In addition to a small bathroom with a toilet and sink, there were three beds which took up most of the room. First, on the right, was Barzuk's. On his little bedside table lay a Bible with a large knife on top. The knife had a slightly curved blade, a bit like a scimitar—the kind of saber most commonly associated with Eastern Europe, but perhaps best known as the signature weapon of the Arabs and Ottomans. The handle was made of beautiful wood, with a large brass knob shaped like a bear's head at the end of it. Barzuk saw that I was looking at the knife.

"Pretty nice, huh?" he said with an accent, but in otherwise perfect English, despite being a Frenchman. "I got that from a Ukrainian Cossack I met at the front this summer. Not the kind of fake Cossack you usually see, but a real one. Complete with that odd hairstyle and a really big mustache."

To the left of Barzuk's bed was a small refrigerator, and then, against the window, were two more beds. The one on the right was Tor's, and the one on the left belonged to Papi. On Papi's bed lay a large wheeled travel bag. It was left open and contained various clothes and equipment that he had left behind.

"Take whatever you wish," Barzuk said. "All of that is considered group property as it is now."

Most of it was rubbish, but he had left behind his uniform, an airy and light British summer stand in MultiCam. It turned out that we had about the same size, and it would come in real handy, as it was still very hot outside, especially during daylight hours.

I went upstairs and knocked on the door to Mikola's and Greek's room. I had no weapon, something I was hardly alone in, now that many new recruits showed up the same day. Tor had said that, at least for the time being, since he was still in

Kyiv, I could borrow his. It was an old AKM from 1958 in 7.62x39, a different kind of ammunition than the 5.45x39 that was standard nowadays. 7.62x39 was a scarce commodity, and I was not allowed to use it recreationally at the shooting range, but with that caveat in mind, the rifle would be otherwise mine to manage until he returned. The AKM was kept in "Båtsman's house," and I asked Mikola if he could guide me there and help me acquire it.

We went out past the dining room through the hole in the wall we entered the barracks area through. There we took a right turn, continuing the same way the horde had gone in the morning, until we reached the small ledge below which the beach and the sea began. The wind from the sea chilled nicely in the heat. As it passed the cliff edge, it took with it the scent of the small pine trees that adorned the length. The smell of the sea that married it from the pines was like feeling the memory of my summer holidays in the Finnish archipelago, and the forestry work in the Swedish Jämtland with my father at the same time.

We continued along a stone walkway, framed by beautifully maintained vegetation. Mikola pointed to the fine brick buildings between the small ornamental pines.

"Yanukovych, the ex-president the Ukrainians booted out last winter, had one of his private residences here," he said. "He spent his summers here, and these big buildings used to be his."

The large brick houses were not really in class to be classified as real castles, but for having functioned as simple summer cottages, it was probably also the best description. Near the entrance stood an old red motorcycle with a sidecar which a couple of soldiers were in the process of repainting to black.

"It's a bit ironic that it's now Azov who controls Yanukovych's old base here," Mikola continued. "When we came here, we could see everything he had collected. The houses were full of gold, expensive furniture, and porcelain. There were lots of animal furs: minks, foxes, wolves. Even more exotic, like lions and such."

The description was that of the luxury life of a stereotypical dictator, especially as it came in contrast to the simple people's village outside the walls of the current military base.

After a short walk, we arrived at a nice log cabin which was facing the beach right on the edge of the cliff. Outside the wall were large boxes of ammunition and anti-tank mines stacked on top of each other. Mikola knocked on the front door while I waited a bit behind. The door opened, and we were let in.

Inside, it was like a mix between a sauna's wooden interior, a luxury villa's wall-hung paintings, and a weapons depot. The boxes and mines outside were obviously just the leftovers which did not fit inside. It was pure armory: rifles, heavy machine guns, mortar shells, RPGs with rockets, mines and ATGMs. All of it stacked on top of itself like a militant game of Tetris.

The man who let us in returned to the living room, sat down on the sofa, and leaned back, seemingly bored. Behind him, AKs and RPKs hung on the wall. One of the rifles was an old AKM which fit the description of Tor's rifle. It was difficult to determine how old the man in the sofa was. He could have been anything from twenty-five like me up to upwards of forty years old; he was extremely broad-shouldered, and rough-hewn. His face was generally very round, and he probably could have looked friendly had he wished to, but the fleshy contours together with an unobtrusive look on his face gave a very strict impression. I held out my hand and introduced myself. He sat up slightly and shook it, but by his body language, he made it very clear that the gesture alone was more effort on his part than I really deserved.

"I am Bonn," he said with a strong Slavic accent, "but you can also call me Misha." He leaned back again and put his thick, tattooed arms behind his head. "What do you want?"

I explained that I was a new guy lacking a rifle, and how Tor had said I could take his AKM until he returned and I was issued my own.

"No," Bonn replied shortly, "there is nothing for you here. You can leave."

Mikola tried to help, but the intimidating-looking Ukrainian was determined. The rifles belonged to Båtsman and this was not a handout welfare store. New guys weren't getting guns just because they'd like a rifle, especially not foreigners. Mikola seemed surprised and sought answers. Apparently, some kind of dispute had been created when Mike created his own company and tried to recruit men from Båtsman. Strictly speaking, the weapons were still Båtsman's, and if people were to change units, there was nothing he could do. The weapons, however, would stay firmly with him. While Mikola tried to mediate in front of Bonn's stone face, I asked if I could at least look around a little more.

"Sure, sure!" Bonn replied, waving his hand. "Just make sure you do not touch anything. If you touch something, I will break your neck, okay?"

Bonn spoke better English than one might expect from just looking at him, and he had made himself abundantly clear.

"Absolutely," I replied. "Look, but do not touch."

I walked around the arsenal and looked around. There were metal containers of belted AGS ammunition on the floor, with several wooden boxes of unbelted 30mm grenades next to them. A bunch of RPG-7s lying in a pile, some with iron sights and others with optics. Next to these, a smaller mountain of associated RPG-7 rockets towered. TM-62 anti-tank mines were stacked on top of each other up to the roof, and next to these towers I saw tin boxes for belted 7.62x54R. Behind these were a handful of PKM machine guns to which the belted ammunition belonged. One of them was painted in green camouflage with light spots. It was one of those machine guns which I had really hoped to be issued.

I bent down to take a closer look at the ammunition. As a conscript, I had previously only used regular FMJs and tracer ammunition. Here, on the other hand, was the entire range on offer: FMJs with regular and lightweight bullets, armor-piercing, and even armor-piercing incendiary. The incendiary rounds came in green casings, unlike the others in copper-colored steel, and the bullet tips were painted in black and red.

As I was kneeling down and looking at the ammunition, a new Ukrainian came up and tapped me on the shoulder.

"You, you," he said as he grabbed my hand, opened it, and turned it with the palm up. With his other hand, he put a hand grenade into mine and proceeded to squeeze my fingers around it and the spoon. He changed grips and put his index finger in the pin, which he pulled out while still holding my hand firmly with the other. He then released the grip on my hand, stood up, took a few steps back, and giggled restrainedly but with noticeable excitement. I saw that Bonn and a few others were also standing there, eagerly watching the expected spectacle about to unfold. Even Bonn's serious look had eased up slightly as a smile appeared.

The first thing I noticed was that the grenade, an F-1, was black. The F-1 was normally painted in some shade of green. They could be brown, but I had never seen black before. I knew that live and practice hand grenades were often distinguished by different colors, so I assumed that what I got in my hand was just a blind, harmless practice grenade. The purpose was, of course, partly to "fuck with the new guy," but also, perhaps only subconsciously, to see how this particular "new guy" would react to quickly and roughly assess the quality expectations.

I looked back down at the grenade in my hand, at the expectant Ukrainians who were still giggling like school girls, then back at the hand grenade again. For a moment, I thought I should just let it go and drop the grenade, as if I didn't care. I discarded that idea quickly. It could end in either of two ways: either, and most likely, they'd interpret it as if I were a jubilant idiot, or completely crazy—a permanent stamp on my forehead with which I could forget to ever get my own rifle issued. The second outcome was, of course, that the grenade was actually live, after which I would blow us all into smithereens—and considering the number of explosives in the small room, likely half the base, too. I chose the boring, albeit safest option: to just play calm but slightly annoyed, to stand up and hand the grenade to the man I got it from.

"Put the pin back," I said.

The prankster looked a little disappointed and upset, as if his hopes had not correlated with the outcome. He grabbed my hand and inserted the pin back into the grenade. Bonn went back and sat down on the sofa—still bitter, but a looking a bit more well-off than before. I approached him and asked if they had gotten anywhere with the weapons issue, but the decision was firm: there was no AKM or other rifle in here with my name on it.

It was getting close to noon. Mikola said that the group, which was scattered between different rooms and buildings, used to gather outside the barracks before meal times and then dine together. We went there to wait, and soon the members started showing up one by one.

A large man in a grayish camouflage uniform appeared first. He was about my age, but both taller and wider over his shoulders, even though he still looked small next to Mikola. His skin was light and eyes light brown. His hair was snagged and contrasted by being very dark on the head and then lightened on the sides down toward the short beard which was so light that it turned light gray. He introduced himself as "Gratch," as his Ukrainian accent cleared out his heritage. His official role was as a platoon commander for a platoon that did not yet exist, so until then, he was something of Mike's deputy commander in practice. He asked me a little about who I was and how I ended up here. It felt a bit like a short, impromptu interrogation but was not something I thought about anymore in the moment.

61

As Barzuk opened the lower floor hallway door, I saw the Greek coming out from the upper, accompanied by three other men. The first two looked to be about my own age, as well. The first was a sorry-looking, skinny guy wearing an oversized, loose-fitting Ukrainian army uniform. Not the old dark green color variant, but the newer, light green one in digital pattern camouflage. His skin was light, and his hair was dark brown, tousled, and uncut. He held his hands in his trouser pockets. The closer he got, the more of a bum impression he gave. The other was smaller. Thin and dark in complexion, with black hair and an equally black mustache and a short beard. I had a hard time placing his uniform. It was similar to both the German Flecktarn and the British MultiCam, but nothing of the sort. It was only when he got closer that I saw a faint national sign which read "ITALIA" on one sleeve, which explained the unusually dark skin color.

The third and last man was a few years older than the others. He was a bit shorter than me but seemed even smaller, being very broad and compact in his stature. He wore British MultiCam trousers and a combat shirt in the same pattern which, due to its tight shape, emphasized a thick belly slightly. He wasn't fat or anything, but not at all as thinly built as the hobo and the Italian. The thick and compact one closed the corridor door behind him, threw his wood-gripped AKS-74 over his back, and lit a cigarette, ascending the stairs down to ground level. As he approached, I could see his face better. He was very light-skinned with narrow, blue, slightly oblique eyes. The man was evidently a proper Finn.

From the barracks on the other side, the door from the downstairs level was soon opened, as well. Outside came a face I recognized: it was the confused Swede I had seen in the filmed interview earlier, Leo. His red beard had grown somewhat since then, and just like the hobo from upstairs out of my own barracks, Leo was wearing a Ukrainian digital camouflage uniform. Shortly after came two young boys at most twenty years old. One was tall and thin and wore sunglasses and white-black camouflage pants. The other was unusually small and a bit square in shape, dressed in a MultiCam uniform. The short description I had previously gotten of the two Belarusians—"the Bulbash boys," a tall and a short one—had been as accurate as it was simple.

With that, the entire "company" was accounted for. As we stood there in a circle and talked as we waited for the doors of the diner to open for admission, I tried to place my thoughts into the reality unfolding around me. I was definitely in the middle of a strange mix of men in uniform. The giant Mikola and the fierce Greek quarreled over yet another deep but strictly unimportant conflict, which the latter of them had recently created out of thin air, only for the sake of argument itself. Leo tried in vain to access a post where he could partake in the discussion, each attempt landing far outside the goal. Gratch and Barzuk tried a little more intelligently to be some kind of voice of reason, but each attempt was stopped immediately by the Greek, who wanted no solution—only chaos. The Italian repeatedly asked in frivolous English what the Greek and Mikola were really talking about, while the bum, who looked very cold even though it was hot outside, laughed nervously as he tried to curl up in his oversized uniform. The Finn did not participate at all but only observed everything quietly from the outside while puffing his cigarette. The Bulbash boys did not seem to speak any

English at all but laughed about something among themselves as a separate entity inside the group.

Was this the foreign elite Azov had been rounding up? Was this the so-called "Foreigner Group" I had been told about? And if so, where did I fit into it?

Inside the diner.

CHAPTER 8: PARANOIA

The diner was even more Spartan on the inside than the outside would suggest. The ceiling lights gave light, but only a very dim one. Without the sun penetrating through the thin white curtains, the electric light wouldn't have been sufficient to illuminate the large hall. Next to the thin white curtains that hung over the windows were large, thicker red ones. The ceiling was high and the thick concrete walls with square pillars in the middle of the room were painted in two kinds of pink. A little darker from the waist level and down and a little lighter upwards toward the white roof. On the pillars hung posters of historical Ukrainian figures. I recognized some of them from the banknotes I had in my wallet. Ivan Mazepa, who fought with the Swedish hero King Charles XII (also known as Carolus Rex) at Poltava was one of them.

In the middle of the kitchen was a thin little wooden partition. It shielded the serving itself from the rest of the hall as it led the food queue around it, and it functioned as a kind of bulletin board. Instead of today's menu, or important information for tourists, it was now filled with simple, colorful drawings made by children with crayons that were to be sent to the Ukrainian soldiers. The motifs were usually centered around the Ukrainian flag, or sometimes the red-black nationalist variant. The girls often depicted soldiers with their families, houses, or flowers in their paintings. The boys were distinguishable by being a bit more action-packed, depicting scenes like Ukrainian tanks smashing adversary ones with the Russian flag waving from the turret, fire and blood spewing out from their knocked-off turrets. The colorful drawings were often accompanied by texts beside them—sometimes shorter and simple, written by the children themselves, sometimes longer in astonishingly beautiful writing style by the mothers. The mothers were saying "Thank you for defending us," "God is with you," "We send all our angels to help you," and such.

The serving area for the kitchen was small, filling the bowl of one soldier at a time. The food was as simple as it had been described to me. I handed out my little dog bowl, and it was filled with three ladles of boiled buckwheat, or *greshka*, as it was locally known. A small chicken leg placed on top of this was followed by a ladle of broth to make the greshka a little more interesting.

One by one, the group sat down at one of several long wooden tables in the room.

"What the hell is wrong with you?" the Greek asked, sitting on the edge.

I looked up, spoon still in my mouth, noticing how he was looking at me.

"How the hell can you eat that shit, let alone with such enthusiasm?"

"Hey, it's not that bad," I replied, still with food in my mouth.

"You're the first Swede I've seen eating that. Everyone else hated the food—Tor, that man-child, in particular. And rightly so! A dog would not eat this shit!"

"I'm half-Finn," I said. "Could very well be because of that."

The Greek dropped everything and locked his eyes on me.

"Are you a Swede, or a Finn?"

"I am both, with dual citizenship. My father is Swedish, my mother is Finnish; or rather, she is Fenno-Swedish, to be precise about it."

The Greek looked seriously upset.

"Fenno-Swedish? Finnish-Swedish? What the hell kind of bullshit is that? That doesn't really exist. You made that up right now," the Greek insisted, confused and annoyed.

Mikola filled in and explained that a large part of the population in Finland, especially historically, were Swedish-speaking and were thus called Fenno-Swedes.

"Don't bother trying that bullshit with me, you stupid brute! Both of you have conspired about this before coming in here. I'm not buying it, you understand? Finnish-Swedish, and Swedish? And Finnish? Ha! That's retarded! You are either Swedish, or you are Finnish. Nobody can be both. Even a small child understands this!"

The Greek stood up demonstratively and walked over to the long table next to the wall behind him. There were large thermos containers filled with hot water for tea and coffee. Next to these were also sugar, freshly cut lemons, and large glass jars of honey. The rest of us followed one by one, depending on how quickly we emptied our dog bowls.

The conversation continued over a cup of tea. Mikola said it was rumored that the separatists would try to kidnap Azov soldiers in the area, and how we needed to be careful if we moved outside the walls.

"Bro, that's complete bullshit," the Greek said again, "and you should know that. You should! What do you think they're going to do? Drive around in a van at night, and then just ambush the first lonely Azov they see, grabbing him into the car? Bro, I am serious, that's retarded."

Next to the freshly cut lemons, there were some white pieces of what looked like butter in a bowl.

"What's that?" I asked Mikola, who looked over his shoulder to see where I was pointing.

"It's called salo. The Ukrainians eat that a lot, and they really like it. It's basically just pieces of pickled, salted, and cut pigs' fat. I don't like it very much."

"Fuck you, you disgusting Swede, or Finn, or whatever you are," the Greek shouted. "You're seriously not contemplating eating that shit, too, I hope? Fucking don't, I'm serious. You white people fucking disgust me. Jesus!"

The pickled pigs' fat turned out to be really good, especially together with white bread. After eating and washing our dog bowls, I went outside, sat down on the stone steps at the dining room entrance, and lit a cigarette. I saw out of the corner of my eye how the group bum approached with his Italian friend close behind him.

"I have one like that!" he said in a distinct Fenno-Swedish dialect. I saw that he was pointing to my tattoo, which was visible when I had a short-sleeved T-shirt on. "Not on the arm though, obviously, but on a green beret!"

I looked a little skeptical at the shaky and nervous-looking bum at first, but if we were both Coastal Jaegers from Finland, there was a common denominator. We started talking, and he soon introduced himself as Långström, originally from Savolax in the eastern parts of Finland. I quickly noticed that he was not as stupid as I had initially thought, but we did not get far in the conversation until the previously friendly tone disappeared in the wink of an eye. Långström's face suddenly changed from one moment to another. The pleasant and open smile sank down, and his gaze went from welcoming to extremely cold and, above all, unpleasantly scrutinizing. From friendly chatter with a new acquaintance, it now felt as if I was sitting in a prison cell with an interrogator reading me through and through.

"Well, then. Yes, then it's all good," Långström said and turned around, walking back to the barracks.

The Italian quickly followed him and seemed inquisitive about the conversation he had just had. I stayed, smoking the rest of my cigarette while I tried to understand the reaction and the sudden change in behavior. Above all, I had a hard time letting go of the thought of his gaze. I had never experienced someone looking at me like that, looking straight through me before. I tried unsuccessfully to understand what it was about until my lungs told me I was simply smoking filter.

I went back to the room. Barzuk had the only key, but he was already there and the door was open. The scars on his face and around his eye were not very large in themselves, but the large blue and green miscoloring around them together with the stitches made it look as if his face was rotten. During the car ride retreat from Shyrokyne, he had tried to push a tree that had fallen across the road to the side when a heavy mortar shell struck just a few dozens of meters away, spraying his left side with shrapnel. After this, he, a Croat volunteer, and Sonic had continued with Papi as the driver until the car lost the road trying to avoid the artillery raining down around them.

According to Barzuk, Papi had told him to take the Croat with him. The young man from the Balkans was injured by secondary shrapnel and, in shock, unable to care for himself. In the meantime, Papi would stop to see what condition Sonic was in. After that, he did not know exactly what had happened. He said, however, that even though Papi had left Sonic there by the road, the Irishman had still been the first to scramble a new car and return to pick him up, once he had finally reached friendly lines.

"Everyone was fucked and they don't remember, but I do. Papi couldn't carry Sonic and left him there. When he got back and got a car, and everyone else was still in shock, he was the only one who tried to get a car and go back to get him," he said. "Papi was a funny guy. Above all, he was good at drinking. He used to smuggle vodka into this room, and we would get fucking wasted in here, having a blast!"

"There's a new story," I replied, "an Irishman who's a professional drinker," and laughed.

"No, no! It was not because he was Irish!" Barzuk responded firmly. "It was because he was a former Legionnaire!" Barzuk looked proud when saying it. I understood that he wanted to connect a character trait he liked in another man to his home country if possible, so I just let him keep that without arguing.

There was a knock on the door; Barzuk got up from his bed and opened it. It was a Ukrainian I didn't recognize from before. He looked suspiciously at me, the new foreigner, then back up at Barzuk, who was a head taller than him.

"Come outside and close door. We must talk. He does not need to hear."

Barzuk did as he was asked and stepped out into the corridor. He did, however, not close the door properly, instead leaving it with a slight gap allowing me to listen in. The Ukrainian spoke poor English and almost whispered while Barzuk spoke in a normal conversational tone. The door, even though it was not completely closed, made it difficult to hear everything, though I did manage to perceive a worried tone and a few words. "Spy" was one of them and a name: "Michelle." The conversation was short and Barzuk soon returned, now actually closing the door behind him.

"What's really going on?" I asked. "People are behaving fucking strange around here."

Barzuk sighed slightly, went to the small refrigerator on the floor, leaned forward, and opened it. It didn't have much to offer inside: a few bottles of water, a small slice of cheese, and what looked like half a salami.

"I'm not really supposed to talk about it," Barzuk began as he poured water into his tin cup. "But...one of the foreigners here was a spy. They only took him yesterday, just before you arrived."

"Spy? How?" I asked, and Barzuk continued the story.

The man they had taken was quite young, French, and had also arrived recently. He had never been in a fight or anything; he had only been staying at the base. Barzuk didn't know exactly how he had been caught. He guessed that it was either through the SBU, the Ukrainian security service, or it had simply been discovered via the base's own Wi-Fi system. After the Ukrainians took the spy, Michelle, they had asked Barzuk to go through his laptop and email conversations to translate them. The email account contained drafts in which detailed information about the Ukrainian military positions was written down. These drafts had never been sent anywhere, but they had been opened from an IP address in Donetsk, the enemy capital. In addition to these, there was also direct communication between Michelle and French voluntary communists on the separatist side.

On top of that, the laptop was otherwise full of communist material, making it evident that he belonged to such a French political organization. It also became clear that he always wanted to fight for the separatists; however, as he had failed to get a visa to travel there through Russia, he had structured a new plan. He was to go to Ukraine, which was visa-free for EU citizens, and from there, initially spy on Azov on behalf of his communist brethren on the other side of the front. He would then, as soon as he got access to weapons, try to take another foreign volunteer hostage. Then he would cross the front line to the Russian side with this as both a guarantee and trophy.

The pieces of the puzzle began to fall into place. Ever since I arrived, I had experienced people's behavior as very suspicious toward me, as well as each other. The Ukrainians had looked at me strangely, and so had several of the other foreign volunteers. The fact that they'd captured a spy so recently explained everything.

"Okay, listen," Barzuk said. "I checked the guy's computer, his emails, hard drives, and everything. He's not really a spy." Barzuk was very serious, but at the same time had a hard time keeping from laughing.

"Like, that plan of his...what the hell even was that? He was going to take a hostage here, okay, fine. Then what? Would he just, like, walk out of the base with this guy, and then over the front? Across the mine fields? And then what? Would he just be received there as a hero? Medals, maybe? The guy was completely retarded, simple as that."

The laughter Barzuk struggled to hold in trickled more and more as he recapped the main parts of the story. "And then he writes about this via fucking Hotmail? On an open Wi-Fi network? Ha! You hear this?" Barzuk was forced into beating his chest after swallowing water down his lungs.

"Seriously. He was not a spy. He was just an idiot LARPing like one. He thought he was playing a live action role-playing game where his character was a communist French James Bond or something. All that 'top secret information' he 'leaked' to the enemy was bullshit, anyway. Lots of numbers he took from open sources or just pulled out of his own ass. Nothing of it is current news. Completely meaningless information for the Russians."

"What happened to him?"

"They threw him in the salt pit," he replied, smiling a little bit.

The "salt pit" was out on the lawn, after the vehicle park under a small pavilion. It had previously been the ex-president's personal wine cellar and now served as a temporary detention center.

"They are still questioning him. Hopefully, they'll be throwing the communist in jail here, permanently."

Barzuk looked pleased, but his tone soon changed again. He looked straight at me, his swollen and scarred left eyelid open, as well, so that the blood-red eye was clearly visible.

"The problem isn't that he's released any secret military intel, or whatever. All that was just bullshit, like I said. The problem with all this shit is that it's got everyone wired up. We're all paranoid as is, but now everyone sees spies and shit everywhere, too. So, you know, just keep that in mind."

There was a knock on the door again. Barzuk shouted that it was open and free to enter, and it was opened. It was the Greek.

"Oh, oh, it's just the two of you here," he said, curled up, and put his hands together in front of him as if to give a prayer. "I hope I did not disturb any kind of…male friendship sessions going on in here."

Barzuk laughed and said that we were just starting to get to know each other seriously, as only adult men can do. He also invited the Greek to take part in this spectacle. The bald and bearded Mediterranean exclaimed excitedly: "Opa!" And he made a quick gesture where he seemed to throw himself into the room.

The live sketch with homosexual undertones ended there, and the Greek returned to his original case.

"What are you really doing?" he asked, and Barzuk replied that he had just drawn up a summary of the spy incident.

"Ah, so he knows, then?" The Greek looked at me. "Did you hear that part about how he was going to take one of us hostage, too? And take him to the Russians? What a fucking retard! I mean, all of that communist nonsense doesn't leave you expecting a lot, either, but that shit was fucking retarded, beyond the norm. Really."

The Greek restrained himself from starting to go off on a rampage about communists in general.

"Anyway," he continued, "we're sitting upstairs in the big room. Come on up. Let's talk organized bullshit as a unit instead of just inside our pathetic little separate cabins."

We followed the Greek out, up the stairs and onto the second floor. He pointed to one of the first doors on the left side of the corridor and reminded them that it was his and Mikola's room. We continued a bit further into the corridor before the Greek opened another of the doors on the left and stepped in with us still following. This room was significantly larger than ours and therefore a great gathering place. There were three beds, two in a row on the right side and one alone on the left. On the beds sat Mikola, the Finn, the Italian, and Långström. Barzuk took a chair, and the Greek sat down on one of the beds. The Greek looked at me, grinned, and clapped his hand in the only vacant seat next to him.

"Come and sit with me, little child."

The discussions were much like the ones that took place in the dining room. Russia and the war were recurring themes, obviously, but so were contemporary world politics in general. History was also a hotly debated topic, and it became particularly interesting in that the many nationalities in the room since school age often learned about it from different points of view. The Greek was the most diligent of the debaters. He was unusually well-versed in all the subjects discussed but also enjoyed provoking everyone else on purpose too much for some sensible conclusion to ever be nailed down in the end.

"Fuck this," said Långström. "I think we all could use something to drink right now."

"To drink?" replied the Greek. "Yes. Yes, you are right. Whose turn is it to go buy beer this time?"

"I can go," said the Finn from the other side of the room with an extremely thick accent. He stood up and asked who else was coming.

"New guy!" cried Greek, putting his hand on my shoulder. "Praise you, son of joy and glory. Lady Fortuna has given you the chance to make your nation proud. It's time for you to become a hero and carry out a beer raid for the care of your infinitely loving brothers-in-arms."

I said that of course I can go and buy beer, but I wanted to borrow a gun in that case, looking over at a vacant AK-74 standing by one of the bedsides. As it turned out, though, nobody was eager to lend me a rifle. It wasn't just that I was the new guy in general, not to mention the history of espionage, which was to blame: the fact was, nobody really had any idea if I even knew the front or back of a weapon. I had already told the others about a fairly simple, but still relatively strong, military background, but for everything they knew, I could just as easily have lied about it all.

"I have a rifle, it's a-good," said the Finn, throwing an empty backpack at me. "Come on, we are a-going now."

"Continue forward!" the Greek shouted after us as we went. "Give us pride, fortune, and glory, you, the bravest of sons!"

It was already pitch-black outside. I followed the Finn via a small path that went over the lawn between the barracks. Although the rest of the battalion area was protected and enclosed by walls, our barracks area was relatively open. There were guard posts at the two entrances, but there were other holes in the defenses where you could get in and out of without being detected.

"Alcohol is forbidden here, you understand?" the Finn said, to which I nodded. "They will have a-beat you, really, if they find you with a beer here. They will be beat you up so you not want alive anymore, understand?"

I nodded again and the Finn showed me the gate which we were exiting through.

"We have to do this a-good, okay? If anyone guard comes, you playing cool, do you understand?"

I nodded again and soon we were out on the main road. The sun had set, and the yellow street lights lit up the village surroundings. Everything seemed closed, but after a while, we found a small shop. The Finn started picking two-liter plastic bottles into a basket, and then a number of smaller one-liter ones.

"Look at this one: it is a Polar Bear," he said, holding out one of the smaller bottles. The plastic and the label were both green with, quite rightly, a white polar bear on it. "This one is really Russian, they get you with that, they beat you worse. I should not a-buy it, but I think it is taste better than the others, so I buy it, anyway."

We paid at the cashier. The Finn unloaded all the beer into my backpack, and we headed back to base.

"We take another road back," said the Finn, leading me on what appeared to be the back road, toward another gate in the fence.

We walked on the left side of the small road as a small orange UAZ van, a Soviet copy of the classic Volkswagen bus from the fifties, came driving slowly on the right side toward us. When it passed, it continued for about fifty meters more before the red brake lights lit up and it stopped. Both the Finn and I had noted it without saying anything. My companion pulled his AKS-74 from the side of his body to his chest. We both remembered Mikola's talk about alleged kidnapping attempts around the base.

The UAZ started to turn on the narrow road to get back toward us, now on our side of the road. Both sides of the road had walls, so there was really nowhere to break off to or find cover. The Finn pressed the safety lever downward and slowly pulled the charging handle backwards before whispering to me.

"We split. I am going forward fast, you slow down but keep a-walking, about twenty meter behind me, okay?"

I did not understand how it would help anyone, least of all me, but before I could understand the plan, my partner had already hurried forward. I was alone, unarmed with a backpack full of beer. The only thing I had was a folding knife I got from my friend Anders. I pulled the right arm out of one of the backpack's shoulder straps, grabbed the knife into it, and unfolded the blade. Should the car stop and someone try to grab me from it, I thought I would try to get them to grab the bag instead of me. After that, at least I had the knife. It wasn't a good plan, but resources were limited and the clock was ticking.

I heard the engine sound and saw how my shadow in front of me moved in the light from the headlights. The UAZ approached slowly. Slowly closing in, slowly, and then it passed. It also continued past the Finn further ahead before disappearing at an exit after a few hundred meters.

I hurried forward to my friend.

"Why in hell you not keep a distance like I said?" he asked indignantly. I had apparently moved way closer to him than he had wished for in my nervousness.

"What the hell kind of that plan was that, anyway?" I asked, but I only got some mumblings about "Swedish fucking idiots" back.

We passed through the gate back to the barracks area without being detected by any Azov guards. Soon, we arrived and opened the door to the room on the upper floor. The others sat closed in a circle of deep philosophical discussion. The Greek stuck his head up and looked at us.

"Ah, the prodigal son returns!" he shouted, striking out his arm toward us.

Långström also looked up at us, cheering. "Beer! Beer! Beer!"

CHAPTER 9: THE POLYGON

I was awakened by the cell phone alarm ringing, completely exhausted. It was 8:30 in the morning and thirty minutes to line up. I recalled some vague memory how I vomited all over the toilet in mine and Barzuk's room loudly enough for the neighbors to bang the thin section of wall in complaint. I was hoping it was just a bad dream and imagination.

"Fuck, man, you were making noise last night," said Barzuk, thus confirming that this was not the case.

In a way, I still felt good. It was as if the final nervousness and stress over the new situation had been drowned and then literally exited my body together with the same drink the very same night. I was ashamed that I had gotten drunk to the point of puking, but other than that, I felt in high spirits—above all, mentally.

The battalion gathered for line up and count on the driveway past the beautiful entrance gates. There were many soldiers at the base, but it was obvious that the battalion was only about half the force at most. Gratch, as deputy commander for our company, stood at the front and waved us in, a company hardly stronger than a squad in numbers.

The battalion commander shouted something in Ukrainian, and all the men in the line smashed their heels together. Unknowing in why we stood in attention, like any soldier I simply imitated my surroundings. The commander shouted as the men tied their right hand, bringing it to their chests above their heart. As I continued to imitate, the commander shouted a line from a poem or song, and the battalion answered—and then another, and another.

I looked around: none of the other foreigners sang along, so I also remained silent, reserving myself to only listening. I understood nothing of the song more than that it felt powerful and fate-saturated. It was strong, and the energy in the battalion's common voice had a resilience that really echoed, even though there were no walls on the open field to echo between. It was only half a battalion of

72

men, but it sounded like an entire brigade. I thought of the young man at the nightclub who, with excuses about his studies and family, explained that he could not take part in the war. He had made me think of the Ukrainians as useless, a people who deserved to lose. The Azov battalion's song now contrasted this.

"Ukraina! Ukraina! Ukraina!"

After the song and the commander taking up the force count, the battalion was disbanded for breakfast.

"That was the prayer of the Ukrainian nationalist, if you were wondering," the Greek informed without me having to ask. "It's quite a hardcore piece of poetry, really."

The days mostly went like this. I was not the only one who lacked a rifle, and without Mike in base to command, the group mostly went around in circles. Time was passed with further beer raids in the evenings and line-ups with singing in the morning.

Långström's Italian friend wished to be called Yuri. He had a background as a paratrooper and suggested that we take the opportunity to pass the time with physical education. CrossFit was something he personally kept close to heart, and he offered to organize and lead such sessions for the rest of the group.

The heavily built Finn was not only a regular army reservist, but also a former French Foreign Legionnaire. He immediately declined the invitation for physical exercise.

"I haven't a-done that before. If I want to do running and shit, I go 'eted make a Legion boot camp again." As he turned around walking away, it was obvious his word was final. Nobody would change it, and there was no use bothering to try.

The rest of us changed out of uniform into PT dress in order to allow Yuri to begin tormenting us physically. After warming up a few laps around and past the courtyard and driveway, we headed into the ex-president's enclosed old tennis court. The group, minus the Finn, lined up. Yuri described and demonstrated how we would follow his example for a beginners' CrossFit session. It didn't seem very beginner-friendly to most of us, though, and I feared what a normal-paced one would be like. The slender Italian probably started from a rather low level, in his opinion, but even those in the group who were most well-trained struggled to keep pace.

The Greek looked up at me, sweat dripping from his forehead. "Fuck! This shit is gay as fuck," he panted, to which I replied with sympathy and confirmation.

Others also agreed about the gayness of it all. Yuri soon took notice of our complaints and began waving his hands.

"Stop, stop! Okay, guys, stop! Stop!" he shouted, giving us a much-needed break. Yuri pointed and gestured like any Italian would. "My friends! My friends!" he began, pausing a moment to struggle in his scarce English. "My training, it is not a…gay training!"

Yuri paused and pondered for a moment how the words in the next sentence would come together, while we were still breathing out. After a short period of reflection, he continued.

"My training is a-very manly! I am not a homosexual! I love a-the pussy!"

The rest of us had a hard time keeping up with the laughter. Yuri's physical ability was reflected in his total inability to understand sarcasm.

The days went on. I came past the entrance to our barrack after lunch. Långström and Yuri sat there, each with an AKS-74 in their hands. The guns were mostly disassembled, and next to them they had geared up with a bucket of water, soap, and a dish brush.

"We couldn't wait any longer," Långström explained.

They more or less left Mike's captain-less ship, instead turning over to Båtsman's group to get weapons. The rifles were fine-looking, with beautifully striped red-brown grips in lacquered, laminated wood. The insides were another matter: well used and dirty, not that it mattered much to the two foreign volunteers and their bucket and brush. As gun oil was in short supply, water and soap worked just as well for the purpose of cleaning out the fouling from the Soviet machinery.

There had been talk that we could go to the shooting range, which was a short distance from Urzuf—provided we paid for the gas from our own pockets. I looked enviously at Långström and Yuri with their new weapons and wondered if I might also move over to Båtsman in order to finally get a gun of my own. Mike, with whom I still had contact, was clear, however, that Båtsman was a fake bastard, and under no circumstances should I trust him. He and Tor would soon come down, anyway, at which point I would not only get my own PKM, but would also finally get started training on the other heavy weapons.

"My eye is still shit. You can borrow my AK if you want," Barzuk said.

It was also an AKS-74, solid green painted with a railed-up plastic handguard which mounted a red dot sight over the gas tube. Below the handguard was a pistol grip. The rifle was light and comfortable to hold as I dry fired in the room, but somewhat rugged and rough-hewn in its construction.

"The rifle really belongs to Homer, though, so don't break it," he said, and I promised to be careful.

As I practiced reloads and general handling, it felt like the exact opposite of the long, heavy, and precisely cut Ak4B I had previously used for several years in the Home Guard. All the controls, on the other hand, were largely the same as from the RK 62 and RK 95 TP I used in Finland seven years earlier. The Soviet gun felt a lot less solid than the Finnish rifle, but on the other hand, it weighed next to nothing to make up for it. I didn't have to practice a lot before muscle memory from three hundred sixty-two days in the Finnish military kicked in. The gun came with a single 30-round magazine and nothing extra, but I was content enough finally having something else than my folding knife to rely on.

Tor came back to the base by himself. He had finally tired of waiting for Mike, who, just like before, still was only one day and one special errand away from coming back to the front lines.

"I was bleeding cash like fucking hell, and Mike kept asking I lend to him, also. I started to feel like fucking social services over there."

We had a cartridge ration for training of thirty rounds per man and day, and as we had been dormant for a long time, we had saved up on ammunition. Mikola and I went to the central warehouse to pick up ammunition, leaving with two wooden boxes full of paper packages of thirty cartridges each—brown paper with a light green line on top. I opened one of them and inspected the cartridges: they were 5.45x39 caliber for the AK-74, marked by their green steel cases wrapped around copper bullets, with a clear green tip painted on top. I understood enough to know it wasn't ordinary FMJ, but I was not familiar enough with Soviet ammunition to know what kind of ammo we had received.

"It's all tracer," Mikola explained. "Apparently, for some reason, there's a huge surplus of this ammunition in 5.45, and we use it specifically for training at the Polygon."

We started to load ourselves into the vehicle that would take us to the shooting range (or the "Polygon," as it was locally known)—it was a small minibus spray-painted in various colors resembling camouflage. It was awful to look at, really, but it drew the least fuel for the purpose of transporting our group to the Polygon and back. A journey that would take about thirty minutes, depending on whether we got stuck in the mud or not.

"Isn't that Homer's AK?" Tor asked after looking at the weapon I carried over my neck.

Tor laughed and said that it was the gun which Homer had carried when he shot the guy behind him in the throat by mistake. "It was me who went forward first. I still do not understand how he shot the guy behind me. It should have been me who got hit."

Barzuk, who was standing at the side, intervened in the discussion: "You know that those two belonged to rival hooligan gangs in Russia, right? That pull of the trigger may have been very conscious and anything but a mistake."

Tor looked surprised, but he admitted that Homer was strangely unfazed after he shot the guy. After thinking for a while, Tor offered: "However, not caring much about life and death isn't an uncommon trait for the Russian. They aren't really the sentimental kind of people."

The discussion had already become a bit too conspiratorial for good taste, so Tor quickly led it back on track to continue telling the part of the story after the shot was fired.

"While we evacuated the guy he shot, he was bleeding like a stuck pig! I sat in the back seat with his head lying on my lap. I was sure he would die. The blood just poured down my dick and legs, onto the seat where it created a big puddle. At first it was warm and nice, but the blood quickly got cold." Tor laughed out loud. "The coagulated blood was splashing like hell in my ass as the car was driving fast on those terrible roads!"

A Ukrainian stepped past and looked at us, as we were getting ready to embark on our journey.

"If go near Manhush and get fired upon, do not return fire," he said.

Barzuk, having just been ready to leave us, met the Ukrainian with a somewhat confused tone. "What? Why?"

"Because is young National Guard. They like to make joke."

"You mean us getting shot at with a heavy machine gun is funny?"

The young Ukrainian looked at Barzuk as if the Frenchman were slightly retarded and responded: "Yes, it is." The Ukrainian was quiet for a moment before adding: "Usually, they miss."

We stepped inside the cargo space. As the driver closed the sliding door on the side, we found ourselves in the dark, without windows that let us see the surroundings—so began the journey to the Polygon I had longed for. The driver took great pleasure in riding hard and fast. It started with violent shakes and turns to avoid the worst pot holes in the asphalt. When these then, after a clear left turn, ceased, and the sound from the tires changed, I understood that we had switched from asphalt to dirt road. At times it was hard and dry. In other places, you could feel the wheels sliding in wet mud.

"Fuck!" Tor complained loudly as the driver made a very sharp right turn. "There are high cliffs above the beach around here. He is going to slip over the edge and kill us all driving like this!"

He yelled and tried fruitlessly to bang against the metal wall separating us from the driver's compartment in order to make contact. In actuality, I, too, ought to have been terrified, being locked in with the others in that small, dark, violently bouncing tin can, yet I was not. To me, it was like a carousel of life and death, and I had no problem with the bus landing in either. On the contrary, I was laughing and enjoying myself immensely, never having enjoyed life to such an extent before, stuck in complete darkness out of control. Ironically, I felt freer than ever before.

We survived the trip, and our ride soon began to slow down. The door opened, and we began emerging outside one by one. The outside light dazzled, and it took a moment before our eyes got used to it. The cliffs Tor described were right next to us; below them, the sandy shores of the Sea of Azov began. Warm, lovely algae and light salt water-smelling winds from the sea blew up the cliffs toward us. I took deep breaths, savoring each one as its own moment. Behind us were a couple of simple dirt walls, beyond which were endless open fields as far as the eye could see. The dirt walls, which had been erected to catch bullets, were full of debris and scrap metal—mostly old ammunition containers made of wood and metal, but also shattered helmets and the like. Littered across was all kinds of loose trash which would move when shot to indicate a hit.

The small group became stationary next to one of the dirt walls. We continued our discussions and nonsense, just as if we had been back in the larger room at base drinking beer. There was no assigned commander around to take command and order us around, so we quickly degenerated into a bunch of simple, leaderless conscripts.

Långström, still the most disheveled and least soldierly-looking one in the bunch, also took part in the discussions. However, he soon started walking in circles and became increasingly impatient with the lack of action. He walked around the group, looking at everyone in extreme annoyance over the lack of someone taking charge.

"Fuck! Don't just stand there like a bunch of idiots! Do something! Open fire!"

Leo, who had received a weapon somehow, looked around in great confusion for a moment. He then lifted his plum-colored AK-74, backed up a few steps, pulled the charging handle back, and opened fire at nothing in particular, not even at the wall intended for target practice. His semi-automatic fire just passed by the group and out into the sea.

"What the fuck are you doing, you fucking retard!" Långström roared at him.

"You told us to open fire?" the now even more confused Leo replied.

After this mishap, the foreign volunteers regrouped. It was decided that we wouldn't be doing anything especially complicated to start. We were allowed to get used to our new firearms and fire as we pleased ourselves, as long as each man made sure not to kill anyone by mistake.

The Greek and Tor, who both used AKMs in 7.62x39, were forced to conserve their ammo. They, unlike the rest of us, did not have a cartridge ration of thirty rounds per day; instead, they were forced to eat out of their reserves. Tor was annoyed with the trouble he had implementing his thirty-year-old techniques with the modern equipment.

"In the mid-eighties, we were taught to fire the rifle in this way," he said and stood up, leaning forward with the side of his upper body facing forward. "This shit doesn't work at all with these modern vests and plate carriers, but I also seem to be too fucking old to relearn anything else at this point."

Långström and Yuri set up diligently zeroing their iron sights on paper targets. As professional a move as it was, I had little interest in doing the same since my current gun was leased. The small Aimpoint micro attached to the rail also seemed zeroed well enough. I was more interested in how the AK-74 behaved in general. The rifle was a far cry from its Finnish-manufactured cousins as far as craftsmanship was considered, and I had little hopes of it being as pleasant to shoot.

Only a few shots in, though, and I was pleasantly surprised to find the recoil very low—next to nothing even—and I was eager to try the gun on full auto next. Full auto on the Finnish RK 62 had been a bit of a struggle, especially with its slightly angled buttstock. The RK 95 TP had been a lot better, but still definitely something the rifleman needed to hold on to when letting go of a longer burst.

The Swedish Ak 4B in 7.62x51 was a gun I had only fired in full auto by mistake on a few occasions. Having so much experience with Kalashnikov pattern rifles, my muscle memory more than once had made me push the selector switch straight to the bottom. This would have been semi-auto on Finnish and Soviet rifles; to my great surprise every time, however, this set the hefty Swedish battle rifle into full-blown automatic fire.

Where the Ak 4B jumped around like a horse in a lightning storm, the AK-74, in its neat 5.45mm cartridge, only recoiled slightly by comparison, even in comparison to the heavier Swedish Ak 5 chambered in 5.56. It was as effortless as to paint the target area with a brush. The light recoil synergized well with the accuracy of the lightweight, rapid-firing rounds. I loved the harsh little Soviet troublemaker from the first burst of automatic fire.

I refilled my single 30-round magazine, rocked it into the gun, chambered a new round, and continued. The ammunition rations started to empty, and the winds from the sea blew both rougher and cooler in temperature. The sun had begun to dive down, and it was time for us to move back, at least if we wanted to have our daily dinner dose of increasingly disgusting buckwheat porridge supplied in the kitchen. We called for a ride and soon the tiny minivan skidded up the road again.

CHAPTER 10: TRAINING CONTINUES

Mike finally returned to base to start organizing everything. He had brought with him a pair of proper sniper rifles for the group: two brand-new modern bolt-action rifles with good optical scopes. The receivers lay in modern, ergonomic chassis equipped with bipods and adjustable stocks, making them a lot more versatile than the battalion's other snipers, which were either basic hunting rifles appropriated for the battlefield, or old and badly beaten-up Soviet-era Dragunovs.

The only issue they presented was their being chambered in a Western cartridge: the 7.62x51 NATO posed some logistical concerns, as our available standard-issue rifle ammunition was all Russian. The only available way to feed these guns was with boxes of commercial Sellier and Bellot .308 Winchester. Unfortunately, the sniper rifles were not the only weapons chambered in Western calibers. The most common weapons facing logistical issues were Fort-model pistols chambered in 9x19mm Parabellum instead of 9x18mm Makarov. Fort, the Ukrainian arms company, also produced, among other weapons, Israeli-licensed Tavor bullpups and Negev light machine guns in 5.56x45 NATO instead of the Russian calibers.

The logistics problems meant that these weapons, useful as they may been, mostly collected dust among the storage shelves, lacking ammunition to run through them. Luckily, the sniper rifles fared better in this case—a hundred cartridges weren't enough for more than a few seconds of entertainment in a machine gun, but the same amount would be enough to last days in the bolt-action guns.

Mike told us that there would be some more Westerners arriving: a pair of Frenchmen, and one British citizen. As soon as he said that a Briton was coming, I got a strange, inexplicable feeling. I thought of the Isle of Man, a small island between Ireland and the English mainland. A very long time ago, predecessors in my own family tree had resided there, and, somehow, I got a strong feeling that Brit would come from that very same island.

They arrived the next morning. The two Frenchmen differed in age. One was young, about the same as me, around twenty-five years old. He was rather small and slender, but his face and hands were those of an older man. His deceptively aged look was enhanced by his thick, perfect black beard. The other Frenchman was older, somewhere in his forties. He was lighter in both skin and hair, sporting a more Norman appearance. Some time had passed since the spy incident, and the worst paranoia had subsided, but the French were still a dubious source of volunteers now. While Mike was showing them around the base, Barzuk and I followed behind—not too close, but within hearing distance. Both the Frenchmen spoke poor English, and between Mike's descriptions of the area, they conversed a lot between themselves in French. Barzuk, who they didn't know was a Frenchman himself, eavesdropped to see if they'd say anything between themselves worth remarking.

After a while, the older Frenchman caught wind that something was amiss. He halted, turned around, and waited for his pursuers to catch up. He confronted us with a hostile look on his face.

"What are you doing? Are you following us?"

Barzuk pretended to not understand any of it, and it was up to me to answer his suspicion regarding our motives.

"Well, we simply took the opportunity to follow because of the beautiful weather outside. My friend and I didn't really have anything else to occupy our time, and the fresh sea winds are lovely this time of day."

The older Frenchman, who introduced himself as Fabien, asked if either of us spoke French. Barzuk continued pretending not to understand either English or French as I continued to play the part.

"No, sadly, not a word," I said, and continuing in a shy manner. "I studied German in school, something I've come to regret afterwards. Had I gotten to choose today, I would definitely have chosen French. It's a beautiful language, and it's very pleasant to just listen to the two of you speaking to each other. Even if I sadly don't understand any of it."

My charms were successful, and Fabien's defensive and previously suspicious facial expression gradually turned into a smile.

"Yes, of course, our language is beautiful," he said with his strong French accent before proudly continuing. "It is the language of love!"

After following them in this way a while longer, Barzuk soon came to the conclusion that they both seemed okay. Fabien had originally come from the French Navy but later worked as a security contractor. In addition to liking the military and mercenary life in general, his political convictions were what compelled him to be here, describing himself as strongly anti-communist. The younger of the two, calling himself Cuix, had been an engineer in the French army who handled mines and explosives during his tours in Afghanistan. Cuix was a fighter simply looking for a good war to fight in—just as the earlier generations of his family had done before him.

When the British volunteer finally appeared, it was easy to spot him in the crowd. Among all the soldiers of different sizes in different kinds of uniforms, there was a lone young man who didn't fit in at all. His trousers were in MultiCam camouflage, like many others, but on his upper body he wore a thick coat in extremely English-looking tweed wool. The top of his head was covered with a flat cap in matching material and color.

He was about thirty years old, and as he cupped his hands around the cigarette in the corner of his mouth, rough fingers appeared that protected the lighter's flame from the strong sea breeze. Like Cuix, he also had a full beard, though somewhat more messy and not quite as perfect as the young Frenchman's. It also differed in color, being somewhat blonde-brownish bordering on light red instead of completely black. As I saw him, I unexplainably became even more convinced that he came from the Isle of Man. I was so sure of it that I wanted to tell him right away as we greeted. However, as I quickly noticed that he was quite reserved toward his surroundings, I decided not to. The odds that he would be from there were extremely low, and I would probably have come off as a nutjob for thinking so. On the other hand, should my feeling indeed be correct, and he actually did originate from the small and sparingly populated island, the risk of him becoming increasingly paranoid seemed great. I knew I would have been, in his position. It would have been impossible to explain how I'd guessed his homeland so specifically correctly, and any explanation would probably not have been enough for the new, obviously cautious stranger.

"The name is Chris, though you can just call me Swampy," he said. "That's how everyone else knows me."

He had no serious state-affiliated military background other than as a junior cadet in the British Army. He had, however, spent a long time volunteering with Burmese rebels in Southeast Asia, where he developed an autodidactic expertise as a minesweeper in the jungles. Outside of that, he was also quite fond of hunting.

"I've mostly been hunting birds, as we don't have much in the way of bigger game where I come from. It's a small island, located in the sea right between Ireland and England. It's very well-known for its TT races, a big annual motorcycle race. Maybe you've heard about those? The place is called the Isle of Man."

As his eye had begun to improve, Barzuk also started to accompany us to the Polygon, forcing me to return my borrowed rifle to him. With Mike returning, though, I was able to get access to Sonic's AK-74 instead. His back injury would continue to torment him for a long time, so his practice gun came into my possession. The AK was iron sights-only, with a plastic handguard and a very beat-up wooden stock. Like my previous gun, it came with only a single magazine, which was, as luck had it, a black 45-round RPK mag. Glad as I was to finally have a personal weapon, I still kept asking Mike where my PKM was—something which he assured me was an issue stemming from the higher levels of Ukrainian command: "The Ukrainians are so goddamn lazy. They never get anything done."

Mike mostly stayed behind at the base while the rest of us continued with training: Yuri's physical education in the mornings, then either tactical training around the base or shooting drills at the Polygon every second day. Långström had taken command and quickly started structuring everything. The more we practiced, the more organized everything became.

The rifle drills started to become more complicated and demanding. It was obvious that Långström, still sporting a haircut that only grew more terrible by the day, had greater military experience than the average Finnish conscript. Exactly where it came from, however, remained a mystery. I had initially been very suspicious of him, partly because of his hobo-esque appearance, but more so because of his nervous tics. The Greek had informed me that it was just a typical case of a junkie going through detox. And the hands in his pockets? Well, that was simply because, despite the nice weather, he was freezing, finding himself north of the equator for the first time in several years. From what I had been able to figure out, he had been fighting in some kind of extremely hot, humid climate before coming to Ukraine. "The jungle," as he himself had called it—that was as far as it could be narrowed down.

"Stress shooting," as Långström called it, was something he was particularly fond of. "Any fucking ape can shoot, hit a target, then reload an AK quickly on a shooting range," he said. "Doing so while exhausted and under stress, however, is something completely different."

While the group waited in a cluster some distance away, Långström took one man at a time to the shooting range, where he would lead and instruct him individually: two 30-round magazines were to be fired at a solitary board about forty meters away, with a reload in between, and the number of hits would be counted. Between shots, the shooter would perform various physical tasks according to how Långström decided. When everyone had finished, the group would lie down on their backs in a row and take turns running over each other's stomachs—the worse the shot, the more times the shooter would be stepped on.

I knew that this shooting would be strenuous, but I had the confidence that I should be able to handle it quite well. Quickly firing at short range was always my thing, and, in addition, I had a hidden ace up my sleeve that I assumed would aid me even further. I readied myself, AK locked and loaded, with the barrel pointing toward the ground.

"Fire!"

I lifted my rifle at the same time as I lowered the selector to automatic fire, emptying an accurate shower of rounds toward the board. With the AK-74's soft recoil, I knew I would land every shot, thereby saving myself from the physical torment by expending all my ammunition more quickly. I would then simply apologize for accidentally switching to full auto instead of semi-auto—an easy mistake to make for an amateur. It was a nifty and great plan, really. Or rather, it would have been, had it not been that Långström saw through it immediately.

"Fuck you! No full auto, you fucking negro! Lie down! Down to the ground!" he screamed at me in fury. "Ten push-ups! Ten push-ups! Faster! You cheap fucking asshole!"

I dropped the rifle and started doing push-ups as fast as I could.

"You're going too slow, you fuckup! That's twenty push-ups now! Faster!"

The pain, the smell of gunpowder, and his strong voice calling me out as a cunt in Fenno-Swedish dialect was a great nostalgia trip, bringing me back to my late teenage years in the Finnish military. Långström understood exactly what I was trying to do, and now he was going to punish me with extra physical torment to even the score.

"What are you even doing? What the hell even was that? Get up! Move forward! Ten meters forward! Faster!"

I grabbed the rifle, got up, and started running.

"Fire!"

I lifted the rifle and fired during my advance.

"Faster!"

I kept walking forward while firing, but faster.

"Ten paces right! Faster!"

I started to walk sideways, still firing while Långström kept physically jostling me around, hard enough for me to almost lose my balance.

"Fire!"

I fired.

"Get down! Ten push-ups!"

I threw myself on the ground and continued with the push-ups but only had time to do a few before Långström roared again.

"Get up!"

I tried to stand back up, but Långström kicked me in the back, pushing my face straight down back into the ground.

"Ten push-ups, I said! Ten push-ups! You don't get up before you're done! Faster! You fucking piece of shit!"

Out of the corner of my eye, I could see the saliva spraying out of his mouth while large blood vessels dilated on his neck and forehead.

"Crawl forward! Five meters!"

I crawled, but just as I arrived, he roared again: "Too fucking slow! Crawl back! Five meters!"

I tried to crawl backwards but immediately got a boot in my back, hugging me to the ground.

"Crawling, I said! Ground level! Fuck! Fire!"

I struggled to grab a proper hold of the rifle.

"Faster! Fuck! Fire! Fire!"

I didn't even bother trying to find the iron sights anymore. Instead, I simply corrected my aim by looking at the dirt being thrown up from the impacts around and behind the target. The first magazine was soon emptied.

"Reload! Reload! Fuck, reload! Faster!"

I used my left thumb to quickly release and swing the empty magazine out of the gun and lifted myself slightly to the side in order to bring a fresh one out of my vest.

"Five…four…!"

I grabbed the new magazine and began the process of inserting it into the rifle.

"Three…two…it's fucking urgent right now!"

The magazine locked in the wrong position; I tore it out to try again.

"One!"

Långström kicking Tor during live fire exercise.

I realized I wouldn't have time.

"You're useless! Ten push-ups! Faster! Faster!"

I struggled through the push-ups and got another try, now with even more shaky hands than before as Långström began to count down once more.

"Five...four...it's fucking urgent right now!"

Suddenly, I heard loud bangs. Hot air hit my face under high pressure, and dirt spewed from the ground around me. It was Långström firing his own AK down into the soil around me.

"Reload! Faster! Fuck! Fire at the enemy!"

I finally managed to insert the second magazine, and the stress shooting continued in the same pace for the duration of thirty more rounds.

Finally done, I used up a few more moments to mentally fight the strong urge to vomit. Once I finally managed to catch my breath, we went forward to count the impacts. Despite the short distance, I had only hit just over half of the shots. Långström counted the holes in the board.

"Twenty-nine, thirty, thirty-one, thirty-two..." He paused for a moment, examining the board intently. "Thirty-three is there. Thirty-four. Yes, thirty-four. Hmm, well, you got thirty-four hits, it would seem. All is good, then. Go back and send the next guy in line over to me."

Fortunately for me, things went even worse for many others. The results averaged about thirty for most. One recruit in particular barely made ten hits at the board, so despite my far-from-great result, I ranked high enough that I didn't have to be run over too many times. The "punishment" of being run over also really wasn't as bad as first expected. We laughed through most of the torment, especially as Gratch's turn came, the physically biggest one of us all. The air pushed through us, but as soon as it returned, the moaning was replaced by laughter.

After a while, Långström decided that we would start marching to the shooting range instead of using mechanized propulsion. It wasn't a particularly devastating change, since on more than one occasion we ended up waiting for a truck that never came. Marching was a great form of physical education in general, anyway. Above all, though, while moving on foot, the group would also get to conduct and practice tactical movement in a natural way, in both urban and open terrain.

"Do you want a PKM for this?" Långström asked.

Completely forgetting how long a walk long it was both there and back, I immediately answered yes. Via the contacts he made in Båtsman's group, he was able to borrow a machine gun for the day. The gun belonged to a Ukrainian called "Radik."

"You can borrow it for Polygon," the Ukrainian said, "but only if you don't destroy it. And, when you bring it back tomorrow, you clean it before. I don't want machine gun full of shit, okay?"

I grabbed the gun and promised to return it in as good or better condition than how I received it. The PKM was painted in the same manner as Barzuk's AKS-74, green with white spots—not just the receiver, but the entire gun. I struggled to understand how it was possible that the paint survived the extreme heat without melting and evaporating from the barrel when firing. I figured they must have used some special kind of industrial paint for the purpose.

The next morning, we got ready. Långström was leading, as usual, while Mike stayed back in his room. Even though we were already well into October, it was still very hot outside. Warm clothing was thus unnecessary. Outside of ammunition, the only things we needed to bring were food and water to last us the day.

We locked and loaded as we exited the gates. The Finn immediately expressed his discomfort with how I let the PKM's barrel lie almost horizontally when carried.

"You know, this thing is bulky enough that I can't simply carry it like the average AK, right?" I told him. "I know this gun inside and out, and the selector switch is on safe. Don't worry about it."

Finn looked at me, still highly dissatisfied. "Well, you do 'et as you a-want to. Just make a sure you do not a killet'ed me when you fucket'ed up."

On the way through the village, we strived to take as many strange detours as possible, where the group would be forced to act tactically to cover transitions and the like. Instead of walking the large main road, we climbed walls between the houses and took to crossing bridges. I was reminded about the Finn's concern more than once as my loaded PKM was pointed at people's faces several times during difficult crossings. Pointing a gun at a friend wasn't anything I enjoyed myself, either, but in some cases the big gun made it simply practically unavoidable. At least the safety was engaged, after all. Even if some of the thick brush of some tree branch would get stuck within the trigger guard, the gun wouldn't discharge.

We passed road crossings, practicing who would cover in which direction and when—things most of us actually knew before, but all in slightly different ways, with everyone coming from different armies and doctrines. In the beginning, as always, exactly everything went wrong. Total confusion covered every attempt like a blinding fog. Soon, however, with practice, things slowly began improving. Once we had eventually learned to work together, we had also managed to combine the best of all our individual experiences and approaches.

Of course, there were some exceptions, such as Leo and Abdullah, the latter of whom had returned from Kyiv together with Mike to also join the group of foreigners. Neither of these two brought any real military experience to the table. In Leo's case, there was nothing more than a little youth activity from the Home Guard in his mid-teens—not much different than the Boy Scouts. Abdullah had been in Azov for quite some time and was wiser because of it, but it was evident that, despite having been at the front line more than once, he lacked the basic fundamentals of military training. These few scratches in the vinyl were, however, the exception. As an entity, we other soldiers began to get to know each other more and more, and little by little, everything started to flow really smoothly.

We soon exited the village and began venturing out into the open fields. This proved to be the most difficult part of the march. The dirt roads were dry and as hard as asphalt. The sun was blazing, and my feet began to boil in my thick boots, aching more and more as time went on. The worst part, however, was that there was nothing to look at. Unlike in the north, we passed no spruces or pines. No rippling streams. No rose hip bushes or beautiful birch tree lines. All you could see on the horizon were yellow, open, endless fields. You walked and you walked, but the terrain looked the same, seeming as you hadn't progressed the slightest bit. There was no goal to focus on, like an intersection or turn in the road. The terrain was a single, boring infinity. If by chance a tree line finally appeared at the end of the field, only a new, equally large identical field followed behind it.

The only break we actually encountered on the march was a small Ukrainian army post on the cliffs above the beach. It was maintained by a handful of Ukrainian soldiers encamped in a simplistic military caravan. In addition to this, the only convenience they had was a small fireplace. None of the soldiers spoke English, so Abdullah acted as interpreter between us.

"They ask if we want tea."

Everyone in the group nodded. Tea would be nice.

After the short break at the army outpost, we continued forward. At times, we changed the scenery and passed down by the beach. Down there you could enjoy the soft sand under your feet and a guaranteed fresh sea breeze. Even though the sea looked the same everywhere, at least the large cliffs on the right side of us were moving as you passed them.

What seemed like an eternity later, we finally arrived at the position where today's shooting drill would commence. We had passed the usual range and continued a bit further until we arrived at a large valley. A small stream flowing into the sea from inside the countryside had broken the high, yet soft cliffs into two parts. Långström's plan was that we would do a combat shooting where the squad would be divided into two parts, where one pushes forward while the other supports the first with fire. The layout was simple, really, but it would be the squad's first more serious shooting.

Långström walked us through everything. The squad would start on a column toward where the valley began. From there, we would act as if we had gotten enemy contact, forming a line. Targets were to be arranged through the use of simple balloons which Långström would tie to the bushes and the vegetation down in the valley.

If all of these were shot to pieces before the exercise was completed, there were a couple of derelict old houses in the valley on each side, of which the one on the other side of the river could also function as a natural shooting target.

When the squad formed a line of fire from above, it would be divided into two. The offensive team, Alpha, which consisted of riflemen led by Långström, would assault down the valley first. This was to take place under cover fire from the supporting squad, Bravo, which was to be led by the Greek. Bravo was smaller in number but fielded me and my PKM, thus attaining equal or perhaps greater firepower than Alpha. As Alpha found positions in the valley, they would in turn give cover fire, allowing Bravo to also move down. The assault would continue forward in this manner, one covering the other's assault, and vice versa, until the fort below was reached. From there, the squad would start pulling back in the same manner as it had assaulted. Should all the balloons be expended before we were back up on the high grounds, there were some old brick building ruins inside the valley which would do for targets.

We started moving forward with Långström in the lead.

"Contact front! Line!"

The squad spread out and split up, assaulting and covering down the steep valley side. The balloons were popped one by one, and soon only the ruins remained for targets. For the first time, I had the chance to actually fire a machine gun into a target that gave proper evidence of where each individual bullet had impacted.

In Sweden, we had fired at Saab targets—human-shaped plates that fall upon being hit, then rise back up. With those, you will know if you've scored a hit, but with the delay from impact to when the target actually falls, they never provided any actual flow. Exactly where the bullet had impacted was obviously impossible to tell.

The ruined brick building in front of us was a different matter entirely. Dust sprayed from where each bullet hit in the wall. Using the bipod for a rest, I was able to turn the PKM during ten-round bursts, let it wander from right to left in the target. The bullets landed exactly as I had wished them to, in a fine line from one side of the house wall to the other. It worked great, unlike how my instructors in Sweden had taught me to handle a machine gun: firing long bursts of fire but never moving the gun during the procedure. It had always seemed like a pointless waste of a belt-fed weapon's capabilities, and seeing the gun fill a several meters-wide area in a single burst made it all very clear. It was both easy and effective to simply let the machine gun cover a large area with fire in a single burst—not at all impractical, as my useless instructors had taught me.

A disgusting stench irritated my nose, and I wondered if it could be the old corrosive Soviet-era ammunition creating it. We started pulling back, and the supporting team which I belonged to was the first one back on the top. I was finally free from incompetent and lazy instructors who insisted on tiresome safety regulations prohibiting me from firing from the hip. There were no safety regulations in Ukraine. Everyone else was busy doing their own thing, meaning if I missed, only I would know, anyway.

I took a stable kneeling position, grabbed the folded bipod legs firmly, and pressed the PKM to my side, pointing the barrel at the ruined house. Just as I had done before, I quickly painted a line from where the gun barrel ended in the corner of my eye, forward into the terrain. Thinking I had it about figured out, I pulled the trigger and let slip a burst. The target showed no impacts.

I'd missed, and the tall grass around the ruins gave no signs of where the bullets had impacted. As the belt was entirely comprised of FMJ rounds, no tracers were around to clue me in, either. Did I hit right or left? High or low? It was impossible to determine. The ruins lay approximately one hundred fifty meters away, much longer than I had previously practiced on the previous range. I was still panting from the run up the hillside, but decided to give it one last go. I pressed the PKM back to my right hip, repainted my mental arrow from the barrel, and once more pressed the trigger, releasing a long burst of a dozen rounds down the valley.

The ruins began to throw off smoke, marking the impacts exactly where I had calculated them to be. To make sure that it wasn't just a simple fluke, I changed positions, redirected the machine gun on target and fired another time. Once again, just as before, the dust billowed from the dry brick wall exactly where I wanted.

"Yeah!"

I heard a loud voice from my left side. It was Långström, who had also just arrived back up on top, his part of the squad following behind him. He waved his clutched right fist in the air while following my firing with enthusiasm.

"Fucking great! Get some! Feed it!" he yelled in excitement while I continued to paint the walls from side to side.

I was filled with an intoxication—the intoxication a man feels when he reaches a certain level of his craft. In simply hip-firing the PKM, I suddenly knew how Mozart felt when he played the piano. The annoyance of every time an instructor had told me to use a firearm like a simpleton was blown away. What I hadn't been allowed to practice during my years of service in Sweden, it had taken a single drill in Ukraine to figure out. Before the belt was emptied, I was able to find and hit specific parts of the building instinctively without any problem, despite the distance. Not having the gun kicking up dust in my face while firing, as in the prone position, even made finding the targets faster.

The PKM had cooled down and I saw how the paintjob had indeed melted away from the barrel, which explained the disgusting stench I had experienced during the drill. It was not at all some kind of special color as I first thought. It was just that the gun had never been heated up properly after its coloration. The sun started to set, and we needed to start heading back since most of the ammunition was gone, anyway. It was already late, and in order to save time, we decided to walk along the beach, providing the shortest route back to Urzuf.

"We need be careful of the Border Guards," Abdullah said, "not those army soldiers by the fire. They seemed to be okay guys. The real Border Guards though…they might be problem."

He was referring to a checkpoint above the beach, about halfway back to base. We had passed it without any issues earlier in the morning. By the time we would pass it the second time, though, it would already be dark outside.

"They like to drink their vodka, and by now they are probably pretty drunk already. They like to shoot first, and ask after shooting…"

CHAPTER 11: MINES AT THE BEACH

Old, badly healed marching fractures from my conscription time began to ache, and the abrasions grew ever bigger inside the humid climate inside my boots. An attempt to change socks made little difference. The thick Gore-Tex padding was drenched in sweat, and the new socks were soaked after just a few minutes. It was getting darker outside, but the temperature remained warm. We had only marched slightly longer than twenty kilometers and had a meager ten left to go. It wasn't very far, and I had gone longer distances many times prior without problems such as these. I wondered if I should blame the heat, or if it was my own fault for having become a simple weakling over the past years. Luckily, for the sake of my own confidence, I wasn't the only one in pain. Walking along the beach also helped ease the pain. Sinking down into the sand was a nuisance, but at least the soft ground was easy on the feet.

We soon arrived below the Border Guards' posting. Down by the beach was a long staircase leading to the checkpoint up on the cliffs above us. There on the outskirts of the village above was the actual roadblock and gate we passed earlier. After some deliberation, it was decided that we would follow Abdullah's advice to just continue along the beach below and hope that no one would take notice of us. The alternatives, moving up and trying to pass through the checkpoint, or using the stairs, could easily create a confrontation, in case the men guarding it were pissed from spirits.

We continued along the beach but did not get very far before a strong flashlight shined down on us. We hastily moved for cover behind one of the many upturned small fishing boats which lay on the beach, just next to the cliffs and the stairs leading up over them.

"What do we do now?" we asked, but Abdullah was all calm.

"It's no problem. I will talk to them," he said and rose up with raised hands, walking slowly toward the staircase to show himself. He shouted something in

Ukrainian to the guards up on the hill top while the flashlight shined on him. The guards roared something back. Abdullah, previously calm, began to sound increasingly nervous in his voice. He soon stopped moving forward, instead starting to back off, slowly at first while the discussion was still going on. Suddenly, he turned and ran back, throwing himself into cover next to us.

"We have to go," he said. "Why is that? What did they say?" Långström asked. "Now! We must go now!" the noticeably nervous interpreter said as the sound of an AK being racked above us echoed in the vicinity.

Långström looked around. The road we had come from was relatively open and without protection, but if we continued toward Urzuf along the beach, there were some more overturned boats, container buildings, and other debris providing varying levels of at least visual cover.

"We continue forward. Quickly now, but low positions!" he hissed. "Don't open fire unless they fire first! We don't fire unless we are forced to! Okay? Come on now, let's go!"

We started our escape, first crawling between the shelters on the beach while the strong flashlight went back and forth across the sand searching for us. The drunken yelling and the clatter of more rifles being loaded at the height continued as more friendly guards appeared by the cliff edge. As we made some more distance, we began moving quicker, soon to be running.

As the distance increased between us and the checkpoint, we began slowing down, still keeping an eye to our rear from time to time, though, to make sure we weren't followed.

"It's a-okay, they do not follow us," the Finn established. "Way down from top is a-long, and they definitely have gone'ted back to a-drinking inside. In morning, they do not even a-remember anything about all this shit."

Admittedly, the theory was fair, but the Greek wasn't taking any risks. "Still, let's try to keep it quiet. In case they, against all odds, have found extra energy and determination to follow us through the booze."

We continued slowly eastwards along the beach. From now on, the situation should be all clear without pseudo-friendly checkpoints and the like. It was only a matter of putting one foot in front of the other and repeating this, and we would soon be back at base. The Brit, Chris, was at the head of the column. He suddenly stopped, holding his right hand in the air to signal "halt" to the men behind him.

"What's the problem?" Långström hissed in an impatient voice.

Chris looked down at the ground and said something vague: "PMNs."

"Huh?" the Greek, who was a bit further back in the line, whispered extra loudly. "Speak up, so the rest of us are able to hear!"

Chris slowly turned around to point at his mouth in front of the rest of the column. "We've stepped into a minefield!" he shouted loudly.

Everyone in the group froze completely. The border guards, whether they followed us or not, suddenly became a much lesser concern, and our sound discipline went right out the window.

"Okay, so what the hell do we do now, then?" asked the Greek.

"I can disarm the mines," Chris replied, "but it will take the entire fucking night, at best. I have no idea how deep this goes. The alternative is for us all to back off and out of it."

Discussions how to proceed erupted. If we backed out of the minefield, we would have to redo almost the entire distance again, which included passing the border guards once more. Waiting for Chris to disarm everything wasn't a great option, either. Abdullah rushed in with an idea of his own he was hoping might save us both these troubles.

"What if we just keep going forward?" he asked.

"What do you mean?" Långström asked.

"We just continue forward like before?"

"Are you serious? That's a fucking minefield right there!" Långström said, seemingly not entirely sure if Abdullah was joking or not.

"But if we go back and around, that's so very far. And my feet hurt," Abdullah complained, looking sad and tired.

"What level of fucking stupid are you, really?" Långström asked as Chris filled in behind him. "You will have a bloody lot more pain in your feet when you step on one of these bastards!"

Abdullah pondered, enhancing his tactics and strategy further.

"Yes, yes, okay, but…what if we walk really, really slow?"

"That's not how mines work, you fucking moron," Långström sighed loudly, finalizing the shutdown of Abdullah's partaking in any further planning while the interpreter looked down on his poor feet in disappointment.

"There's no other choice. We simply have to go back and see if there's some other way around," said Långström to everyone's agreement.

"Careful now, careful steps!" Chris shouted. "Make sure to step in your own footprints in the sand! Not one single step outside!"

The squad turned around and slowly began making their way back. After a while, Chris judged that we were out of the immediate danger, but he decided to remain walking in point and keep his eyes open for more buried surprises.

By having to turn around, the distance needed to be covered had doubled. The nearest ascent for the cliffs was the stairs at the Border Guard checkpoint, and since our last visit with them, it was obvious enough that we weren't going to be using that route. We would have to leg it almost all the way back to the shooting range, and then find some other path from there.

Next to us, the cliff slopes soon appeared to flatten somewhat. Unlike in most areas where they were completely vertical, these slopes actually looked as if they would allow someone to climb them.

"It could work," Långström said, and the Foreigner Group was at least willing to give it a try.

The cliffs were not made of rock and stone, but consisted of porous sand. It was difficult to ascend, as the soft soil gave way beneath us. We reached about halfway up, but then the cliffs became too steep. Instead of moving upwards, we only dug ourselves into them. When someone finally managed to gain a meter of height, he would only slip two meters further down the next step he took. We tried to climb in several places with no better results. The sand was too soft to carry our weight.

Just as we were about to give up, a thick rope was thrown down from the edge. It felt like a literal miracle from above. At the height of the cliff, the silhouettes of some figures could be seen against the bright night sky. They waved at us and shouted something in Ukrainian. Not understanding or knowing who they were, I grabbed the rope, corrected my PKM over my back, and started climbing upwards.

Reaching the top, I saw a Ukrainian flag on a lone pole on the right which I recognized from before. We had been climbing just below the Ukrainian army position where we had previously stopped for tea, and soon, with their help, everyone had reached the top of the cliffs. One of the soldiers approached me, presenting a tin mug.

"Tea?" he asked, a gift I happily accepted. We accompanied them by the fire for a second time, resting a while before continuing.

"You're a fucking machine, you know that?" said a panting Barzuk, looking at me.

"Machine?"

"You're carrying that fucking PKM, and just keep going like its nothing. I've only got this AK, and I'm dying over here."

Barzuk lay back in a supine position. My body was aching, as well, but it was very pleasant to hear that my pains and aches weren't visible to the others.

The march continued again. The pace was slowing down more and more, but later that night we finally arrived back at the base. It had started to rain, and the water was a lot warmer than that which was offered by the only shower at the base. Being covered in sweat and stink, I took the rare opportunity to wash outside.

The next morning, I cleaned my borrowed PKM properly to return it to Radik. As I had reassembled the weapon after maintenance, I did a routine functions check to see if everything worked properly. After pulling the charging handle, I then pulled the trigger, and the bolt carrier struck home just as it should. I charged it back again, switched the selector over to safe, and pulled the trigger again, expecting the mechanism to stay back. To my great surprise, the bolt slammed shut again just like when in fire mode. I checked the selector and tested again with the same result. And again. And again. And again. The selector switch had apparently broken down at some point, no longer working to prevent negligent discharges.

As I had promised to return the machine gun in good condition, I felt pretty stupid, but there was nothing that could be done about it. I took the weapon back to Radik and explained that it had been cleaned and everything was fine, but unfortunately the safety mechanism seemed to have broken down somehow.

"Ah, yes," Radik replied, "it has been broken a long time. I just forgot to tell you."

CHAPTER 12: SHOT AT FOR THE CAMERAS

It was noon, and the company, still only sized as a larger squad, had once more ventured toward the diner, eagerly waiting to see and experience whatever kind of material was on offer this time. The food hadn't been an issue on my part at the beginning, but as time went on, the sheer mundanity had begun to take its toll. Day in and day out of the same buckwheat porridge with a chicken leg on top had also made even me start to poke around in it like the others. Constantly hungry yet unable to eat most of what was on offer, I sustained myself mainly on the pieces of pickled pigs' fat, bread, and crackers. In order to scratch out whatever protein was available, I broke into the chicken bones in order to consume the marrow inside.

"You're a fucking animal!" Tor commented in disgust, finding supporting opinions from the Greek.

"You know, by all civilized people's standards, the Ukrainians are a bunch of savages, but not even they eat like that."

"I can take yours, as well, you know, if you're just going to throw them in the trash, anyway," I replied. Tor grimaced in disgust as he dumped his chicken bones on my plate.

After finishing off Tor's chicken bones, as well, I went out to enjoy my after-meal cigarette. As I stood there smoking, Mike came by and grabbed me.

"You," he said, "come here. You're going to be a part of something."

"What?" I asked, confused.

"Bah! Don't ask so many questions, come on back inside the diner, and you'll see. It'll be good, I promise."

I put out the cigarette and followed. Mike pulled out a chair at one of the tables.

"Here, sit down" he said, and I proceeded as instructed.

Surrounding me were people unknown to me, rigging up light screens, lamps, and cameras. The white light from the lamps dazzled my eyes. I heard them speaking proper English (i.e., not American), and as I was trying to figure out what was going on, a man came and sat down next to me. He looked to be in his early fifties. His face was rough but honest and seemingly felt familiar. His chin was clean-shaven, just like his head. It was like I knew him—I was sure of it. But from where and when, I couldn't place. He stretched out his hand toward me.

"Hello, I'm Ross."

The handshake was firm and proper, but before the handshake was finished, it hit me.

"Wait a minute…Ross? Ross Kemp?" I asked, and he replied that, yes, that was indeed his name.

He was a famous British documentary filmmaker, and I had seen several of them over the years. After doing series on criminal gangs and violent prisons, he and his crew were working on a new series, titled "Extreme World." The first episode would apparently be about Ukraine, the revolution, and the current war. Apparently, I would now be an interview subject in it together with Mike, despite being completely unprepared for it.

Ross asked what we as Western volunteers were doing here. Mike replied that Russia was also a threat to Sweden in the long run. When asked if I shared the same opinion, the best I could come up with was how our massive neighbor was even closer to Finland. There was not much more after that. The interview was quite short but straightforward. Both Ross and his camera team seemed generally decent. There was a suggestion that they ought to join the entire group to down a beer at Urzuf's only open bar, but alas, due to a busy schedule, they were forced to decline and head back to Kyiv.

The next morning, we ate breakfast in the dining room as usual after the line-up and the strength count.

"Some reporters from the magazine *Café* will arrive here today," Mike informed us. He was definitely good at courting the media, and it was, after all, through this that I had gotten to know about Azov in the first place. In this particular case, *Café* was a well-known Swedish men's magazine.

"They will stay a few days to do an article about the group," he continued, but by this time we had all probably begun to understand that his main objective was to have them write a story on *him*.

The correspondents arrived shortly after. Outside the gates stood a reporter and a photographer. The reporter, who would handle the writing itself, was a young woman with long blonde hair. Her photographer was a rather fat man in his early forties, who with his dark beard and his body size made the female look even smaller than she already was. The young woman, Annica, seemed rather inexperienced and nervous. This, combined with the fact that she also came equipped with a rather sweet-looking face, turned Långström on.

"I had just started considering fucking one of those fat, disgusting kitchen babushkas, but she's a lot better," he said. "You can see that she is turned on by soldiers, too. It's obvious. She will be mine before they leave, I'm telling you! Just wait and see."

"Jesus!" cried the Greek, holding out his hands and pointing in the direction where Mike had gone to give them a tour. "She was fat, I'm serious. Disgusting, downright contemptible! Like a fucking elephant!"

"Shut up, you dirty Turk!" Långström replied. "Sure, she's not your ideal anorexic teenage dream girl, but I bet she's got a nice body under all those clothes. You're just jealous because you know you won't be getting any of her, because she is mine!"

The Greek mumbled something incomprehensible about Aristotle and how Western women had completely lost control of their figure, the rest of us enjoying the verbal shit-fest going on in front of us.

Mike spent a lot of time at command trying to get us some missions, and as a rule, he came back after meetings royally pissed off, waving his Makarov around in the air, shouting insults over the senior Ukrainian commanders.

"They never give us any missions! They just want to stay in base! Fucking cowards! One day, I'm going to shoot them!"

It was an increasingly big issue. The group had started to get restless, because we were never allowed to do anything serious. In order to do something other than just sit at base, Mike decided that the group would go out on their own exercise overnight, near the front line. The young reporter and her photographer would be allowed to come along.

We loaded up food and equipment for the night. A truck was requisitioned from the battalion's transport group for most of us to ride on. I had been ill with a fever for a while and would not participate in the overnight stay. Therefore, I rode in the group's only car together with Mike and another Swede, Lynx, who also recently arrived. After leaving the group and the journalists on the truck at the overnight place, Mike, Lynx, and I would return back to base to sleep in our cozy beds.

We embarked in our respective vehicles and began the mechanized "march" toward the front line. After passing the large city of Mariupol, we began to come out near the dividing line between enemy territory and our own. The fields were as open as anywhere else, and soon the road led down into a broad, shallow ground where water flowed over it. Mike engaged the four-wheel drive and rolled through the water. We were close to getting stuck, but the car managed to get up on the other side.

We stopped and got out of the car. The truck with the others remained on the other side. The driver was unsure if it would work, and he did not want to get the heavy, blue-colored civilian ZIL stuck in the mud. Just as Mike and the driver argued about this, gunfire erupted some distance away. Soon, loud cracks from bullets flickered in the air above the slump we were in.

It was automatic fire, and I tried to figure out from what. The sound of the bullets suggested that they came from the direction of the enemy, and the cracks were loud—way higher than the usual small arms munitions. It could have been an NSV or a DShK, two Soviet heavy machine guns in 12.7mm caliber. The rate of fire was too low for the NSV, which averaged eight hundred rounds per minute. This was rather approximately in the five or six hundred range, more like the DShK. It could also have been a KPV, the 14.5mm heavy machine gun mounted primarily on BTR and BRDM armored cars.

I yelled to get Mike's attention, laying out my suspicions about what was firing at us. Mike heard and waved the truck back the same way we came. We, on the other hand, worried that making a new attempt to drive through the water would lead to disaster, and instead continued forward along the road to find another way back.

I was sitting in the back seat. Mike was driving with Lynx next to him in the passenger seat. Lynx had been Air Force Ranger before and seemed to stay fairly calm. Mike, on the other hand, was increasingly agitated, especially as the road ended and we were forced off-grid. I took it for what it was and kept an eye on the sides and through the rear window. We had one single shot anti-tank weapon, an RPG-26. It was very easy to use, even easier than the earlier variants, and even the American-made LAW they were based on. Only thing needed to put it into firing condition was to pull a pin and lift the sights up. If my suspicion about a hostile armored car in the area was correct, we might need to use it—worst case scenario, from inside the car. It would, of course, be unpleasant for all of us inside, especially in the front seat if I had to shoot through the rear window, but the situation was what it was.

Mike was shouting and screaming uncontrollably, hitting the steering wheel with his fists as he brought us out on a large, plowed crop field. The car bounced up and down, the engine screaming as the RPMs went haywire during the jumps. I thought about the situation with the anti-personnel mines on the beach we had encountered before but soon calmed back down. If there were mines such as those around here, as well, they would just blow a tire or something. We would still survive, at least momentarily. I kept my eyes looking backwards and saw how we passed some large signs, which were, of course, completely out of place in the middle of the field.

"I think it said 'mines' on that sign we just passed," I heard Lynx state in the front seat. In a field like this, I realized, nobody would place anti-personnel mines. We had run into a field of anti-tank mines. An AP mine intended to blow up someone's foot or something was one thing; an AT mine that would stop a 50-ton tank was something completely different. If we hit one, this cheap Chevy knockoff we were riding in would turn into a pile of distorted scrap in an instant, with us crushed and scorched into mangled flesh inside.

"Well, whatever happens," I mumbled to listen to my own voice, "I bet it will be pretty quick, anyway."

Mike screamed and cursed as the car bubbled over the mined field. Lynx held on as best he could, and I tried to keep looking back, ready with both AK and RPG in case anything should appear outside.

After a few hundred meters more, we exited the field out to asphalt road again, saved by Lady Fortuna from being turned into burning kebab. Mike started to calm down slightly as we began to drive around, searching for the lost truck.

"Ukrainian fucking assholes!" he shouted as we swung around without finding it.

"They probably returned to the closest Ukrainian army checkpoint, by the main road," Lynx said, suggesting we should also head there in our search. It was reasonable enough that as they retreated from the fire, they would have sought safety among other units along the major road instead of waiting out here alone. Mike turned the car around, and we headed for the closest checkpoint.

"I wonder what those journalists will have to say about all this shit!" I laughed.

"Yes, it's a shame we didn't get any bullet holes in the car to show them!" said Lynx, laughing.

In the rear-view mirror, I saw how Mike's eyes started shining. "Fuck! Guys. We should pull over and put a few rounds into the car before we get back to the checkpoint!" he said.

Both Lynx and I approved of the idea. It would be great fun to scare them that way, and I for my part was very curious to see how they'd react to something like that. Mike's AKS-74 was fitted with a suppressor, and as we stopped by the car to get out, he put two rounds through the right-side rear door. He made sure to place them both carefully in the middle so as not to damage the window. The left window was already broken and taped up, and since the internal heat wasn't functioning, another broken window would mean a very cold ride home.

I took a look at the bullet holes. He had placed them only an inch or so apart, making it seem a lot less credible.

"Bah, they're journalists. They won't take notice of such a thing. They will see the bullet holes and wet their pants immediately," Mike said.

He was probably right. The rest of the group would understand it was all fake, but the joke was never intended to fool them, anyway. As long as the journalists were gone, we could tell the others, I thought. We got back into the car again and continued. We soon reached the army checkpoint, and, just as Lynx had predicted, the blue-colored ZIL truck was parked right next to it.

We got out of the car and gathered together with the rest of the group and the journalists. Mike told an exciting story, full of the colors of life about how we got out of the car to search the terrain for the others. Then, suddenly, we had found ourselves ambushed by enemy infantry up close—Russian special forces, to be sure. We had vigilantly responded with our own fire and driven the enemy off before quickly returning to where we stood now, ever victorious. The engagement was evident enough by the two well-placed 5.45mm bullet holes in the rear door of the car, Mike explained while pointing them out. The young female reporter, Annica, gasped for air seemingly close to tears, finding comfort in Mike's confidence. Her photographer held his emotions back somewhat better, though still visibly shaken by the action-filled story.

Immediately afterwards, Mike went over to the soldiers at the checkpoint. I saw him unfold a large map on the hood of a UAZ jeep, and I moved closer to see what he was up to. Upon getting there, I realized that apparently not only were the reporters being subjected to this story, Mike was taking the chance to also inform the regular army about his heroic deeds. The checkpoint commander took action at once. Combat vests and equipment were put on and readied; the soldiers took positions among the sandbags. The BTR-70 at the checkpoint's disposal fired up its engine. In an instant, this little joke meant to pull the legs of two novice Swedish journalists had escalated into something completely different. I struggled to understand what in all available fucks he was thinking, but Mike seemed quite content having alerted the army to chase ghosts. I returned back to the car, seeing how the Greek was examining the bullet holes in it.

"He did that himself," he said, turning over to me. "Right? To give the journalists something to write about?"

I didn't know what to say. The whole thing escalated quickly to become much bigger than I had anticipated. Not knowing what to do, I simply stuck to the official story.

"I know you guys are making this up, it's okay," The Greek said. "I know Mike does these kinds of things. Everyone knows it."

Tor also joined in: "Mike makes a lot of things up—it's well-known. He does it for the attention. He fired those bullets, right?"

The sun had set. It was dark again, and the nights brought colder and colder temperatures with them. The truck bed was cold, so the journalists were offered to ride in the car with us, squeezing into the back seat together with me. Mike continued to tell in detail about our fictional ambush while Annica continued to struggle against tears.

It was soon the journalists' last day in Urzuf. After spending the bright hours on the shooting range, mainly for the photographer's sake, we decided to visit the local pub to have a drink together. Mikola and I went ahead. It was a large, circular building relatively close to the base. The inside was basically a large dance floor with tables all around. In a slightly smaller, adjoining building section, there was another room with a bit more casual pub-based decor. Next to the bar, there were comfortable sofas, and next to them, a pool table.

As soon as we entered, we were greeted by music. The volume was comfortable as the dance floor was void of Ukrainian holiday-goers, but the number of speakers in the roof suggested that they could increase it a lot more, should the owners have wished. It was nineties pop music of some kind, and the notes sounded eerily familiar. I looked at Mikola, asking if he didn't also recognize this terrible music, which he did. We stood there for a while, still, both digging into our inner record archives. I began to recognize the text and the voice.

"Fuck," I said, "isn't this Dr. Alban?"

"Yes!" Mikola shouted. "That's fucking Dr. Alban! Fuck."

The year was 2014. It was cold outside. We were two lone Swedish volunteer soldiers in Ukraine. We were involved in a war that arguably shouldn't be of concern to us. At the present moment, our involvement itself was very unclear. Now we were going to a bar to drink beer, an hour away from the front line, still armed and in uniform. And the fucking bar was playing Swedish summer torment tunes from the mid-1990s. On top of that, it was Nigerian-born joke of an artist Dr. Alban, of all things possible.

After shaking off the music and the absurdity of the general situation, we headed toward the bar and each ordered a beer. The nervous bartender placed two bottles on top of the counter and backed away without charging us. His eyes were focused on the AK-74 on my chest, muzzle pointing upwards.

"You know, we would like to pay for this drink, too. We're not here to rob you," I said. "How much is it?"

The dark-haired young man behind the counter, hardly even above drinking age himself, stammered in poor English that it was fifteen hryvnia per beer—a very affordable sum for Belarusian import, which both Mikola and I happily paid. The guy at the bar exhaled and collected the bills while we headed to find the perfect seating arrangements in the otherwise nearly empty bar. Soon, the rest of the group also appeared together with the journalists.

Everything was off the record—no notes being taken down for any articles. We just had a good time over a few beers and relaxed. The Greek enjoyed playing around with the journalists, both of them desperately trying to play his games with zero success. The Greek's provocative but logically unassailable questions and examples only led them deeper into traps he had set from the very beginning. When they tried to defend their failures, the Greek only went harder. It was like watching a cat play with a mouse.

"You're sitting here, claiming that you believe all people to be equally valuable. You are lying. You do not think so, at all. Even a child knows the world doesn't work this way."

The idea that everyone is of equal value was a common concept and often considered the mainstay of Swedish political correctness. It was based on a sloppy mistranslation of the UN Declaration of Human Rights. Where the original English text read: "All human beings are born free and equal in dignity and rights," the Swedish version had translated "dignity" into "worth" or "value," something the Greek was quick to target.

The female journalist had long stayed silent and found Långström's charm more interesting than the Greek's brusque rejection of all her ideas: "You're a female, anyway. Your opinion isn't important."

Her photographer, however, continued steadfastly (but still in vain) to defend his beliefs from an increasingly tenuous position.

"Of course, all people are equally valuable!" the photographer said. "It's a fundamental value and human right!"

The argument might have worked on most Swedes, but as he tried throwing it at the Greek, it stuck like drops of water on a goose. The photographer's last ace up his sleeve, that of human rights, was like a ricochet on an angled steel surface as the Greek struck back.

"It has nothing to do with human rights—something like that doesn't even exist, anyway. That's dead text on pale paper, not reality. You simply do not understand it because you are an imbecile and a Swede."

"But—" the photographer tried again, but the Greek interrupted him before he got started.

"Listen here, for once. I'll paint a picture for you to imagine." The Greek let his hands float in the air in front of the photographer. "You're on a ship. It's sinking. Everyone is going to drown. Around you, you see many people in the water. You can save one of them. You had children, you said, right?" The photographer tried to answer, but the Greek cut him off immediately so as to enforce that he did not care about anything the photographer had to say. "One of them is your child; all the others are strangers. You can save one—who are you saving?"

"I...I would try to save everyone, or whoever was the closest, or—" the photographer tried to start, but the Greek hit back with the rules previously laid out.

"You can save one: your child, or a stranger."

"I would try to save both," replied the photographer. The Greek laughed and reminded him again how only one could be saved. He needed to choose one: his own child, or a stranger.

The photographer leaned back, noticeably upset by the situation.

"Then I will save no one," he said.

The entire group burst out in laughter. Not even Långström, who otherwise focused his full energy on trying to get his hand higher up the reporter's thigh, could hinder himself. He laughed so hard there were tears in his eyes, while his female objective looked at him with disgust.

"So you would rather let your children die than admit that not all people are equally valuable?" asked the Greek.

"Yes," replied the photographer angrily, crossing his arms, avoiding looking at the Greek.

"*That* was extremely Swedish of you," the Greek replied before waving his hand in front to signal that his work here was complete. The discussion was over.

A couple of local Ukrainian drunks approached our table. They looked erratic, so I put my hands on my AK, pulling back gently on the charging handle to do a press check. After confirming there was a round chambered, I returned to study the shaky advance of the experienced alcoholics who were quickly but erratically closing the distance. Obviously, as they didn't speak a word of English, the language barrier made it difficult to understand exactly what they wanted. With a mixture of primal sounds and hand signals, however, it soon became clear that they were mostly just wondering who we were. We looked like Ukrainian soldiers, armed and in uniform, but we obviously spoke a foreign language. We managed to explain that we were Western volunteers in Azov living at the military base nearby. One of the drunks lit up in a toothless smile and swayed off to the bar. He scratched around in his pockets, collecting some banknotes and small coins which the bartender helped him count. After a while, he came back, with a bottle in one hand, and the bartender's arm firmly gripped in the other.

The drunk went down on his knees next to the table and twaddled about something even the bartender had a hard time deciphering. After asking a few control questions, the bartender turned to the rest of us and began staggering to translate what the toothless man had been trying to say.

"He say he want thank you for what you are doing for Ukraine. That you are soldiers, that you have come long way. That you are fighting." The drunk twaddled on and handed out the bottle he had rattled together with his last remaining pennies. The bartender translated the twaddle. "He not have much, but want give something, and it's vodka."

I received the bottle and unscrewed the cap, which did not click as if it were from the factory. As it seemed, this was the restaurant's homebrew—of course, the only thing the drunkard could afford. A quick inspection of the liquid's scent also revealed that it was far from first-class homebrew. I lifted the bottle toward the light in the ceiling, which revealed light strokes of a rainbow in the otherwise clear liquid.

I passed it on to Tor while whispering "rainbow" to him. Långström snatched the bottle from his hand immediately and put his nose to it.

"Fucking hell. You'll go fucking blind if you drink this shit, that's for sure!"

Tor and I looked at the still-kneeling drunk. He smiled humbly at us, nothing malicious in his gaze. He wasn't deliberately trying to poison us. His gift was honest, the best he could muster with his meager means. We had discussions within the group on how to handle this while the bartender came over with a number of small glasses. The bartender poured the dubious contents of the bottle into the glasses. Then, together with our wasted benefactor, they each grabbed a glass after lining up the other ones around the table, and the rest of us.

The drunk twaddled something further before emptying his glass. The bartender translated that we would drink for Ukraine before emptying his, as well.

"Fuck it. We only live once. Free booze is free booze," said the Greek, downing his own rainbow drink, followed by the rest of us.

CHAPTER 13: SEARCH AND DESTROY

The summer had ended as the winds from the sea were slowly bringing in a rugged Eastern European winter to replace it. Increasingly thick layers of frost soon began blanketing the ground at night. The thin walls together with the lack of heating meant our summer barracks were nigh uninhabitable during the colder season, which was fast approaching. My disgusting morning breath created visible white clouds above me as I regained consciousness. I turned on the small heat fan Barzuk had bought from a local store a few days earlier to heat up the cold room. Initially, we had left it on throughout the night, but as it had almost caught on fire at one point, we decided to play it safe and only use it while under direct supervision.

The more or less zombified foreigners started arriving one by one at the morning lineup, every man having taken a serious beating from the local rainbow vodka. Långström, in particular, was sporting noticeable bags under his eyes.

"How did you fare yesterday?" I asked as Långström looked up at me, with great effort to produce a smile.

"Yeah, I was faring alright. Had to get up and puke my guts out in the middle of the night. That fucking Urzuf vodka's something I ain't giving a go ever again."

"Yeah, same thing here, but I wasn't talking about the vodka."

Långström looked confused. "About what, then?"

"Annica, that reporter. Did you bang her?"

Långström's headache seemed to worsen. "Ah, that bitch. No, it was no use. I'm pretty sure she was a dyke, anyway."

The day proceeded as usual, with us going through physical exercises with Yuri—maybe differing with me having to go puke in the bushes as early as during the warm-up. After lunch, we conducted some tactical training at the base, first practicing tactical movement outside buildings, after which Långström held a course in basic medical procedures.

"You there!" he shouted, pointing to Leo. "You've been shot in the stomach! What are you going to do about it?" Leo looked confused, but Långström couldn't be bothered to wait for him to get with the program. "Well, obviously, you'd lie down on the ground, crying about how you got hit in the stomach, right?"

Leo now understood what Långström was implying, so he performatively agonized in pain as Långström demonstrated first aid on him. He first quickly ran through everything from scratch, namely the use of tourniquets and blood stoppers in case one sustained an injury in the arms, legs, or neck areas.

"But for today, we will focus on a gut wound, as I said, so we'll move on to that."

The basic combat vests we had received came from Poland. They were probably good for what they were, but they were big, clumsy, and heavy. They had both Kevlar over the chest, sides, and back, as well as large, heavy bulletproof plates on the front and back. For mine, I had moved the plates over to my plate carrier to reduce the empty weight from twenty-two kilograms down to about ten, but most men in the squad still used the vests they were handed out.

"If you quickly need to get the vest off yourself or someone else, you just pull on this," Långström said, pointing to a bright wire located at the front top of the vest. "You do that, and all the panels will fall apart immediately, and you don't have to bother with buckles, Velcro, and all the other shit. However, as it's a fucking bitch to put back together, for the sake of time we will just skip that part now," he continued, telling Leo to remove his vest like normal.

"Well, then, do we see any blood on the shirt underneath? Well, not now, obviously, but it's not necessarily that you'll see any at first, even if it's a real wound. We need to quickly get the shirt off, as well as cut up the shirt to see for sure," he said, pulling up Leo's T-shirt over his stomach, where a rather large, not yet properly healed scar appeared.

Långström looked at the scar, then at Leo. "Fuck, man, who the hell stabbed you in the belly?"

Leo explained how he had run into some problems at home in Sweden recently, which also explained how he'd ended up in Ukraine. He had no military experience or background, something his embarrassingly lacking performance had made abundantly clear. His nervous nonsense while participating in the previous interview also made it clear that he could not have been motivated by ideology or political convictions.

The only baggage he carried was one of several, less successful petty criminal endeavors. His name appeared in a thread titled "Gangland" on the Swedish internet forum called "Flashback," where I had studied the stories about these criminal adventures of him and his friendly associates. They were all basically treated as a joke by the users, and their grand screw-ups were posted as a comedy, garnering both laughter and applause.

What was a literal hole in Leo's gut was, in turn, for me, a big piece of the puzzle. He, a failed petty thief, had apparently managed to put himself in a situation he couldn't handle at home. Without money or refuge, Azov was the best option he could think of, and rather than a new deadly puncture wound, he had chosen the war in Ukraine as a safer bet.

Långström continued his lesson, and after we'd practiced some evacuation techniques for a while, he began wrapping everything up.

"So, let's say we have someone get hit during fight, or a reconnaissance mission in no man's land—how do we proceed?"

Leo took a confident tone. "We'll just leave him behind. Fuck, it's his problem if he gets shot! Ha!"

Långström looked unimpressed at the unsuccessful criminal, whose reddish hair had begun to grow after having previously been clean-shaven. It was not the first time Leo tried to slip something in as if his opinions held merit, without him having a single clue what he was talking about. I used to stay silent, but my nerves were strained, and I asked him politely to shut his mouth before an even more useless, lowbrow idea ran out of it.

After dinner, I went up to Greek's and Mikola's room. Mike still said he was pestering the Ukrainians daily about getting us assigned some missions.

"Something will surely come our way soon," the Greek said. "Sure, this nonsense cease-fire has slowed everything down, but the war isn't over yet."

"In that case, though, our primary role will be to carry out recon missions for the regiment, right?" I asked, which the Greek confirmed as the likeliest scenario.

"I'm not going out on a recon mission against the Russians with Leo," I said. "The guy is a complete fuckup. You've seen him at the shooting range, and you've heard him talk. He's a naive, cocky retard who has zero idea what he's doing. He's going to get us all killed without even understanding how or why," I said.

The Greek argued it wasn't that bad. He agreed that Leo was definitely on the retardation spectrum, but also stressed that worse material had turned into completely fine soldiers.

"Leo just needs time. We give him that, and he will turn out well enough," he said.

"Okay, sure, whatever. But if we go on a recon mission tomorrow, I'm going to trip with my AK and kill him by mistake," I replied.

The Greek laughed as if I was joking, which I emphasized I was not.

"You may not understand and see it the same way," I continued, "because you're not a Swede, nor do you understand Swedish. I can see more of him than you do. He's a walking disaster, and he will get us all killed. If we go on recon, I will kill him. I'm not kidding."

I could read from their facial expressions and tone of voice that the Greek and Mikola both got the message, so I felt a lot better and asked if they would tag along to the bar for a drink instead.

A few more days passed when Mike suddenly gathered the group after a meeting with the senior commanders.

"Finally! We have a mission!" he said. The group lit up immediately. Finally, we would get something to do.

As Mike began briefing us, the hurried feeling we all felt soon began to subside. It became more and more evident that the objective itself—a search and destroy mission, as he called it, would be purely about searching for ghosts. The goal was to find the fictional characters who supposedly ambushed our car earlier. Having been there myself, I knew it was all made up, and even though I'd never confessed to it, most of the others knew, too. It was not just us who were moving out—the entire battalion would be involved in the hunt for this huge nothing-burger. Mike, on the other hand, spoke of it as if he himself believed it.

Be that as it may, it would still be fun to do something other than sit around the base. Even if the enemy soldiers in the terrain were made up, we did, after all, experience some kind of contact before all that allegedly occurred; I had, in fact, heard the heavy machine gun open fire on us at the river crossing. It was still possible that something exciting might come our way. We got ready early the following morning. I had reminded Mike that I still only had one magazine for my AK and asked if I could take some of his instead.

"You'll be carrying the sniper rifle, anyway, right? Then surely you won't be needing of all of your six to eight AK magazines, as well?"

Mike assured me there was nothing to worry about, because although most of the group was poorly equipped in general, the battalion would make a short stop at the Mariupol airport for supplies. Mike and Lynx got in the car while the rest of us would wait for other means of mechanization. The battalion's fleet of vehicles began to rumble, and the leaded scent of burnt Russian gasoline spread through the cold morning air, creating a mist over the assembly area. The old BTR-60PB IFV coughed its rough-hewn engine, and the Ural trucks began to slowly line up at the parking lot. "Mad Max," the big, black-painted armored garbage truck, was also rolling out, and I thought it would have been fun to ride it at some point.

"Here's our battle taxi," said the Greek before climbing up the welded steps at the back of Max.

Tor moaned a little with pleasure. "Fucking rock and roll! I've wanted to ride this bastard truck ever since I got here! So fucking cool!"

"You enjoy yourself," said the Greek. "You see all those rebars they spotwelded on the sides? You do understand that they aren't really bulletproof, right?"

Tor just laughed. He had no issues with that. Everything here was life-threatening, armored or not. The only thing that mattered to him was having a blast before dying.

When everything and everyone was loaded up, the convoy started to roll. We shared the open dump truck platform with a group of Ukrainians from Båtsman's group, the RDSh—Recon & Sabotage. It was extremely cold, and the body of the armored garbage truck was covered in frost. We huddled close together partly because there was a lack of space, but mainly to keep ourselves warm. Mad Max bounced up and down on the bad roads. All the extra armor, ineffective or not, added significant weight to the vehicle, and the suspension was struggling to support it.

After about an hour's vehicle march, the convoy arrived at Mariupol Airport. We climbed out of Max's freezing steel casket to stretch our legs. The sun had begun to rise, and its rays warmed our cold bodies. Soon we'd have mild, comfortable autumn temperatures outside again, and we could start taking off our thicker outer layers of clothing. I grabbed Mike and reminded him I still had only a single AK magazine.

"Calm down, I'll fix it," he replied before disappearing.

We waited for a while. The number of foreign volunteers had increased, though we were still very few. Mike's company was still far away, and we could only muster enough men for two squads at best—and even that was including the Ukrainian interpreters. One of the Frenchmen, this one with a special forces background, stood some distance away enjoying the rays of the morning sun with a cigarette. On his head he wore a gray cap flipped backwards. It was covered by small autographs from various soldiers he had served with during his years soldiering. His helmet, lacking a cover, hung from his side. Its green surface shined in the sun, and it looked like something had been written on the back of it. I went closer to look. It was a black ace of spades with a short text written below it: "I LOVE THIS JOB!"

Someone shouted "Machina!" and the command soon spread like wildfire, being repeated again and again among the soldiers. It was time to mount the vehicles and head out again. Neither Mike nor any of the extra ammunition I'd asked for materialized.

"Where the hell is the ammo?" I asked the others aloud.

"What ammo? What do you mean?" the Greek asked. "How much do you have?"

I patted the one magazine already inserted into my AK. "This is it."

"Fuck," the Greek sighed. He carried eight magazines in his chest pouches, but all his ammunition was in 7.62x39, so it did me no good. "Anyone around here have a spare mag or two of 5.45 for Carolus?"

Most men only had three or four magazines each, hardly enough as it was. The French special forces soldier with the autographed cap cursed both Mike and Ukrainian logistics in his mother tongue and flipped his cigarette away with one hand, opening one of his magazine pockets with the other to present me with another 30-round magazine. The extra ammo was significantly less than I had hoped for, but it was better than nothing.

The convoy began moving again. As we approached the front, I took the opportunity to lift the thick canvas which hid something large in the middle of Mad Max's troop compartment. I had assumed that Max was armed in some way but had waited for someone else to take responsibility for that bit. Since we were getting close and no one seemed interested in uncovering the mystery, I decided to assign the job to myself.

An NSV, a heavy machine gun in 12.7mm, hid under the veil. It was mounted in a modified platform looted off of a tank turret and could be swung around to fire in 360 degrees. Next to it on the floor lay a pile of long belts full of armor-piercing explosive ammunition, about three or four hundred rounds in total.

"Very nice!" Långström shouted to overpower the engine noise as I uncovered the beast.

I had been instructed on this huge machine gun during my conscription, but only in a very hasty fashion. Since then, the only time I had crossed paths with the NSV was during a NATO exercise in Norway called Cold Response, over five years earlier. I quickly started to regain acquaintance with the gun, checking for all the controls. The gun was unloaded, and with a firm grip and some force, I got the top cover to open. Based on my knowledge of Ukrainian maintenance, I was expecting to find completely rusted-out insides, but the gun was clean and seemed very well-maintained internally.

Mad Max halted together with the column at the entrance to a small village, and I began the process of loading the NSV. Most other automatic weapons fire from a closed bolt, meaning the bolt is tensioned backwards and drops forward to automatically insert a cartridge into the chamber. Upon pulling the trigger, the hammer or striker is released, which in turn ignites the cartridge already in the chamber. The vast majority of machine guns, on the other hand, for a number of reasons, instead operated from an open bolt, and the NSV was no exception. Here, instead, the bolt would be pulled and then locked back in the rearward position. Then, as the trigger was pulled, the bolt would release, push a cartridge into the chamber, and then fire it immediately once the bolt locked forward. Thereafter, the procedure would be repeated until the trigger was released, and the bolt once more locked back in its rearward position.

Before I inserted the ammo belt, I wanted to get a feel for the trigger. I positioned myself behind the gun and gripped the large charging handle on the right side firmly, pulling it back toward me to tighten the powerful return spring. When I released the handle, however, it simply struck forward again without locking back in the rear position like it should. Thinking I must have fumbled by not pulling it all the way back, I tried again. The result was the same: the bolt immediately struck back again the moment I released the handle, which shouldn't happen until the trigger is pulled. I attempted again and again—I tore and slammed at every possible control, but the moment I let go, each time the bolt slammed forward again. Manipulating the safety between attempts, fumbling with the trigger, striking the receiver with my AK-74—all resulted in no change. I asked around for someone who knew the gun better than me if I had simply forgotten something.

"I don't remember any of that," said Långström, shaking his head.

By all accounts, something in the trigger group appeared to be broken, and there was nothing to be done about it.

I saw Mike pass outside Max with his sniper rifle on his chest.

"Hey! Mike!" I yelled to grab his attention, pointing at the large machine gun. "The NSV—sure as hell it's supposed to fire from an open bolt like any other machine gun, right?"

To some extent, I wanted further confirmation I wasn't getting something completely wrong; mostly, though, I asked as a form of sending a report up the chain of command, informing them our vehicle-mounted armament was broken. Mike seemed to not understand what I was talking about, so I repeated my question in a simplified way.

"Ah, no, no," Mike replied. "It works just like an AK. The bolt moves forward after you've loaded it, then you just pull the trigger to fire!"

I looked at Mike continue forward on his business, then at the gun, puzzled and unable to fit everything together. There were exceptions to this rule, but I was absolutely certain that the NSV was not one of them. Unsure if the Urzuf vodka had broken my brain, I realized there was an easy way to find out for sure. If the gun actually was striker or hammer-fired, as Mike claimed, pulling the trigger would be followed by a clicking noise inside the receiver. I pulled the charging handle again, let it strike forward, and pulled the trigger. The trigger was dead, and the gun remained silent—no click from either a striker or a hammer falling. Had I done as instructed by our commander and inserted an ammo belt inside the NSV and charged it, the gun would have immediately started firing uncontrollably by itself.

I looked back at Mike, who was standing somewhat in front of us next to some Ukrainians. When we had talked about company formations, platoons, and the like in Kyiv, he seemed to know nothing about even the most basic things. After that, the oddities only continued. Just today, it began with the ammunition shortage, which he couldn't solve, and now this issue with the NSV, where it seemed he didn't even understand the difference between a gun firing from a closed or an open bolt—something even the simplest soldier used to learn early. How was it possible for someone who had worked as a professional soldier for years not to understand this? Had I also done as he said and charged the gun "like an AK," I would have, best case scenario, revealed our position to everyone on this and the other side of the horizon via hundreds of rounds of heavy automatic fire; worst case, I might even have put the lives of people nearby at serious risk.

I looked through the ammunition on the floor. The 12.7mm belts were about ten links each and assembled into a single, longer belt. I divided them up into a number of 50-round belts to make it easier to load and ease the wear on the gas pressure's ability to lift the belt and chamber each new round. Afterwards, I pulled every ninth cartridge from the belts. I also left a few empty links at the beginning of each belt and locked the first one into the gun with the first 12.7x108mm round on top of the feed tray. Now it would only be a matter of tightening the mechanism and releasing it to open fire. Upon that, the gun would fire until it hit the empty link, lock forward on top of it, after which the charging handle would need to be pulled back again—a less than elegant solution to the problem, but a solution nonetheless.

The convoy split up. Mad Max continued along some smaller roads, stopping every now and then to drop off soldiers to search different areas until it was just me, Gratch, and the Bulbash boys left in the back. Finally, it was also our turn to disembark to go on foot. With Max left on the road about a hundred meters away, I calmly began to make my way through a tree line. Gratch stayed back to keep visual contact with both me and the Bulbashes, who were clearing a nearby building.

After carefully looking for trip wires and mines during my slow advance, I reached a position where I had a good view over the fields in front. I turned to see where Gratch and the Bulbashes were staying. Everything seemed fine. A bit further away, I saw a couple of other Ukrainian soldiers moving up the road, beginning to climb into Max's troop compartment. I turned back forward to keep watch across the fields until I realized the broken NSV remained on the platform. I turned around again, seeing how the two soldiers who had climbed Max were now standing there, looking at the volatile weapon of mass destruction.

We were supposed to keep a low profile and stay quiet, but the situation demanded otherwise. I shouted at Gratch to quickly establish contact with the two men, making sure they kept their hands away from the gun. He himself had remained completely oblivious to what was about to happen. Only after a fair amount of hand waving and primal howling did he finally understand what I was trying to communicate. Gratch started running toward Max, the heavy machine gun, and the men around it, shouting at them to not, under any circumstances, pull the charging handle.

The men backed away, narrowly avoiding calamity, and calm returned. Gratch, the Bulbash boys, and I soon embarked again, soldiering on with our meaningless search and destroy mission. As expected, no enemies ever appeared. We aborted the venture later in the afternoon, and the column began the hours-long return to base. I manned the NSV on the way home, mostly to avoid someone else doing it and causing an accident. Standing there, bouncing up and down while the overweight garbage truck's suspension struggled with the bumpy roads, I had more time to think about this with regards to Mike.

I had previously asked him about his time as a professional soldier. The media had clearly written about it, how he had been employed at something called "NB21" for several years as a sniper. It was strange, as Sweden did not get a professional army until many years later.

There were some professional units around during that time, as well, he had told me, and about how they were positioned mainly at the border with Norway. As strange as I thought it sounded, I had simply assumed that when the media more or less confirmed it, it must have been true. No journalist could possibly be so incompetent as to let such a statement slip through if it was untrue?

I noticed that bouncing behind the NSV really wasn't that bad at all. The gun platform was sturdy and offered a good handle to hold on to and, unlike my friends sitting on the floor, I never hit my head against the metal walls. The sun had been given ample time to heat the air back up, and the fresh airflow was a lot more pleasant than it had been during the early morning hours.

I thought of a story Mike himself had told several times. The details seemed to change from time to time, but the very core of it was about how he had shot an enemy at 1400 meters. It was a kill which would have placed close to or among the ten longest ever at the time. I had myself tried shooting at extreme distances close to this and knew how difficult it was. Obviously, I wasn't a trained sniper myself, and I knew very well the power of luck. However, there were other details that made the story all the weirder, details I hadn't really taken notice of before now.

The story was a little different each time he told it, especially depending on who the audience was. Something which always recurred, though, outside the distance which he obviously boasted well about, was the details and description of the shot itself. The enemy he shot had been a truck driver, and he had shot him through the side door window. Mike had many times described with great empathy about how he had heard the glass being smashed by the bullet upon impact. However, as the target remained in his seat for a moment afterwards, he had initially thought that he had missed. It was only when the door was opened and the man fell out from the driver's seat down to the ground that he saw the blood on the seat and knew the bullet had struck home.

At 1400 meters, obviously, you wouldn't hear the sound of a glass window cracking. Anyone ought to understand that, but until now, that part had completely passed me by. Being able to see blood on a car seat through the scope also sounded very strange the more I thought about it. Even with the best optics available in Ukraine, a target the size of a man would be very small at that distance. The shot itself was extremely difficult, though still far from impossible—the rest of the description, however, made no sense. It sounded like how someone who had no experience in long-range marksmanship would picture it, based on what they had seen in films and video games. I began to doubt Mike's background as a professional sniper, or if he had any military experience at all.

As we arrived back at the base, I removed the belt from NSV and grabbed Gratch to remind him to let someone in charge know about the issues with the gun. A weapons technician needed to take a look at the gun before an accident occurred. I then hurried to the dining room. It was nearing closing time, and there were many mouths left to feed. Just as we were queuing up to get our daily dose of material, I saw a familiar face passing by. A tall, slender young man in a black uniform who had just finished his plate. It was Bear, the guard from Kossatskiy in Kyiv, who had assisted me on my first day in Ukraine while applying to volunteer at Azov.

"Hey! You finally got to come here, as well!" I shouted toward him.

Bear explained how he had arrived the same day with a new batch of fresh recruits. He had waited a long time and was eager to finally see some action, excited to take part in the war. Once here, however, he had only been met by further disappointment.

"They have put me in a unit that only do guard duty," he said. "They will put me in a block post, and I will just stand there. Look at cars. Reading documents."

The war was in a deadlock. Outside Donetsk at the airport, active fighting was still going on daily, but here, in the south outside Mariupol, it was mostly quiet. Bear's previously happy tone was covered by a veil of growing depression. He asked if it was true that we had our own group, specifically for foreigners.

"Yes, that's right," I replied.

"Could you talk to them and ask if I can join your group? I know I do not have much military experience, but I promise to try my best. My English is not so good. I know that. But I speak both Ukrainian and Russian. I can help as a translator if you need one. Please, try to help me transfer to you guys. I want to do something, and not stand in a fucking block post all winter."

I explained to Bear that, apart from the pointless excursion we had been on during the day, we hadn't been going on any missions ourselves. He probably wouldn't get any action in our company, either. We could, however, use an extra interpreter, and I would ask for a transfer on his behalf.

"Thank you! Please, do it. I will do anything. I just do not want to stand in a block post and rot all winter."

CHAPTER 14: QUALITY EQUIPMENT

There was a knock on the door to the room.

"Come in!" I yelled.

Mike opened and turned to enter through the doorway. "I've got something for you, but only if you've been a really good boy," he said.

I immediately saw the reason why he entered the room in such a clumsy fashion. He was carrying a PKM machine gun from the carry handle. Behind him, Gratch followed with two 100-round cloth cassettes and a large 200-round Maxim-style metal box in his hands.

"Finally!" I shouted straight out, getting issued the belt-fed machine gun I'd been waiting for.

Mike and Gratch only stayed a short while before it was time for them to go to the commanders meeting again. It would, like before, result in no new assignments for us, not that it mattered to me. I began to get acquainted with the weapon I had been missing since I learned to love it during my military service in Finland. We had mainly been issued Finnish-produced, old, worn-out KvKK M62 light machine guns. The few times we were allowed to use the newly acquired PKM machine guns instead, the contrast was day and night.

The joy, however, did not last long. The closer I got to examining the machine gun, the more details I reacted negatively to. Some were just odd, others downright worrying. The receiver seemed fine, but the top cover was strange. It was stamped 2014, obviously being newly manufactured. It looked to be of a milled construction instead of pressed sheet metal like a normal gun, though, and the quality in general seemed off. The gun had been presented to me with two barrels—one mounted, another a spare, and they were both complete nightmares. The insides where chrome-lined and seemed mostly fine, but the outsides were a completely different matter. Huge potholes of rust covered the surfaces, on some occasions several centimeters long and millimeters deep. The barrel attached to

112

the gun was missing part of its front sight, having been completely eaten away by the rust. The spare one was missing its carry handle. Both barrels were missing their gas regulators entirely.

With some help from Mikola, I was able to gain entrance to the armory of one of the infantry companies. There, I asked if they had any PKM barrels of their own to spare. They did have a few, and the commander gracefully allowed me to pick one out of the stash. Unfortunately, their barrels were in just as bad of shape as the two I had gotten, full of rust and missing parts. Luckily, at least I was able to find one still retaining its gas regulator.

The next morning, we went out to the Polygon again. While the rest of the group performed the usual shooting drills, I embarked on a mission to see if my machine gun would work. I had already understood that it consisted of some newly manufactured parts and others looted from old, broken weapons, a Frankenstein's monster. The barrels remained the greatest concern, as I was worried whether they would hold together, or just outright explode when fired. As luck would it, no accidents occurred. All three of the scarred barrels held up fine when firing, even though they all left a lot to be desired precision-wise. The biggest problem, however, was that they could barely fire fifty shots without jamming—a criminally bad performance.

After experimenting a lot, I was able to slowly decrease the number of stoppages. I removed the ejection port cover and filed down the extractor some. Having found the barrel which seemed to have the fewest issues, I replaced its missing carry handle, flash hider, and gas regulator with parts from the others. Soon, I had a machine gun that would run at least something like two hundred rounds between stoppages—not great, not terrible.

Low-quality PKM with the two spare barrels.

Instead, other problems erupted. The weakly constructed and already damaged bipod broke off. The only carry handle was also weak and soon broke off, as well. The only remaining solution was to wrap a piece of cloth around the barrel, in order to have something to grab onto. I was already missing the much heavier and bulkier, though never-failing Ksp 58B I had used in Sweden. Because of all the troubleshooting, I had already used up a lot more of my belted 7.62x54R than I had initially anticipated. Not knowing when I would get a resupply, I needed to find some other way to kill time.

A small minivan came driving along the road, halting not far from me. A handful of Ukrainians opened the rear doors. I saw them pulling something big out of the trunk and decided to leave my piece of shit machine gun in order to investigate. It was a PTRD, a 14.5x115mm World War 2-era single-shot anti-tank rifle. The receiver markings explained the gun had been manufactured in 1942, and even though it was obsolete against modern tanks, the huge cartridge was still extremely potent in its own right. It could be used to gain advantage over both lighter-armored vehicles, as well as men who foolishly tried to take cover behind walls or other kinds of obstacles.

PTRD bolt.

In order to make use of it, though, the first step was to find out if it worked. For this, a brave (or stupid) soul was needed to test fire the small cannon, something the men apparently hadn't decided on before heading out. I heard a voice volunteering; seeing how my hand had lifted into the air, I realized it was my own. While cursing myself, I lay down behind the giant piece of metal rod, grabbed a shiny and nice brass cartridge from the open tin container next to me, and slid it into the chamber. I closed the giant bolt, locked it into place, and took aim. I inhaled, prepared to die, and pulled the trigger.

Click.

After a few seconds of extreme tension, I unlocked the bolt to extract the cartridge. Where the firing pin should have left a proper dent in the percussion cap, there was only a slight dimple to be found. After some trial and error, I was able to disassemble the stupidly simple bolt and extract the firing pin. It looked fine enough, just as the firing pin spring sending it forward. Everything seemed to be in order, but despite several attempts, using ammunition from different tin cans, the gun kept making simple clicking sounds instead of huge ka-booms. The Ukrainians returned to base, disappointed in not having been able to get the gun up and running. I decided to take a walk around the firing range, blissful in not having had my face torn off by an exploding anti-tank rifle.

There were a lot of shards from RPGs and the like around, which I familiarized myself with. The covers were light, in thin aluminum. It was easy to see why these weapons proved ineffective against human targets, outside the psychological aspect, at least. Even though the explosion on impact was powerful, the low amount of lightweight shrapnel they created indicated a weak area of effect. My eye detected a slightly larger piece of shrapnel. Unlike the others, it shined in a bright pink color. Instinctively, without giving it time to think, I lifted it up to take a closer look. It seemed to be the back of an RPG head, and out of the dented and twisted aluminum casing hung what looked like a broken gyro in some sort of wire.

I realized what I had just done. The pink color was the explosive which largely remained inside the broken projectile.

Carolus test-firing the PTRD.

115

Never touch stuff you come across at the range. I had no idea how many times I had heard it said over the years. All I knew for sure was that I had fucked up. Badly.

I carefully placed the dud back on the ground, being exceedingly careful not to change the angle in either vertical or horizontal position during the process, fearful of the gyro lighting a spark. I memorized the location through a small bush a few meters away and then went to find Chris.

"Dude, I found a thing over there," I said, pointing to the bush a few hundred meters away. "I think it's a dud, an RPG which failed to detonate."

The Brit sounded unimpressed, like he had heard this before. I got the feeling that he had previously been distracted from more important business by nitwits on several occasions to investigate suspected mines, only to discover the person alerting him had mistaken a simple rock for a bomb.

"It was pink on the inside," I said to pique his interest.

"Did you say pink?" The color description had succeeded in alerting him, as Chris' tone had become significantly higher than before. "Where? Show me!"

As we got within ten meters, I could point the object out. Chris told me to wait and carefully continued forward to take a closer look. He waved at me and shouted.

"Back off! Go back! Don't come any closer! Go get my backpack and bring it over here!"

I did as I was asked, then watched Chris from a distance fit a small explosive charge next to the culprit in order to blow it in a controlled manner. With a wire cable between us and the charge, we moved away, seeking shelter in a small depression in the ground. Chris took out his flashlight, unscrewed the battery cover, put the wires in between, and screwed it back on. He raised his head, scouting the open field one last time to make sure nobody had wandered off into the danger zone.

"Alright, I'll turn the lights on now," he said, then looked at me to confirm that I'd registered the warning. He clicked the flashlight, which was immediately followed by an explosion. A black cloud of smoke rose into the air, soon being picked up by the sea breeze. Following its route inland, I reminded myself not to go around touching shit anymore, even if I was bored.

The sound of the explosion was nice, though, and the next day, we were set to practice explosives in combination with urban combat. Although most of the Polygon consisted of open fields, one part of it consisted of several large structures. To save time, we had requisitioned an old, worn Ural truck for the purpose.

The driver took a handkerchief from his pocket to clean his narrow glasses. He was much older than the rest of us, even Tor, who was approaching fifty. He was not the only elder in the battalion, but his appearance was special, which made him stand out from the crowd already during the morning line-ups. He was a bit on the short side, and his compact body type gave him a somewhat square shaped appearance. His hair, which had likely been black at some point, had now mostly turned silver-gray. What stood out the most, however, was his thick, wide black mustache. The driver didn't speak English, so Bear, who had now become part of the group, acted as interpreter. The driver called himself "Inkasator," which could be translated as "cash transport"—the kind of vehicle he drove before joining Azov in the war.

"He's really funny. It's a shame you do not understand what he's saying," Bear said.

As a young man, Inkasator had been a "liquidator," one of the many Soviet conscripts who was called in to clean up after the Chernobyl nuclear accident in 1986. According to Bear, it sounded as if Inkasator suspected that he was already living on borrowed time because of it. Many, perhaps even most, of the liquidators had already perished from cancer, and their numbers were thinning out more and more with each passing year. Therefore, even though others thought him too old and gray for it, he decided to join Azov. For him, it didn't matter much whether he died in the war today, or if death came to him a few years down the line while lying in his bed.

"He says it's better that he dies than a young man with his entire life ahead of him," Bear explained, "but he may not be able to accompany us on any recon missions. He is radioactive and luminous at night."

Inkasator's large mustache spread out under his nose and a broad smile showed. The old man looked at us and stretched out his hands to paint a round field around him.

"At night, I am big, shining, green lamp!" he forced forward through poor English, ending with a hearty laughter.

We embarked onto Inkasator's old Ural truck, and Bear continued to tell us about Inkasator.

"You know, he's a real troll, too. He's lying and fucking around with new recruits. In the kitchen, he managed to trick a young soldier that if you sound like a chicken, then you get an extra chicken leg with dinner. The idiot did it! Standing there, cackling in front of the babushka handing out the food! The babushka just looked annoyed, while Inkasator was next to it, looking like he would die from laughing!"

The old man fired up the engine, which deafened us all to further possibilities for Bear's stories. The one about Inkasator's unjustified lying to the recruits was a bit funny, but I had completely forgotten about Bear's complete lack of ability to be quiet for even a moment. The loud roar of the diesel engine ironically offered a feeling of silence, which I enjoyed immensely as we skimmed off toward the urban area of the Polygon.

CHAPTER 15: A GUARDIAN ANGEL

The truck began to slow down and soon came to a halt. We disembarked through the back. Chris brought with him a black sports bag full of explosives, which I was looking forward to training with. If there was one thing I missed during my time in Sweden and Finland, it was that explosives had been extremely limited in my experiences in both countries. Most of my blast training actually came from when I had gotten to blow up tree stumps out in the field with my uncle as a child.

This part of the Polygon had previously served as a military garrison. Some scattered documents and debris which had survived the weather and wind suggested that it had once been a base for anti-aircraft artillery, still active post-Ukrainian independence. This was also confirmed by a large Soviet mosaic on one of the short sides of the house walls. It depicted a large Red Army soldier, anti-aircraft guns, and red Soviet flags. Outside the occasional trash and accompanying communist artwork, there wasn't much left in the area other than the gray bricks which made up the buildings themselves. Most of what could be looted had been already—furniture, windows, toilets, everything. Only large, empty, and white brick skeletons remained, and their dark window openings gaped like hundreds of empty, dead eyes gazing over the fields outside.

Most of the buildings inside the base were between two and three stories. Just outside, a few hundred meters before the entrance, there was a lone, similar-looking five-story building that towered over the others. The area outside this provided a better turning radius for the truck than inside the confined base, and we proceeded the last bit on foot. We soon passed through what previously must have been the closed entrance gate. All that remained was the small gatekeeper's building and some sticks where the gate itself, which once guarded the base, must have been.

We soon got company. From ahead, a UAZ jeep quickly closed in toward us at full speed. Behind us, a division from one of the infantry companies appeared. They came in a brand-new Kozak-2, a Ukrainian-made, four-wheel drive, 5-ton armored car. It was painted in gray, with black and white camouflage stripes. The roof sported a turret equipped with a 12.7mm machine gun and smoke dischargers. It contrasted heavily with the battalion's otherwise archaic vehicle fleet, especially as it stood next to our dilapidated old Ural truck, in particular. The arrival of the Kozak made us hopeful that better equipment was on the way. I looked down at my rusty piece of shit PKM, thinking that maybe, with luck, I could get a new one of those, as well.

One of the buildings at the Polygon.

The UAZ reached us first. A low-ranking army officer stepped out, hurrying toward us shouting something in Ukrainian, which Bear began translating into rudimentary English.

"He is asking if we are going to fire RPGs on the buildings. We aren't, right?"

Långström confirmed that we weren't planning anything like that, which Bear relayed to the officer. The man breathed a sigh of relief and continued talking as Bear proceeded with the translations.

"He say army have position behind the buildings, near the beach. Sometimes people here fire RPG on the houses, but they often miss, and the rockets land inside their position. He say they don't like that, because they have to stay in the trenches when it happens. They can't go outside, take a shit. He ask us if, please, we could stop firing rockets at them. That would be nice."

Seeing as we had no RPGs, the officer turned back to his position very well satisfied. Långström and Bear, meanwhile, turned back around to talk to the head of the armored car group. Obviously, there had been some form of double-booking on this part of the Polygon, but they were confident it would be possible to work something out where both parties could thrive. The armored car commander explained that they didn't really have anything special planned. They would practice maneuvering their Kozak on the fields outside while letting the gunner practice firing while on the move. The infantry would also practice working together with the vehicle. They weren't planning to enter the abandoned base, but would only use the buildings for target practice.

There was a row of two long barrack buildings closest to the road and the field. For our part, we really only needed a single building for our purposes. Långström suggested that the mechanized infantry would use the building to the east as a target while we held up in the one to the west, near the road and guard hut.

"We take the building to the left side then, and you guys shoot as you wish on the right one, okay?" The other commander nodded in agreement, returning to reconnect with his group waiting by the road behind him.

"Ha, oh fuck...I just hope that guy knows the difference between right and left now," Långström joked as we entered the western building, stopping in one of the large rooms next to the mid-section stairwells.

Chris opened his sports bag and began preparing his explosives, detonators, and other equipment which would be used for today's lesson.

"Eyes here now. This is how Soviet or Russian explosives usually come pre-packaged," Chris said, showing off a small, reddish-colored block of TNT, similar in size and shape to a paper-wrapped stick of butter. "The detonators can come packaged in several different ways, but they usually look something like this," he continued while holding a small metallic stick the size of a cigarette.

I oftentimes kept a camera nearby to document both our exercises and our everyday life, and today was no exception. I turned its dials for aperture and shutter speed according to what seemed appropriate for the dimly lit interior of the ruin. When satisfied, I lifted the camera to take a photo of Chris.

"Stop that!" Chris cried out, holding up his hand. "Mate, you don't have a flash on that thing, I hope? Take pictures however you'd like, but be careful around explosives. These particular detonators are probably okay, but there are some that are extremely sensitive. A camera flash may be enough for one to go off in some cases, and I'm quite fond of my hands, you know, right?"

I always made sure the built-in flash wasn't engaged, as it ruined most photos, anyway, but I double-checked to be extra sure we didn't all blow up. Chris resumed his lecture but was quickly interrupted again—not by my camera this time, but from something hitting the other side of the wall next to him.

"What the hell?" he said, backing up surprisedly.

"That sounded like a bullet impact," I said, looking at Långström. "We did enter the right building, didn't we?"

"Yes, it's all fine," said Långström, "but obviously you have to assume the worst and be ready for that one idiot in each unit who shoots in the wrong direction, you know."

Chris tried to continue, but it didn't take long before another bullet hit the wall, and then another one. Within moments, the occasional knock on the wall turned into a continuous clatter of bullet impacts.

"Fuck! I told them to fire on the building on the right, didn't I?" Långström cried out in frustration. "I fucking knew it! I fucking told them to fire on the building to the right, but they didn't fucking know the difference!"

Picture of the explosives at the Polygon—taken without flash, of course!

A bullet flickered through the air as it passed us through the doorway and out one of the window openings on the other side. The whipping sound echoed inside the room, striking our ears as each foreigner began hitting the deck. It started with only AKs and other small-caliber weapons, but soon the heavy machine gun joined in on the show. The armor-piercing explosive rounds hit the brick like sledgehammers, throwing debris around and kicking up dust indoors. Occasionally, bullets passing through the windows could be seen by the naked eye, as they created small tunnels of air behind them as they sliced through the dust.

"Fuck! Långström!" I shouted while Chris scraped his explosives back into the sports bag. "You were looking at him when you spoke—you both stood opposite each other! When you said 'left-right,' you meant *your* left-right! He saw his! He thought you told him to shoot at *this* house!"

Tor looked around confused as the bullets hit the outer walls and bounced between the inner ones. The Greek was simply laughing at the stupidity of the situation.

"Shit!" said Långström. "We have to get out of here. We're moving outside, behind the building further into the base. Quickly! Low positions! Follow me!"

He began crawling out, with the rest of us following. The bullets continued chasing us out on the other side, as well, whipping both the air and the overgrown bushes around us. We soon reached the second row of buildings, behind the one being used for lively target practice. Some bullets managed to travel unmolested all the way through the first building and hit the second, but the intensity of crackle had decreased markedly. On the second floor, we soon found a part of the stairwell offering no window openings facing north toward our friendly antagonists. The group gathered around in the staircase as Chris opened up his bag of bombs again, ready to give his lecture yet another go.

"Well, good thing we didn't get too far into the lecture, at least. We'll just start over from the beginning."

The gunfire soon died down, as the armored car had expended all of its ammunition, and calm set upon the abandoned base once more. After Chris had run through the basics again, he proceeded to show us how to get through doors and walls during combat with the help of explosive charges. He applied a decent amount of TNT to one of the brick walls on the ground floor to blow a hole large enough for a man to climb through. We backed away to another room and, just like before on the shooting range, he connected a wire to his flashlight.

"Fire in the hole in three, two, one," and click. A loud explosion shook the building. As the smoke settled, a surprised Chris noticed that the brick in the wall barely moved at all. The younger Frenchman, Cuix, who was also an explosives man, seemed equally surprised.

"It barely damaged it. It should have blown a hole right through," Cuix said in his thick French accent as Chris scratched his head.

"Yeah, and I even put a little extra juice in it, just for the hell of it. We need to get more bricks and stones to put behind the charge as a ballast, and maybe use some extra explosives."

This time around, the bang was markedly larger, making the entire building shake. The dust took a lot longer than before to sink to the ground, but when it finally did, the brick wall still remained intact, outside of a few bricks having moved around.

"Well, fuck me, mate. These Soviet walls are tougher than I thought."

As it turned out, the explosive recipes both men had learned were tailored to blowing up huts in Afghanistan, or simple plywood at military training facilities—both far cries from brutal Eastern European architecture.

"Fuck this. I'm making a bloody hole in that bloody wall," Chris said, lifting a TM-62—a ten-kilogram anti-tank mine—out of his bag. Using a long wooden plank to support it, he managed to place the mine against the wall at chest height.

"This ought to sort things out," he said firmly while using electrical tape to attach an additional TNT block and detonator onto it. He then went outside and strapped his GoPro camera to a tree outside to record everything. The rest of us waited on the other side of the building until Chris came back with an expectant and excited grin on his face, wire in one hand and flashlight in the other.

"Cover your ears now. This is going to be loud. Fire in the hole in three, two, one."

Click.

The entire vicinity shook. A loud roar was followed by a crash and, for quite a while, the sound of bricks and debris landing all over the area. We hurried out past the edge of the house to see the destruction. A large cloud of dust lay over the courtyard, bits and pieces of building scattered everywhere. Chris had not only finally managed to create a hole in the wall, he had torn down and collapsed the entire eastern section of the large two-story brick building.

Långström (left) and Chris (right) admire the explosive aftermath of demolitions training at the Polygon.

"Bloody hell…" Chris mumbled, looking at the badly damaged tree where he had tied up his camera. "That was a proper new GoPro…now I have to buy myself a new one."

Suffice to say, everyone's appetite for large-scale explosions had been sated for the day; we would henceforth continue with a gunfire exercise between the buildings. The conditions were simple: we would move between different buildings with the squad split up into two units, just as we had done before in the ravine. As we moved throughout each room, we'd fire upon the targets spray-painted on the walls. The goal was partly to get individual soldiers with less experience to practice routines, but it was primarily to solidify the group's cohesion in general. As many parts of the buildings had collapsed even prior to Chris' anti-tank mine shenanigans, it was also a matter of practicing rapid movement through ruined rubble without getting badly injured or killed by the environment itself.

We quickly learned that firing indoors was simply too dangerous. The PKM and AKM worked fine with their 7.62 caliber rounds, but the AK-74, using the fast-moving and slender 5.45mm bullet, created a hail of ricochets inside. The tracer bullets bounced erratically between walls, floors, and ceilings (and frequently back at us), forcing even the hardheaded Långström to reconsider the wisdom of this exercise.

Before anyone lost their teeth, an eye, or worse, we decided to replan the procedure. The shooting would mainly take place outdoors during the transition between the buildings. Firing indoors would be avoided other than on a few occasions where the layout of the rooms were deemed less risky. In order to make the drills a bit more interesting, it was decided we would also throw live hand grenades.

As with everything else, we also suffered from a shortage of grenades. We had, however, recently received a couple of boxes of mixed hand grenades from Kirt, one of the company commanders in Azov's Mariupol battalion. We would keep the RGD-5 and F-1 grenades for use in combat, but the box also contained a number of RGO grenades. These were small and easy to throw, which made them suitable for practice. A rumor going around about them being unreliable duds also made them less fit for serious business.

What otherwise mainly distinguished the RGO from the other grenades was that they had a different ignition device, which worked both with time delay and impact detonation. RGD-5 and F-1 both had a simple delay, and they both exploded about four seconds after being thrown. The RGO had a similar delay function, but this was activated only if the grenade hit the ground or an obstacle shortly after it was thrown. If, on the other hand, it was allowed to fly undisturbed through the air for more than one or two seconds, it switched to impact detonation, meaning it would explode the moment it hit something.

I had never thrown a live hand grenade before, just practice grenades, so I was admittedly a tad nervous. I told myself that it was no different from throwing a rock but still found it a bit difficult not to distinguish that I was, in fact, holding a live, volatile little bomb in my hand.

The firing drills continued, and soon it was my turn to the spearhead the squad and throw the grenades. We had moved from the first building and were now going in through a window into the second. I lobbed the grenade inside, and immediately upon it detonating, we entered through the large window opening. The room was full of smoke and dust, which made it as difficult to see as to breathe. I was the first one in, but as I entered, I got stuck on something sharp protruding on the inside of the wall, below the window. Whatever it was, it cut deep into one of my buttocks. If there had only been something to stand on inside, it would have been fine, but the floorboards closest to the window were missing. I ended up hanging there, like a pig at the meat packing facility hooked by its ass. I tried to lift myself up with my hands on the windowsill behind me. The vest and PKM on my chest, however, made me front-heavy, and the movement was unnatural—I had a hard time getting power to my arms. At the same time, the rest of the group was still following behind me, trying to enter the room. The fact that my hooked-up ass was blocking the entrance proved no issue as the others continued instinctively, as if it were for real. Someone put his hand on my shoulder, using me to climb on top, which only pushed me down further on whatever devilish contraption was cutting into my ass. After everyone had already passed through the window, I finally managed to get free, rejoining the drill until we were done.

"What the hell happened over there, after you threw the grenade?" Långström asked me.

"Oh, nothing, I got stuck on something by the window," I said, rather than actually admit to having been injured in the ass during a simple drill. I put my hand inside my pants to caress my sorry buttocks in an attempt to determine how serious it was. The back of my trousers was torn up on the side, and so were the long johns underneath. Visually inspecting my hand, I could see that I was bleeding a lot, but not as dangerously as if something vital had been cut. The still-clear blood flow from the wound led me to think that the puncture was probably not too deep, most likely in the fat reserves primarily. I went back to the window we entered through to seek an explanation for what had happened.

Just below the window sat two large, sharp, rusty metal hooks that protruded from the wall on each side. In size and design, they were reminiscent of a bear claw—a tool used to clamp timber together for transportation. I was still bleeding, but increasingly less, as luck would have it. It was probably a good thing, anyway, I thought—less risk of getting an infection from those rusty pieces of shit. I then kicked and bent them inwards to avoid being further raped by them in the future.

We continued the exercises, and soon it was my turn to throw again, now in another place. The grenade was going in through a doorway this time, and I stood as close to it as possible, my back against the wall. As I had learned earlier, if possible, you should always try to peek quickly into the room you intend to throw into. The reasons were several, but above all, to check there was no debris in the way, which the grenade could bounce off of back toward you.

I peeked in quickly. *Fucking shit, that's a lot of rubbish*, I thought.

Inside was a narrow entrance hall with stairs on the right side. Apparently, someone had thrown down tables and chairs from the upper floors, leaving a chest-high pile of junk in front of the entrance. In order for the grenade not to bounce back at me and my friends, I had to throw it rather high.

I stepped back with my back to the wall with the doorway on my right. After bending the pin to make it easy to pull, I pulled it and prepared to throw it into the room. I'd throw with my right arm, high up so that only my hand and wrist would be visible from the inside. It was a kind of throw I made many times before, but never with a live grenade, and above all, never as high up.

Something went wrong as I lacked the muscle memory to make this kind of throw, and I lost momentum at the last moment. I threw the grenade like an old lady. Inside, there was a loud "clunk" as the grenade bounced off something. It hadn't cleared the pile of junk and was now coming back toward me, and as it hit, its time detonator had fused, now counting down the seconds until detonation. In parallel with the ticking seconds, my understanding of reality began to slow down enormously. I heard the grenade bounce in the hard stone floor. First once, then getting increasingly closer with each bounce.

Expecting it to get hot shortly, I pressed myself as hard as I could against the wall, even pulling my toes together in my boots. I tried telling myself that I, after all, was a person gifted with luck, and this would probably be fine, as well. As my life started to pass before my eyes, though, I understood that might not very well be the case this time. The grenade closed in toward the frame of the entrance. My friends were standing behind me to my left, and I found some comfort in knowing that at least, hopefully, my body would take the force of the blast. They'd probably get some bruises, but at least nobody would die on behalf of my fuckup. I was never afraid though, not for a single moment. As I saw my life pass by in review, I felt a lot of things, mainly disappointment over dying in such a very unfashionable way, but never fear. I was completely calm. Death could come however he pleased. I was fine with it.

Then, suddenly, I saw a strong light in front of me, outside the building just next to the doorway. The outside was otherwise gray in tone, but the light gave life to the dead, colorless surroundings. I stood completely still, blinded by the light, turned my gaze toward it, struggling to see. In the middle of the light, a woman stood just a few meters away from me. I didn't have time to look at or memorize her face carefully, only seeing that she was young and beautiful. Her hair was long and dark brown, bordering on black. Her figure was slim and delightful, and over it she wore a long, dark green dress covering her entire body except her neck and hands, which she held up to her stomach and chest.

I closed my eyes and let darkness embrace me, but just a moment later the compact black changed to an incredibly strong orange glow. The temperature, which had been cold on this cloudy day, suddenly changed to that of extreme heat, like in an already hot sauna where someone had thrown too much water too many times in too short a time. I felt the pressure go through my entire body, and I waited for a moment of pain, but it never came. I opened my eyes and saw only smoke.

"Go! Go! Go!" Långström shouted, and I entered through the doorway with the rest of the group behind me.

We cleaned the building until the drill was finished. We recapped the execution—what went well, what could be improved, and so on. After reviewing, we started moving back toward the truck. On passing the door, I glanced down trying to see where the grenade had detonated. There was a small pit in the entrance where wood and granite had been rubbed by the explosion. Just a few more centimeters would have been enough. During our review, I mentioned how my throw had been a bad one, and how both I and the others should exercise caution not to fuck up the same way in the future. However, I never told anyone exactly how close it was, or what exactly saved us—the first part because I was simply too ashamed of fucking up that badly, and the second part because I simply could not explain it, anyway.

CHAPTER 16: TOO MUCH OF TOO LITTLE

Frosty temperatures came down as soon as light turned to darkness outside. A couple of us were returning back from the Polygon in the group car. It had started raining outside, and the car still lacked internal heating. Normally, it wouldn't have been any worse than a cold seat, but the freezing rain caused extra problems. As soon as the drops hit the windshield, they froze, creating a thick layer of ice impossible to see through. Bear, Tor, and I sat in the back while the Bulbashes were in charge of the vehicle in the front seat—the tall one in the passenger seat, and the tiny one at the wheel. They enjoyed driving when the opportunity arose and no one had any complaints about it really, even if they were far from the most experience drivers. Lacking winter tires, the car was sliding around on the surface of the road. Tor tried to argue that he should take the wheel instead, but the Bulbashes, exhilarated from the dangerous ride, refused to stop the car.

Soon, however, the situation became completely unsustainable. Even though we stopped and knocked the ice off the windshield, the gained visibility would only last a short while when we started driving again. Having to stop every hundred meters or so to break the ice away would never work. The solution was as simple as it was elegant: having the driver hang out of the side window while driving. The tiny Bulbash simply lacked the reach, so the lot fell on the tall one, clad in double balaclavas and large protective goggles, so as not to freeze his face off. The awkward driving position and harsh weather might have slowed anyone else down, but the big Bulbash liked speed as much as the tiny one. We were skidding around a lot on the icy roads until we finally returned to base alive, to enjoy a disgusting dinner that made us wish we weren't.

The food situation was becoming unbearable, and I was losing weight. Having finally reached the same stage as the others, I now left the gretchka untouched on the plate, unable to consume it despite my hunger. Our commander, who still wasn't offering us any missions, was as tired of the food as everyone else.

"Let's take the car to Mariupol tomorrow and get a really good steak instead of this shit," he suggested. It was a literal case of first come, first serve, as the transport had limited seating. After breakfast the next morning, the Niva hauled a full load of starving soldiers eastward. Upon reaching the city, Mike took us on a tour of the streets where Azov had done battle during the early summer, overpowering the separatists holding Mariupol and reclaiming it for Ukraine.

"Along this street, we rolled up and pushed the enemy back", he told, with one hand still on the steering wheel and the other one pointing at one of the large brick buildings in front of us. "One of the Russians was holding up in that window over there, but I killed him with my sniper rifle."

The story was as lively as the one about his kill at 1400 meters earlier, but I had trouble believing anything he said anymore.

We soon arrived at a restaurant Mike had chosen, where we were showed into a private dining room–- small and cozy with nice windows and a piano next to the large table. Mike translated the menu and showed me which steak was the biggest and grandest, but when it later arrived, it looked nothing like what he had previously described. On the plate in front of me lay some kind of colorless lump of processed lard, seemingly with vegetables of some sort mashed inside to increase its size without adding more meat. The lump was large, something he was correct about, but nothing else fit the description. It became evident that his claim to read and speak Russian was also exaggerated.

Despite the initial disappointment regarding the so-called steak, whatever it actually was, it clearly wasn't made from buckwheat, at least. Like a hungry wolf, I devoured the mysterious lump of flesh in front of me like the others, and it wasn't long before all of us were leaning back on the chairs, empty plates in front while tapping our unusually swollen bellies. Chris, who in addition to being a mine clearer, was also a musician. He sat down at a piano in the hall and played a few tones, accompanied by our burping and farting noises.

With our primary objective completed, we would make a quick stop at the eastern outskirts of the city in order to complete the secondary before we returned to base. The front line began just outside Mariupol and went from north to south in an almost vertical line, east of which the Russians were controlling most of the villages. Mike had recently come across a large batch of boots and thick winter socks donated to Azov by civilian volunteers, and we still had a lot left after the group had acquired what we needed. The plan now was to try to trade the rest of it for other equipment we were lacking in, namely ammunition—especially for my machine gun. According to Mike's experience, the regular army soldiers manning the front would be well-stocked on ammunition, but they would also be freezing their feet off manning their checkpoints and pillboxes. After getting lost in the outskirts of the city for a little while, we finally found a small bridge which led us out toward one of the bigger block posts outside of the city.

It was an odd feeling getting out of the car. We had been near or at the front some time before, but the invisible line had never contrasted so clearly before. Behind was the large city, with all of its familiarity and relative safety. In front of us, on the other hand, was a completely hostile country with armed enemies waving different banners. Behind us, the cloud cover burst open and the sun shone over the city, but over no man's land lay a compact and dark gray cloud cover, drowning all light from the skies. There was nothing friendly over there on the other side of the large fields and forest groves. Even though we weren't going there, the unpleasant feeling from merely seeing this was palpable. I wondered how it would feel to actually cross the line and continue forward along that road, straight toward a historical and hereditary enemy of which I had heard and read so much about but never before seen myself.

We heard artillery shells firing in the area, both incoming and outgoing. I tried to listen to how it sounded, depending on where and how it landed. Suddenly, it sounded as if some of the shells fired from the enemy side also landed back at the same side of the front line, only a bit further south.

"Ha, yes! That is right, they fire at themselves!" the block post commander laughed and pointed. "In north is Russian Cossacks. In south, there is Chechens. They don't like each other. Most of time they shoot at us, but sometime they shoot each other, too."

Mike talked to the soldiers at the block post, doing a show and tell about how comfortable his socks were: "These things here, you know, these are great socks!"

It was like seeing a professional salesman in action. The soldiers themselves looked at the products on sale, coming to the same conclusion as the salesman laid out for them. The boots and socks both seemed comfortable, and they wanted them, but they were unsure about how much ammunition they could spare. They had to report their ammunition expenditure to their commanders, partly because of the cease-fire agreements, but mostly to stop them from selling it for booze and whores.

"Come on, now, that's no problem," Mike told them confidently. "Just say you had an enemy contact over there by the trees, and you were forced to open fire on them. How will your senior commanders, who aren't even here, know if you expended five hundred rounds or five thousand? Come here and count empty casings? No, they won't. They are way too comfortable in their offices to come out here, and you will be comfortable with your new socks and boots out here, as well!"

Mike was close to finishing the final negotiations regarding how many rounds of small arms ammunition a pair of thick woolen socks was worth when the ground suddenly shook. Thick, dark clouds of smoke rose up north of us, about where we had crossed the small bridge just prior to finding the block post.

"Fuck, those were Grads," Mike stuttered.

The heavy enemy rocket artillery had hit the small suburb area, and the previously sleepy soldiers at the block post hurried to take defensive positions. Mike became noticeably nervous and suggested it was time for us to leave, quickly, before anything else fell down from the skies. We dumped the socks and boots, leaving them for the soldiers while throwing the ammunition they had brought forward into the car. We embarked and quickly made our escape.

Everyday life at the base continued as before. Tor was so restless that, without thinking twice, he agreed to adopt a small puppy. The dog was black with short, rough fur and white paws. Nobody could quite narrow down the breed, as with all stray dogs, but it seemed likely that it would grow up to be some kind of rather large Sighthound. The base was littered with stray cats and dogs, but the Finn had found this particular one on a mission and brought it here.

"Claymore is not like normal dogs, you can see to that. This dog is really smart," he had said. "I have raised and trained dogs before, and I see it very clearly."

The Finn was planning to bring the unusually intelligent mongrel back home with him when he returned to Finland. However, owing to long assignments and missions, he didn't have time to take care of the dog at the base himself. Therefore, the black and white puppy got to stay in our room, as we were unfortunately far from as busy.

It was true that the dog was unusually intelligent, and Tor, who had never had a dog before, immediately took a strong liking to the little puppy. I had never owned a dog myself, either, but at least I saw it as a pet animal, not a child. I tried to explain, as pedagogically as possible, how he needed to be more determined, to little or no use. A few moments later, Tor would be back at it, rewarding bad behavior with praiseworthy tones, or exacerbating the dog's already hectic eating behavior with sweets. He trembled before the day he would have to give it back to the Finn, to the point that he almost cried at the thought of it.

The reason the Finn was partaking in missions and assignments was that he switched over completely to Båtsman's group, as was the case with Yuri and Långström. The missions they partook in were usually rather uneventful— patrolling a front section here, protecting an area from nothing there. However, it was undeniably more interesting than just sitting around the base, at best waiting to go to Polygon for some live fire exercises.

Yuri poses next to the shot-up car after returning from Pavlopil.

After the morning line up, I sat down on the bench outside the door to our barracks. It was cold in the air, but the sky was clear, the late summer sun still fighting to provide warmth, something I wanted to take every chance to enjoy while it was still offered. As I sat there smoking a cigarette, some civilian cars drove into the parking space in front. As the sun's rays shone over the cars' dark, lacquered bodies, I saw how there were several bullet holes scattered over the thin metal hulls and through the windshields. It was part of Båtsman's group returning after conducting vehicle-borne reconnaissance during the night. Together with Ukrainian soldiers, the Finn, Yuri, and Långström stepped out of the shot-up vehicles.

Långström looked pleased and laughed. He had been involved in combat several times before, and it was obvious that he liked it.

"I love it when people shoot at me!" he shouted happily and patted the front hood next to the bullet holes.

Yuri was equally excited. "My friend! My friend! Take photo of me here!" he stammered forward. He got down on his knees and posed smiling next to the 7.62mm caliber airholes next to him.

They had been driving through the village of Pavlopil, in no man's land. Hoping it would not have been occupied by enemy forces, their plan had been to conduct recon on foot a couple of kilometers further east. Flares from the village had, however, proven that the enemy positions were closer than expected, forcing them to abort the mission. On their way back, still in their cars, they had been engaged by enemy infantry nearby, forcing them to drive-by themselves out of the zone of contact, firing through the windows.

The Finn was happiest of them all as he tried to describe the events of the night in careful detail: "They started shooting like fuck from behind. I just turn around in seat and fire back, and I think, fuck, I am not a-wearing an armor plate tonight! Road was full of tracers in different colors! It's crazy that no one got a-killet-ed."

He calmed down a bit and lit a cigarette before showing where he had been sitting and how he had lifted his weapon.

"You don't know a-how you react before it happen, and I thinket about it many time. You never know first time, but I just turn around and fire back. Without a-thinking about it, I just do it, on instinct like I was trained to do it. Not a panic, just very calm and good. It's only later now, I get a lot of the adrenaline, and it feels very good!"

He rolled the sleeve of his uniform and pointed to the inside of his forearm. "Here I will make tattoo of this date, with big numbers, and if I have some children, then I will also tattoo in their dates below, though smaller."

It was difficult not to be happy for him for finally getting to experience a real, albeit brief firefight, that he finally got to know what every soldier who hasn't experienced combat can only ask themselves—if they're really cut out for it, cut out to be a fighter. I tried to let off a smile without showing how disappointed I remained in my own situation. I had left home to go to war, but despite being so close to it, I felt as far away as ever. It seemed as if my first engagement with the enemy would never come, and I grew increasingly depressed because of it.

Mike was talking more and more about returning home to Sweden, and the thought was tempting. I was as tired of the food as I was of doing absolutely nothing. Barzuk had already left a while back, and several others were on their way. My entire adventure into Ukraine was a great big failure. Whatever kind of small career opportunities I had in the Swedish Home Guard were lost forever, as were all chances of doing anything similar in the future. I also knew I had already been branded as an extremist by both politicians and the media. All my mental preparations, and the suffering they had caused, seemed to have been for nothing. The price tag for this misadventure was incredibly high, and what did I actually have to show for it? I hadn't even been allowed to fire my rifle once at the enemy. Could I even say I had participated in the war? Sure, I had experienced some dangerous situations, and had even been fired upon—once, allegedly, even by the enemy. That was more than many others claiming to be war veterans, but did it make me one? I certainly didn't feel like one.

Earlier, I had tried to keep the opportunity open to join up with Gia, the Georgian I had previously met in Kyiv alongside the Hungarian journalist, Sandor. However, the latter had informed me that despite going to the front, it had been very obvious that the charismatic and funny Georgian had not been in the country to search for battle at all. Once he had emptied out his own (and later Sandor's) pockets—money spent on hookers and booze—he had left Ukraine back to where he had come from. There was no opening in some other unit or fighting that way, either.

It was dark and cold outside. Tor and I had retreated into our room where the puppy, Claymore, had just planted a small but extraordinarily disgusting stink bomb on the floor. Tor hurried to care for the accident, but Claymore, proving his Sighthound speed and constant hunger, was faster. Tor yelling "No!" did little to reverse the situation. I was laughing well at the humorous situation, stopping only when Tor in his frustration tried to dig the shit out of the puppy's mouth. I felt compelled to step in at that point to let him know he was only making it worse for himself.

"The dog eats shit, sure. They often do. That hardly means you have to get shit all over your fingers yourself, though, does it?"

There was a knock on the door, and the Finn entered, sitting down on Barzuk's now empty bed.

"I've got a little change in situation. I am going to home soon, but I will not be able to bring Claymore," he said, disappointed.

Tor lit up immediately. "I can take him! I can take him with me to Estonia!" joyous as never before. The Finn looked at the hurried Swede bitterly and asked if he even knew how to train a dog, even more so if he could get it out of the country.

"Eh, no?" Tor replied, confused, obviously having never thought about it.

The Finn sighed again and began to explain the most important parts he needed to think about, both in terms of vaccinations and passports, as well as future upbringing.

A few more days passed, and soon everything was cleared up. Mike, Tor, and I would return to Kyiv and then back home. The Finn would also come a few days later. I missed my friends back home but did not look forward to the trip. I had lost a lot of weight and felt like a wreck, heading from nothing back to nothing. I was ashamed to show myself in my condition, but seeing as the Ukraine adventure was a failure, staying only seemed to make it a prolonged one.

We got a ride to the western neighboring town of Berdyansk, from where we would first take a bus to Zaporizhzhia, and from there the night train toward the capital.

Bear was disappointed, closing in on depression. "I think now, when you guys go, you will not come back, and then the Foreigner Group will disappear. I will have to stand in a block post and rot."

I told him I might come back later, trying to cheer him up. Deep inside, I was still hoping I would find a good reason to return, but at the very moment of me saying it, I was far from sure it would happen.

Another foreigner, a Slavic ex-Legionnaire joined us on our journey. We sat around the table on the train. He barely spoke any English and had been serving in another unit, but he seemed just as dissatisfied as we did—apparently, especially with his commanders.

"I have gift for veterans! It's a big dick!" he joked loudly while laughing beneath his thick, dark brownish beard while swinging his hand repeatedly up and down, stroking an invisible giant cock at Tor's head. In front was a pair of salt and pepper shakers and a small jar of toothpicks. The Legionnaire picked up one of the toothpicks, held it in front of him, and pointed at it with his other hand while vividly explaining to us what we were seeing.

"Here! Look! This is new kommandir! He eat, he shit, he sleep twelve hours each day!"

Once we finally arrived in Kyiv, we headed for the same hostel we had stayed at earlier. As I approached the gate, some young people were standing outside the small bar next door. One of them was the young Russian bartender I had met before leaving for the front.

"You're alive!" she exclaimed, dropping her cigarette to the ground, and she ran toward me to give me a hug.

As I still wasn't sure if I'd be coming back, I had brought most of my equipment back with me. The only things I left behind were smaller, easily replaceable articles I thought the men who remained might find use for: magazine pouches, water bottles, my sleeping mat, and other such items. I brought everything else back with me, including the heavy Polish "bronik" I had been issued, adding another twenty-two kilograms to my already heavy load. The guys on the front already had protective gear covered, and I didn't want to risk coming back only to find my own looted and stolen. I would, however, not be able to bring everything on the plane, which is why I asked the kind girl at the hostel reception if I could store it below the stairs next to her desk. Somewhat confused, she agreed that the large black bag could remain there until I might come pick it up at a future unknown date.

Tor stayed in a hotel that allowed animals and was preoccupying himself with how to smuggle the puppy into the EU. Mike left soon after arriving in Kyiv, and I was once again on my own. It got very lonely very quickly, and I used the time to ponder my situation and life choices—often at the Gorky Pub just below the hostel. The Finn, however, soon arrived at the same hostel, and to my great joy he also took a liking to the cozy, quiet bar. He had long been very rigid toward me, "the Finn who doesn't even speak Finnish," but had started to drop the tough exterior bit by bit. Especially after he had finally gotten his firefight, he had been in an extraordinarily good mood.

"How did you end up in Ukraine, anyway?" I asked him while ordering a new round of beers.

"It is a-pretty simple, really. I always wanted to see and experience a real war, and then, when it started here, it was a-perfect: the right enemy, the right kind of war, and a good reason to fight. A perfect war, at least for my generation. I can help Ukraine in the war, Ukraine can help me get to fight in a war. It was a win-win."

"The Legion, then?" I asked. "How did you end up there?"

"There was a book that was written by a Finnish Legionary," he began, and I immediately guessed which book it was. The Finn continued: "It is only published in a Finnish text, I think, so you probably have not been reading it. He was called Sahara, and the book has same name. He was also a mercenary in Congo and South America, too. He was a-really hardcore—" he continued before I interrupted him.

"Per Andersson, right?" I asked.

"Yes!" the Finn looked up in surprise as he dug the book out of his bag. "You have heard about Andersson?"

"Yes, I remember that book. He was a good friend of my uncle in Finland and lived right next door on the countryside when I was young. He looked after me and my cousin when we were staying there over the summers."

I began recollecting some of my childhood memories of the man who apparently made us both want to be soldiers, how he used to sit on my uncle's terrace, where he went through several packets of Camels a day, a habit and brand he brought with him from the Legion. Often together with a bottle of cheap whiskey. Or two if he was in the mood that day—maybe even three. There were also other, more well-known stories like how he had once gotten tired of his wife's nagging and just cycled to Spain one day without any notice.

"Every time I do something stupid, my mother asks me if I did it because she left me with Uncle Andersson over the summer all those years ago."

Uncle Andersson had been dead for a few years at this point. The bar took music requests, so we asked them to play Jean Sibelius' *Finlandia* in Uncle Andersson's honor, and we toasted to his memory. It wasn't long, however, before the bartender, a big, bearded hipster, shut the beautiful classical music down.

"Your music is too depressing. People in here will start suiciding themselves."

The next morning, we grabbed a taxi to Boryspil, the same airport I landed at a few months earlier. We shook hands, wishing each other good luck before we boarded our flights to our respective homes.

CHAPTER 17: A NEW YEAR

The rockets were detonating all over in the pitch-black skies above, explosions echoing between the buildings and rolling through the streets. The colorful burning orbs slowly fell toward the ground, illuminating all surroundings. The clock was just past midnight in the northern Swedish city of Sundsvall. 2014 turned into 2015, and we were watching the spectacle above from our booze and black metal-infused frenzy below.

One of my friends said I seemed a bit jumpy, and he was probably right. Even though I still enjoyed the fireworks immensely, there was a distinct feeling of terror lurking behind the celebratory explosions. It wasn't necessarily in a bad way; at least here, the fear could pass through me in a controlled environment. However, I couldn't quite stop myself from recoiling slightly. I remembered back to that day by the front line block post, as the Russian rocket artillery had leveled the area near the bridge beside us. The sound above was very similar, and I wondered how it would have felt had we driven through that area at the wrong time. I allowed it to occupy my mind completely. Would it at least resemble something like this, like powerful fireworks overhead?

I had been home for two months already. At the end of November, on my birthday, I got an unexpected call from a large Swedish evening newspaper while on my way to the store to buy cigarettes. The young man on the other line said they were doing a big report on the Swedish volunteers in Ukraine. He sounded rather low-key, and despite his insisting that he wanted to represent us fairly, it was obvious he was far more interested in fishing out personal information about the foreigners, rather than learning anything regarding the war itself. The questions he asked were mostly about politics and other tedious nonsense, questions which he repeated again and again with the intention of slipping me up, to entrap me into saying something by mistake, whether true or false, that he could take out of context to smear me.

Against my better judgment, I wanted to believe him when he said that his goal was to report everything accurately, not to create sensationalistic political propaganda. I wanted to believe he was just a simple moron. Therefore, I spent about an hour outside in the freezing cold, until my poor fingers felt like frozen knives were stabbing through them, trying to explain everything so simply even a child should have been able to understand. When the article finally came out, I saw the headlines and the preface—the parts most readers actually register in their minds. "Nazis" and "war crimes" were the strongest and most recurring words. What I had actually said was cut out, with the exception of a few short quotes.

I obviously should have known better than to have talked to that journalist at all. Not because I left a topic hanging, or hurt myself or someone else, but just because my fingers hurt for nothing.

An old acquaintance I hadn't met in several years suddenly gave me a call. He laughed his way through most of the conversation.

"Fuck, man, I saw you in the newspaper. I went to school with the guy who wrote that article. He was a no-good retard back then, and it seems he never really grew out of it, either. What a load of nonsensical bullshit that was! "

The journalists who had visited us down east had also released their article. The photographs were nice, but the text itself was all the more mediocre. There wasn't much to suggest the female reporter bought Mike's story about our car getting shot up—that was, after all, the intention. More depressing was the fact that she had, in such a short timeframe, still managed to confuse both events and people's names. She even described my name, Carolus, as being made-up. Sure, most people used *nome de guerres*, so it might have been forgivable had she not added that my own baptized name was in actuality something I had taken from a series of Swedish white power records from the nineties; this merely added another level of incompetence, intentional lying, or both. The fact that the article was praised by other media and reporters was pathetic to witness.

Bear had been partly correct regarding his fears about the Foreigner Group splitting up. Several more members had left Azov—or Ukraine altogether—in the past months. Mikola, like me, had returned back home to Sweden. He hadn't initially intended a long stint in Ukraine, but he wound up staying six months. Now back in Sweden, he was finalizing his arrangements to relocate to Ukraine permanently. The extremely fit Italian paratrooper Yuri had, with his dark skin and absence of any kinds of subcutaneous fat, found difficulties with the cold. Like the migratory birds, he had gone on to warmer latitudes searching for other wars. Tor was not only busy trying to figure out his new dog ownership situation, his young adult children simply wouldn't allow him to go back anymore.

The only ones left now were Bear, Greek, Långström, Fabien, Cuix, Chris, and Leo. Some new members had also been added. Among others, the radioactive old driver Inkasator had asked to be permanently transferred to the group. Långström, the hobo, had taken control over the Foreigner Group in Mike's absence. At the same time this shift in command occurred, they had also started to be awarded missions—actual military tasks and assignments to complete. It was mostly reconnaissance against enemy positions during nights, and usually rather uneventful. However, missions were missions, and they were constantly active, sometimes several times a week.

I kept in touch with the rest of the team still in Ukraine to keep myself updated on current affairs. The conversations mostly comprised of small talk, racial slurs directed at one another, and the regular bullshit, but one day the Greek contacted me with a completely different tone in his voice.

"We are all going to die here," the Greek said. I had never heard him make a joke beginning with that sentence, nor with such pessimism, so I asked him what was going on.

"Långström—he is fucking crazy. You know this, right? And now we are all going to die."

"Why?" I asked once more.

"He has a new plan. He wants to steal a tank from the Russians."

The Greek slowly begun explaining the premise of the situation. The Foreigner Group had been doing recon on an enemy block post east of Mariupol. It was manned and guarded by an enemy force of somewhere between fifteen to thirty men and two T-64BV main battle tanks. The T-64 was an old machine, and the current Ukrainian army standard tank. Once the pride of the Soviet armies in the early 1970s due to its innovative design, it had been plagued by technical problems from the start and was now closing in on becoming completely obsolete. Old as they were, the almost 40-ton beasts of war remained well protected and dangerous, boasting powerful 125mm smoothbore high-velocity cannons and machine guns.

As winter arrived in the eastern front, the foreigners noticed that the tanks were mostly unmanned. However, the crews regularly ran their engines at certain times each day to keep them ready at a moment's notice, before retreating back into their warm quarters. As if by chance, the Foreigner Group had, back at their own base, managed to scavenge a T-64 tank ignition key from a pile of scrap. Långström had even managed to convince one of the Azov tankers still in training to man the vehicle, once the Foreigner Group had gotten rid of the guards. They still hadn't received any vehicles of their own from the state, and the opportunity to acquire one from the enemy was tempting enough for one young man to take a chance.

The terrain in front of the enemy position was open ground, with the exception of a thin tree line and a ditch next to it. Långström's plan was simple, elegant, and arguably insane. The Foreigner Group would sneak up on the enemy positions under the cover of night, with Chris and Cuix clearing out any potential mines. After the tank crews warmed up the engines and left the tanks unattended, the bored and sleepy defenders would be stunned by hand grenades and violent automatic fire spewing from inside their own positions. The enemy would obviously outnumber Långström's small reconnaissance unit by at least two to one, but his team would have the advantage of surprise.

"What are you supposed to do about the second tank then?" I asked the Greek.

"Ah, we aren't going to steal it, if that's what you're implying. We'll just blow it up."

The tanks were lined up one behind the other. A hand grenade through the turret hatches should be enough to ignite the main gun ammunition in the unprotected autoloader carousel on the bottom floor. With the defenders suppressed and still confused by the second tank exploding, the Foreigner Group would jump in the first tank and speed it straight back toward their own lines. If everything worked out, they would be gone before anyone, living or dead, knew what had actually happened.

"It's a pretty good plan, really, if you think about it," the Greek said, having calmed down a little bit while going over it himself. "The thing is that it's also insane. That kind of bothers me, you know? If something goes wrong, we're all dead." He went silent, pondering the situation for a moment before continuing.

"But, if you think about it, it would be pretty hardcore to steal a tank, too. I don't think many other guys have done that."

I had nothing to disagree with. I had never heard about such an operation before—at least none which ended successfully.

"You need to take some info down for me, okay? In case I die, you know? Just some formal bullshit, okay?"

I took his information down and wished him and the rest of the guys good luck, waiting to hear from him again. A couple of long days passed, bringing only silence.

Then there it was, a sign of life.

"How did it go?" I asked, partly excited that at least all my friends hadn't died, but mostly to hear their gallant and glorious stories.

"Well, nothing happened," the Greek responded in firm disappointment. "The Russians obviously moved the tanks at some point, because when we came back, there wasn't anything to steal."

"I'm very sorry to hear that," I told him. "I know very well how much your people enjoy stealing shit. Must be tough."

"Bro, that's the Turks, not us. And for the record, I would steal your virginity if I could. Anyway, regarding the tanks, the more I think about it, the situation probably turned out best for both us and the Russians, though Långström is, as you might expect, pretty disappointed. He was really looking forward to this."

Greek informed me that the group had changed bases and accommodation. They left the freezing summer barracks in Urzuf and moved to Yurivka, a slightly smaller village located a bit further east on the road toward Mariupol.

"It's bloody well much nicer over here," Chris said. "No bar around, but there is still a store selling scotch and beer. The group has a pair of rooms to share among ourselves in a simple, but solid hotel building. The walls are made from brick, and we have heating indoors. There's even a shower in each room, with hot water!"

It was good news, indeed, because I was, in fact, finalizing my own plans to return south and join back up. Looking back, the fears I fixated on at the Gorky Bar turned out to have been legitimate, after all: Sweden didn't have anything to offer me, and I had left Ukraine only to return home to absolutely nothing. My yearning for combat never dissipated, anyway, and Greek made it sound like things were on the upturn over there. Whether this was the right decision was hard to say—I hadn't planned this meticulously, instead relying on instinct. Truth is, Mike's antics were one of the primary reasons I waited so long to come to a decision, but I did not want to go back, and try it one more time. If not, it would all have been for nothing.

Mike, as it turns out, also made the decision to come back. He wasn't going to the front anymore but said he had some good things brewing in Kyiv, a stop I was making, anyway. Even with this news in mind, my mind was already made up. We booked the same flight from Stockholm-Arlanda airport: a Ukrainian Airlines Boeing planned to take off in the morning of January 9th.

The Greek contacted me once more bringing news, both good and bad.

"Which one would you prefer first?" he asked. I chose the good news.

"Well, okay. So, two guys from the battalion died today. Do you know what they did? Supposedly, they had been part of a column of vehicles or something, and the car in front of them drove out into a minefield, right? As you'd expect, the car exploded, so what do you think the guys in the following car did? Right? Come on! Guess! This is the grand question! What do you think they did?"

I hardly had to think about it. His tone and the fact they were apparently dead made everything rather obvious.

"I assume that, instead of stopping the car and trying to rescue their friends on foot, they kept driving toward their friends and blew up themselves?"

"Exactly! Just think to yourself for a moment now! Ponder this thought! How fucking arduous it is to walk a hundred meters or so, when you can just take the car! Of course, you keep driving, quickly, like a brave knight on his white horse— despite there *obviously* being anti-tank mines buried in ground in front of you, right? Well, anyway, yeah, they died. Just shit, ash, and distorted pieces of metal left."

Filled with curiosity, I interrupted my friend's description regarding the scene. "If that was the good news, then what's the bad news?"

The Greek was silent for a moment before continuing in a much calmer tone.

"Well, before going on that mission, they needed to borrow a PKM, and, well, seeing as you weren't using yours…"

"What the fuck?" I asked. "I'll be back there shortly, and you just blew my gun up?"

"Calm down, calm down, calm down now!" urged the Greek. "The weapons situation is a lot better now than last time you were here, okay? There's an AK-74 here no one uses, anyway, and we also have an RPK-74 for you. That's basically a machine gun. It's okay, you won't go unarmed, calm down. I'm sure at some point we can find a replacement PKM, as well."

The RPK-74 was basically just a slightly bigger and heavier AK-74—a bridge between an assault rifle and a proper machine gun. The barrel was longer, thicker, and equipped with a bipod, allowing for somewhat more accurate and sustained fire before overheating. However, it was still magazine-fed just like the AK-74, bringing its firepower down significantly from my old belt-fed and now melted-down PKM.

"Sure," I said, "does the RPK have any mountings for optics on it?"

The Greek replied that he thought that Tjeck, one of the new guys, had one he didn't use.

"Alright, good, then see you in a few days."

"Yes, we do. Tjeck and I will be in Kyiv then, so we will meet there."

I walked over to my gun safe, opened it, and lifted out my .300 Winchester Magnum moose hunting rifle. Hunting season in the north was over, and the rifle didn't need its scope anymore. It was an old piece of optics, a Weaver K3-W, easily forty years old. Its magnification wasn't anything special, either—only a simple 3x. It was, however, very versatile at different ranges, and the wide-angle lens made it great for shooting at moving objects. It even had an interesting history already, as its previous owner had been a prime suspect in the assassination of Swedish Prime Minister Olof Palme in the mid-late 1980s. Now it would go on to take part in the ongoing war in Ukraine, mounted on top of an RPK squad automatic weapon.

I contacted Chris the evening before my flight to ask how everything was going.

"Not great, mate. I'm at the hospital," he replied before apologizing for being loopy, as he was only powering through on account of morphine. I asked what had happened but received only a short answer.

"Leo is dead."

CHAPTER 18: A DARKER KYIV

According to Chris, part of the group was returning from the Polygon in the group car, the 4x4 Chevrolet-made Niva. Leo had been behind the wheel with Chris in the passenger seat to his right. In the back seat sat Fabien on the left and Bear on the right, with the old driver Inkasator stowed in between them. Leo had been going quite fast—not insanely so, but enough that the summer tires had lost their grip on a patch of ice, skidding the car off course at the single, lonely tree on the empty roadside. Leo tried his best to avoid it, but the course was set. He crashed the car into the lone tree trunk, demolishing the front completely.

As per usual, none of the occupants wore seatbelts, ironically, as a safety precaution: these particular seatbelts were very difficult to get out of quickly should the need arise. As the car ground to an abrupt halt upon meeting Mother Nature, the men in the back seat were thrown forward. Bear's body pushed Chris forward, who hit his head so hard into the windshield that his skull cracked the glass, and his ribs broke when his torso hit the glove compartment. Bear survived on account of his skinny frame, and with a bit of help from Lady Luck. As for Leo, things didn't go as well: he got squeezed in between the seat and the steering wheel.

Chris was still doped up on morphine as he described the ordeal to me, but he managed to describe first hearing a series of cracking sounds followed by something he could only compare to a balloon losing all its air. As he had recovered himself initially, he looked around at the others to see if they were okay. Everyone in the back seat responded, but Leo remained silent. Chris got out of the car to the other side, pulling Leo out of the driver's seat and onto the ground. Leo's eyes opened for a short moment but soon went out again. Chris, still dizzy from the impact, pulled the cord on Leo's vest, disintegrating it into pieces. He threw away the stiff and rigid front plate from his chest and started cardiopulmonary resuscitation.

Placing his hands on top of Leo's heart and lungs, he readied his upper body and pressed his weight down, completely shocked that his arms sank right into Leo's chest cavity. Light red, oxygen-rich blood bubbled from Leo's mouth and nose, and Chris immediately understood what just happened. As Leo was squeezed into the steering wheel during the crash, his front armor plate had broken all his ribs and then continued inwards, smashing his lungs and heart, which is what caused those cracking sounds. Leo's face was already pale as his lips turned blue. He was dead, and his life could never have been saved. His fate was sealed the very moment the tires lost grip on the icy surface of the road.

Chris' concussion together with the opiates flowing through his blood were taking their toll, and he had to hang up the phone to continue resting. As I boarded the train toward the airport, I contacted Greek to ask for more information.

"How do you know this? Who told you?" he asked immediately, and I informed him that I had talked to Chris.

"Fuck!" the Greek shouted. "You didn't tell Mike about any of this, I hope?"

"No, I haven't talked to anyone about it."

"Good! And you shouldn't. Needless to say, first chance he gets, he'll run to the media about this shit to feed his attention disorder, and then there will be fucking chaos to cash in."

I explained that I was well aware of Mike's narcissistic personality traits at this point, and I, too, could imagine the shitstorm that would ensue if Mike blabbed to the press.

"Good," said the Greek, "we must try to attend to this matter very carefully now. The fewer who know right now, the better. Do not tell anyone."

I confirmed that I understood, then immediately broke my promise. My mother knew that I was on my way back down again and had no major issues with it; however, I knew it was only a matter of days before the newspaper headlines would be reading "Young Swede killed in Ukraine," and I wished to save her from that moment of anxiety. I picked up my phone and gave her a call from the train.

"You cannot talk about this with anyone, as it's not public information yet," I told her, "but another Swede just died in Ukraine. When you see it on the news in a couple of days, you can stay calm, knowing it's not me."

The train entered the underground station at Arlanda, and I met up with Mike at the airport above. As he never mentioned anything about Leo, it was obvious he knew nothing of the accident. Instead, he was preoccupied with showing off two large duffel bags full of equipment he had collected for the group: medical equipment, German Kevlar helmets, Swedish army-issued winter gloves, etc.

"Take a look at these chest plates. I was given them by a Swedish SOG guy, a special forces operator," he said.

The plates were old and worn but seemed as if they would still be in working order all the same. His story about how he acquired all this wasn't impossible, but there was no way to verify it. Plates like these had also been used by regular forces in Afghanistan, and were often discarded after a single use. It seemed a lot more likely that they had simply come from someone who had cleaned out the garbage bins outside some regiment. Whatever way they had ended up in his hands, the plates still seemed good, and as they were molded to the shape of one's chest, they would be more comfortable than the flat, square-shaped Polish plates.

The weight of Mike's two equipment bags was well in excess of luggage limits. The woman at the check-in counter was hesitant at first but found herself quickly overcome by Mike's charm.

"I'm a volunteer soldier in Ukraine. It's possible you may have heard about me. This is equipment donated from Sweden, and I'm bringing it with me to give it to the brave soldiers fighting at the front."

Not having the heart to interfere in such a noble cause, the woman let the bags fly with us without argument.

Once down on the ground in Kyiv, we met up with Greek and one of the new guys, Tjeck. He was a young kid, around twenty years old, without a single hair on his chin. His young and almost childishly nervous behavior contrasted with his very fit, muscular, though somewhat short body type. He wasn't completely new to Azov but had previously been stuck in a unit mainly performing guard duties. His great interest and profession had always been firearms, but he lacked all other forms of military training. He radiated a naivety and innocence. His joyful youth and carefree aura, always being close to laughter, made him difficult to dislike.

We checked in to the same hostel as before, where I picked up the gear I had stowed away next to the reception last time. In our new room, this time equipped with a window offering air intake and a view of the buildings outside, Mike began explaining his plans. Kossatskiy, the current Kyiv base, was far too small to train recruits, yet the fresh soldiers arriving at the front had to be trained from scratch. He and Gratch wanted to address this issue by creating a proper boot camp for new recruits in Kyiv. The Bulbash boys and some friends of theirs had already signed up, and Mike tried to persuade me and Greek to do the same: to stay in Kyiv and work as military instructors for Azov's new soldiers, with him as our commander.

Training people to fight was a cause as noble as it was vital, and I told him I would gladly help in whatever way I could. I refused, however, to stay in Kyiv. I had come to Ukraine to fight, and I would travel east to do so, as would the Greek.

It was getting late, and we all needed to fuel up with dinner. Mike was very generous, offering to pay the tab more than once. This was in great contrast to what we had seen from him before—he had constantly asked to borrow money from Tor last time. Suddenly, he had a lot of cash to spend. I remembered that he had previously been soliciting donations in Sweden, supposedly to buy winter clothing for the group. He had yet to show us these warm winter jackets, and I couldn't help but wonder if it was this money he was using now to pay for our hamburgers and steaks, in order to grease me and the Greek into working for him through subconscious bribes.

Regardless, I wouldn't say no to a free meal, nor the following beer rounds accompanying them. Time flew by, and the clock struck close to four in the morning as we were arriving back at the hostel. Just as we entered the room, I was reminded that we lacked any bottled drinking water. I had once previously made the great mistake of, while royally hung over, rehydrating myself from the tap in the bathroom—a chlorinated experience I wished to never live through again. I told the others I would head back outside to buy some water from the nearest street kiosk.

It was dark outside, save for the street lights, but even though it was early January, the weather was still quite temperate and nice. It started to rain, but despite the winter season, it was warm and resembled the nicest of summer rains. I decided to take a moment to enjoy the warm waters and the fresh scents they released when hitting the soil and pavement, taking a seat at one of the benches in the small esplanade outside the hostel gates.

I knew for sure that this time in Ukraine would be very different from the previous visit. The cease-fire agreement signed in September the previous year was fragile and collapsed more and more by the day. Ukraine was on the verge of losing the battle at Donetsk Airport, where the fighting never ceased. Fighting also seemed to heat up north of Donetsk, especially near the strategically important junction town of Debaltseve, situated between Donetsk and Luhansk. In the south, outside of Mariupol, the group's reconnaissance missions into enemy territory discovered that enemy activity was steadily increasing.

It was getting late, and I decided to return to the hostel across the road. Just as I had reached the gates and started to tap in the entrance code, I was suddenly grabbed by someone, pushing me up against the banging steel doors. Both my arms were locked tight, making me realize there were two assailants behind my back. I managed to wriggle out of one of the men's grips, turning around to face my attackers. To my surprise, a third attacker rocked back the charging handle of his AKM, pointing it straight at my face, almost knocking my teeth in with its slanted muzzle brake.

Meeting the barrel of a gun had a very calming effect on me, and I gently lifted my now freed up hand up, allowing it to be recaptured once more. The rifleman's friends twisted my arms behind my back and handcuffed my wrists. After squeezing the cuffs properly, they led me to a dark minivan parked next to the stairs of the small basement bar, pushing me up against it. They seemed surprised I couldn't respond to any of their questions in Russian or Ukrainian. One of them turned out to speak a bit of English, and he began interrogating me to the best of his ability, all the while his rifleman friend kept his AKM trained on me.

"Why are you here? You sell drugs? You have drugs? Show me where drugs are!"

I slowly realized that these men were actually police—better that than Russians or crazed rapists. I denied all the charges levied against me, but the men began frisking me down, still hoping to find the drugs they sought. Finding half a can of Probe Whiskey in my pocket and my favorite brand of Swedish snus made them initially hopeful. I was, however, forced to explain that their finding was merely upper-lip-tobacco, an explanation they were soon regretfully forced to accept. The knife they found tucked in under my pants, though, threw more fuel on the fire.

"What you do here? Who are you? Why you have knife? And why do you need knife this big?"

The knife was big, indeed—a seax. I bought it figuring it would serve as a useful multi-tool in the field, filling the roles of both a knife and an axe. As such, it was much bigger than what Ukrainian law allowed people to carry.

The police officers, all three looking like Slavic white gorillas dressed in dark blue coveralls, weren't interested in waiting for me to offer some kind of explanation.

"Get in," the English speaker said, pushing me into the middle back seat of the unmarked police car, himself taking a seat to my left. His friend with the rifle stepped over on the other side of street, getting in on my right side. The third man took the wheel, started the engine, and drove off. I had never ridden in a police car before, at least not because I broke any laws, so admittedly the situation was kind of exciting. It was very late, though, and I was both tired and pretty boozed out, making it as annoying an experience as it was thrilling.

"I have my passport in my left back pocket," I said, explaining I was a volunteer in Azov on my way toward the front again. The English speaker dug up the passport and proceeded to flip through it.

"You are lying. Who are you? Are you a spy? What do you do here?" he asked, remaining unconvinced.

I told him to open my chest pocket, as well, where I kept some documents in my name with official Ministry of Defense and Azov stamps on them.

He gave them a quick read-through before passing them over to his Kalashnikov-armed colleague. Their gazes met with shared unease. The driver continued to stroll aimlessly around the city, judging from the many random turns he made without any kind of planned destination. The handcuffs were cutting into my wrists, making the turns uncomfortable.

"Calm down, don't make a problem," the English speaker told me as I tried to find some less uncomfortable way to sit.

"I'm very calm, as you can see," I replied. "These damn cuffs simply hurt."

I was drunk, tired, and just wanted to go to bed. Based on the driver's still seemingly aimless journey through randomized small streets, I concluded we weren't going to a police station. It was obvious they were driving in circles, clearly in an attempt to scare me, but I simply wasn't having it. After all, I was prepared to go to war—getting kidnapped by some nitwit police officers didn't seem very frightening compared to that. It was at best a discomfort and nuisance.

The only ones in the car who seemed nervous at all were the police officers themselves. I couldn't understand what they were saying to each other, but the tone of their voices was nowhere near as confident as it had been when they had initially shoved me into the back seat. They seemed to be arguing among themselves, disagreeing on how to proceed. I remembered that the chief of the Kyiv Municipal Police, a tall and broad-shouldered man named Troyan, was a former Azov member himself. I entertained myself, trying not to giggle too loudly, that it might have been this realization they were arguing about. What would their boss say if it turned out they kidnapped a soldier from his old battalion? They looked at me as I sat there laughing to myself, their eyes flickering through the documents again, making phone calls and sending messages over the car's radio system. Their tone became increasingly nervous, and I could soon hear a voice on the other line yelling at them.

After maybe around half an hour of touring around the inner city, the car stopped back outside the gate where they had previously nicked me. The English speaker uncuffed me and changed to the front seat.

"Where's my knife?" I asked him. "I want it back."

The policemen in the car exchanged a few words among themselves before the English speaker turned over to me once more.

"It is criminal to have knives like this," he said. "There's a fine you must pay first."

I understood very well that after the night shift, the police officers obviously needed a drink themselves. I was also far too tired of the entire situation to make any fuss about it all.

"Yes, yes, whatever. What will it cost?" I asked as the police continued to deliberate among themselves.

"It's five hundred hryvnia," I was told, which was approximately fifteen US dollars. I opened my wallet, gave them an appropriate banknote, and finally accepted my knife back, glad to just be done with everything.

"Fuck, man, how long does it take to buy a bottle of water where you come from?" the Greek asked when I finally returned to the room. "I bet you went to the brothel or something instead, you fucker!"

"No, truth be told, I was just kidnapped by the police and driven around in circles for half an hour for carrying a knife" I said, explaining the delay.

"And they wanted five hundred hryvnia for that?" the Greek asked.

"Yes," I replied. "Better that than having to argue further, or spend the night in some dirty Ukrainian jail."

The Greek laughed. "What fucking losers. Five hundred hryvnia? That's fucking nothing. If it was me, you would have had to give me five hundred US dollars, or more. Did they at least force themselves on you, in a dominant, sexual way?"

"No," I replied, "not even that."

"Too bad," replied the Greek. "I would have."

The next afternoon would be a management meeting. The founder of Azov, Andrei Biletsky, was to receive progress reports on the course of the war, his regiment, and the general situation. This included Gratch and Mike presenting their plans for their new boot camp program and its base of operations, to be called ATEK.

"He doesn't speak English, but he understands what you're saying," the Greek whispered.

The main commander entered the room, dressed in casual civilian clothing to greet all of the fifteen or so participants.

"He is kind of quiet, doesn't say anything if he doesn't have to," the Greek continued. "He is also a connoisseur of history. I know I've made jokes about him previously, but that was mostly bullshit, okay? He isn't stupid. Remember that."

Biletsky was a broad-shouldered man with thick cheekbones. He was only thirty-five years old but carried himself with the confidence of an older man. His hair was brown, shortened on the sides, and he sported a half-length dark beard.

He was probably the most famous radical nationalist in Ukraine and boasted a colorful history to come with it. Even at a young age, then in the Soviet Union, he had refused to join the communist youth organizations and instead hoisted a Ukrainian flag on top of his school building. After the breakup of the Union and the subsequent independence of Ukraine, he continued his political career in a number of different organizations and parties—a career which included over two years in prison after bloody clashes with political opponents, which left him injured from gunfire. At the beginning of the war, he had been leading Azov from the front lines himself, including during the recapture of Mariupol, with great success. However, after having been elected into parliament and his battalion having grown into a regiment, his leadership role had increasingly become an administrative one.

The meeting focused primarily planning the boot camp, ATEK. Mike introduced me as a future instructor, even though I hadn't agreed to this—quite the opposite, actually. I, therefore, had nothing to add on the subject; my only real role in the meeting instead shifted to answering questions on the possibility of setting up an amphibious unit inside the regiment. As Azov's main area of responsibility was the southern front and its long coastline, the use of landing crafts to outflank the enemy had been suggested several times.

"You have experience in amphibious warfare," Biletsky said to me, with Gratch translating. "Can we set up such a unit and use boats, for example, to attack Shyrokyne or Novoazovsk? If so, can you work as an instructor and help create this unit?"

"No, I'm afraid I must advise against trying something like that," I had to say. No matter how sexually titillating I found the idea of violent landings from fast attack crafts in Ukraine to be, I didn't wish to be responsible for how such adventures would end. I could see in Biletsky that he, just as the Greek had explained, understood what I was saying.

"The sea outside is completely open, and the beaches are long," I continued, explaining that without an archipelago providing cover to maneuver quickly between, any boats would be extremely vulnerable from land. "Using divers or surface swimmers could have merit, maybe, but offensive amphibious infantry would require huge numbers and heavy artillery support in order to have any kind of chance. The resources needed for such a unit would be huge, and I believe it would be much wiser to focus these on the land war instead."

I was sort of expecting to see Biletsky disappointed. Usually, he conducted himself very seriously, but I almost thought I noticed a slight smile as he asked the question. This gave me a feeling that he was as intrigued as I was regarding the idea of amphibious infantry—an idea I now firmly shut down, borderline classifying it as stupid. The commander's reaction to my negative review was, however, not emotional at all. He gave small nods during my explanation and never questioned my reasoning, despite me killing his pet idea. Biletsky simply thanked me for my input and moved on from the idea of amphibious landings at Shyrokyne. I imagined many other leaders, especially military ones, would not have acted in such a calculated and fair manner while having their plans criticized so thoroughly. Prior to meeting him, I had avoided prejudging the commander based on what others said about him, but my short impression of the Ukrainian was that of a confident, calm, competent leader.

After the meeting, we all went back to the hostel. Once inside, I opened my laptop to quickly catch up to speed on Swedish news. The top headline told of a Swedish volunteer fighter had died in a car accident in Ukraine. Reading the actual contents of the article, it was also plainly obvious that the information came from another Swedish volunteer.

"You told him, anyway?" the Greek asked. "You did! Didn't you?"

I insisted that I hadn't. "I don't think Mike knew, anyway, or he would have said something about it already. Someone else talked to the media."

"Fuck, no. Of course it was Mike. It's always Mike," the Greek responded, increasingly frustrated.

"No, look here," I said. "Look at this—look at the article on the screen! The only source they mention is referred to as 'a Swede.' Had it been Mike, you can be fucking sure that his name would have been in there, probably with a picture of himself he sent, as well."

The Greek took notice regarding the evidence, but remained unsure and full of questions.

"Who the hell did that, then? It wasn't Sonic. He hates the media. Tor, as well. Naive as a child like he is, he at least wouldn't contact them himself. And if it wasn't you, then who else?"

Mike, who had been outside, came back into the room. "Fuck, guys, I have bad news. Leo is dead," he said.

"We know," said the Greek. He continued, faster paced than usual: "Did you contact the media about this?"

Mike looked completely taken aback. "Eh, no? I just read it in the newspaper myself. A dead Swede, and he was the only Swede left there."

The mystery of who talked to the media remained, but the damage was done. Leo's death was official. Tjeck and I decided to take the bus the next morning to rejoin the group, while the Greek would remain in Kyiv a while longer. Mike told me to at least consider the offer of working as an instructor, and then come back.

"Well, we'll see," I said, having thought about it already.

"Bring one of the bags of stuff I collected with you, anyway," he said, handing it to me to take with us to the front.

CHAPTER 19: A NEW HOME

The bus stopped in Berdyansk at the same station I had left the front from a few months earlier. It was early morning, and hunger compelled us to stroll into a neat little cafe nearby. The women at the cafe didn't speak any English, a language which became increasingly rare the further east you traveled. Tjeck lacked any kind of mastery in the Russian language, but he understood a little, and together with some hand gestures, he managed to order us some breakfast. It was unclear what exactly the dish was called, but it was basically some assortment of blinis—pancakes served with some sort of sweet and sour cream. Simple as it was, it was a breakfast worthy of a prince.

I called up Chris to arrange for the group to come pick us up and taxi us to the base. He was still quite rickety after the car crash just a few days earlier, but was back at the base, regardless.

"Right, mate, it will be a moment, I'm afraid, as we no longer have a car," he said. "We'll have to try to requisition another one, but give it an hour or two, and we'll be there to come pick you up."

The morning brought a warm sun to cancel out the cold breeze. During the previous month of December, the area had been covered in a generous amount of snow, but by now most of it had already melted away. The ground was free from snow, and the air already smelled of spring. With our bellies filled and satisfied, a few hours of waiting in such conditions was no worries whatsoever—quite the opposite. We sat down on a vacant bench outside the station and enjoyed the rays of sunshine.

However, the pleasant atmosphere was soon interrupted by some random angry woman shrieking at us. I turned to examine our sudden antagonist. A crooked old babushka armed with a walking stick—halfway to absolute toothlessness, to complete the picture—had targeted us for a proper scolding. Her exact age was difficult to tell, as it wasn't uncommon for Slavic women to turn from young and beautiful to antiquated-looking in a single harsh winter. The extremely dissatisfied lady could have been anywhere between sixty and sixty thousand years old.

"Fascist" remains the same in most languages, be it Swedish, English, or Russian. Hearing the word in her vitriolic tone made the core of her messaging easily received.

"What is she yelling about?" I asked Tjeck, who listened to her never-ending hysteria with full concentration.

"I'm not really sure. It's very fast, and difficult for me to understand the dialect, but I think she has a son on the pro-Russian side. I think she says he will cut our throats, because we are fascist murderers. Just like they did with the Germans during the war, or something. She is not happy, and she knows many bad words."

The antique angry lady kept going, waving her stick around.

"At that age, her son has to be a pensioner, as well," I said. "How is she still going on, anyway? She has to stop to breathe at some point, at least."

She didn't, however, and our attempts to tell the angry lady to go on her merry way were also unsuccessful. Surely, proper fascists and murderers, as she insisted we were, would have found other solutions to the problem at hand. We, however, instead chose to find some other place to continue waiting for our taxi. The inside of the bus station proved quiet and deserted, and we hauled our heavy bags indoors. The paint on the inside walls was peeling off, and all the lined-up old payphones were broken or vandalized. The blue-painted benches were all dirty, many in a sad state of disrepair. Even the rough stone floor was full of cracked plates and large holes.

"This place must have been pretty nice when it was newly built, about seventy years ago," I commented.

"Yes. Communism did very well for our countries," replied Tjeck, laughing.

The signal from my cell phone echoed in the waiting hall while the broken payphones on the wall kept their silence. It was Chris, announcing they were closing in on the station and would pick us up in a few minutes. We grabbed our bags again and headed outside, soon seeing a black pickup truck coming in from the left. It stopped next to the street, and Chris pushed the passenger door open from inside the car

Chris wore full combat gear. He got out carefully, distinctly grimacing in pain with his AKS-74 in one hand, supporting himself on the door with the other. As he finally stood up, he held his arm out as a sign it was time for a *bromantic* embrace.

"Easy there, mate, easy. Ribs haven't healed up yet."

About an hour later, we were in Yurivka. The descriptions I had received prior to seeing the new Azov base lined up with the picture before me. In the middle of a small, poor village, a gated complex of tall hotel buildings in red brick towered over us—a luxuries, high-class gated community in the middle of a village stuck in Soviet times. Everything inside the walls was clean and tidy. Parks with fountains and ornamental bushes and trees all nicely cut. Everything seemed so new and fresh that it was questionable if the hotel had ever housed tourists, or if the war shut it down before it could open.

Just like in Urzuf, our quarters were a bit more simplistic than the others, but still not half-bad. The brick building was elongated and had two floors, each with a long corridor running through it. The stairs connecting the floors were located right next to the entrance on the left side. Close to the entrance on the first floor, the group was split between two rooms. I would stay in the smaller of the two, together with Chris, Långström, and eventually Greek, who still remained in Kyiv.

Chris opened the thin wooden door, and I followed behind him. Directly on the left was the water closet accommodations. I peeked inside, and, just as had been described to me, the small bathroom was deluxe-equipped with both a sink and a functioning shower. The rest of the room was absolutely cluttered with stuff, probably built to house a maximum of four bunk beds for a total of eight people; it now had three of these. Next to the balcony door, where the fourth bunk bed should have been, stood a large pile of ammunition boxes, mines, and explosives.

Out of the six beds, two of them were also functioning as storage. Next to Chris' ammunition depot on the floor was a single bunk bed whose lower part also functioned as storage. On it lay a bunch of leftover magazine pouches and similar load-carrying equipment, kneepads, and spare clothing. Next to it on the right was a small refrigerator with a kettle on it. On the top bunk sat Långström, his face sunk down into his laptop. Upon hearing us enter the room, he looked down from his elevated position, smiling down toward us.

"A-ha. So you came back, after all? I hadn't expected that."

"Well, I did tell you I'd be back, remember?"

"Sure, but that's just bullshit people say to one another before they leave." He sat up on the bunk, his feet hanging down, stretched like a cat.

"Ah, Finland!" he moaned, whereupon he jumped down, his feet banging loudly into the floor as he hit the ground. He had been letting his beard grow around his chin, but he had shaved away the mustache, making him look even more like a hobo than before.

"What the hell have you done to your face? You look like a goddamn Chechen terrorist or something."

"I know, right? And that's what everyone will think, anyway, in case I end up being spotted by a camera, or whatever," he answered, smirking for a moment before continuing. "Anyway, you came from Sweden just now, so that brings us to the most important point of our current discussion," he said, stepping closer toward me. "I hope you brought snus with you. You did, right?"

I told him I did, in fact, bring a fair amount of the Swedish tobacco goodness, both the bagged variant and the traditional loose kind, the latter in the flavor of Probe Whiskey, my personal favorite.

"Whiskey snus? I've never tried that kind before. I believe I need to investigate this right away," Långström said excitedly, stretching out his flexing fingers like a pair of pliers toward me.

On the left side of the room were two more bunk beds lined up against the wall. Chris had finally managed to painstakingly remove his combat vest and tossed it on the top bunk closest to the balcony, sitting down on the bed below it.

"We have some storage space on top there, and the lower bunk is Chris'," said Långström, pointing onwards toward the last bunks. "You can share that one with the Greek. He sleeps on the bottom, but since he isn't there right now, it's your choice, top or bottom."

I lifted my backpack up onto the lower bed, claiming it for myself, and continued outside to see what the balcony was like. Chris had been continually finding and storing ammunition and explosives since he arrived, and what couldn't fit in the corner inside, he stacked outside. There were several TM-62 anti-tank mines and more ammo crates. In the middle stood a 9K111 Fagot, a wire guided anti-tank missile.

"It's just an empty tube, I'm afraid," Chris said from inside the room. "I was trying to figure out some interesting use for it, but it'll likely end up merely a nice display item instead."

The room next to ours was slightly larger but equally cramped, housing the other group members. It was Tjeck; the two Frenchmen, Fabien and Cuix; Bear, and his new Ukrainian interpreter colleague, Richter. The driver, Inkasator, also usually lived in this room, but he was still in the hospital after the car accident.

"Well, then, looks like it's lunchtime," Långström said.

Chris stood up, holding his chest in agony. "The kitchen is just a little way down the road, past the tall hotel buildings in the middle. It's a lot better than the one in Urzuf. Yurivka has the best kitchen out of all the bases around here, better than the one in Mariupol even."

"You'll probably grow sick of this food, too, though, after a while," Långström said. "Anyway, let's go have lunch, then we'll go to the Polygon. We have some time left before it gets dark. You haven't practiced in a while, so it might be good to let off a few rounds, right?"

The shooting range for the day was the old air defense base. Urban combat was on the schedule today. The RPK the group had saved for me was old and worn but seemed all solid—no parts falling off, no malfunctions. The only thing I took issue with was the long bipod legs, which rattled a lot when folded.

The weather was overcast with light rain, which continued to melt what little snow remained on the ground between the buildings. I looked over at the new interpreter, Richter, who was wearing a thick, green winter jacket under his vest. He was unusually tall for a Ukrainian but as young as most of them, around his early twenties. His gear seemed fine overall, and it was obvious he knew how to equip himself, but the jacket seemed too warm for the task at hand.

"Is that your only jacket?" I asked him. "I think you're going to get overheated moving around in it like this."

"No," he replied calmly, "it's good."

Clearly a young man of few words, his face would probably have given a childish appearance if seen in a simple photograph. In real life, however, his serious demeanor added an extra layer to him that a picture couldn't convey. He was often expressionless, and the few times he did smile, there was something strange about it. It wasn't fake per se, but there was some human quality lacking in his smile. He seemed almost mechanical. His narrow eyes were always attentive to his surroundings, and the few times he spoke, the sentences were short and well-thought-out.

The shooting exercises went well, and it was clear that the group had become even more professional and cohesive since in the few months I was gone. After a while, we expended our ammunition for the day and began returning home, walking back through the skeleton buildings.

Chris, who was in the lead, suddenly called the group to a halt. "Hold up for a moment, mates. I have to investigate something in front of us."

Next to a building lay an RPG-18, a disposable light anti-tank weapon. A short tube which, just like the American LAW it was copied from, extended by pulling it apart. The tube on the ground was in the extended position, as if the rocket inside had been fired. Just as Chris had suspected, though, the rocket was still inside.

"It's a fucking dud they just left behind! Look at this shit!" he said, pointing at several bullet impacts having punched through it. "They apparently decided to try and detonate it with small arms fire, and when that didn't work, either, they just left it! Jesus, the bloody fucking wankers!"

"So, what do we do about it?" I asked Chris, who was already opening up one of his hand grenade-pouches.

"Well, I wasn't expecting any controlled demolitions today, so I didn't bring any of the gear. A grenade should do the trick, though, I reckon."

The rest of the group took cover behind the corner of the building while Chris stood next to the dud RPG, preparing to blow it. "Fire in the hole!" he shouted, dropping the grenade straight onto it before running for cover. The seconds passed before a tiny bang went off, more like that from a fire cracker than a hand grenade.

"Well, fuck me up the arse!" the Brit sighed. "That's the fucking third time this shit has happened!"

It turned out the old F-1 hand grenades, frequently having WW2-era dates of manufacture stamped on their bottoms, were frequently duds themselves. While the detonators would go off, the grenades themselves simply wouldn't explode.

"Fuck it, I'll use an RGD instead," Chris said. He repeated the procedure with a more trustworthy offensive hand grenade instead. This time, shortly after dropping the grenade and ducking behind cover, the RGD lit up both the RPG-18 and F-1 in a great ball of flame and black smoke.

CHAPTER 20: FORWARD OF THE FRONT LINE

Much to my satisfaction, it turned out to be true that the group finally started getting missions. I didn't have to stay more than a couple days in Yurivka before the group received another assignment. Båtsman was gone, but his group remained and now functioned as the Recon & Sabotage platoon. They were heading out to conduct reconnaissance against enemy positions east of Mariupol, and the Foreigner Group would tag along to assist.

"How come you started getting missions all of a sudden, anyway?" I asked. "We were just sitting here with our fingers up our asses for weeks, months, without anything happening. What changed?"

"You mean you haven't figured that out yet?" Långström asked, looking at me as if I were a sluggish retarded person. Seeing as I couldn't find a good answer, it dawned on me I probably was.

"It was Mike, the fucking coward, who cockblocked all the action, because he didn't want to go on any missions," he explained. "The Ukrainians were suggesting things for us to do all the time, but he always found excuses for why it couldn't be done. We were apparently busy with a lot of other things, he told them, when in fact it was just him being a goddamn pussy, too scared to fight."

The Greek, who had arrived back from Kyiv, filled me in on the story: "He was always a charlatan—I thought you knew that? You guys knew each other in civilian life, anyway."

"Knew him?" I asked. "I had never even met him before coming to Kyiv last year."

My friends looked surprised.

"What?" the Greek asked. "He said you knew each other well, that you served in the army together, or something."

I explained that even though we originated from the same city, he was, after all, ten years older than me. We did serve in the same unit of the Home Guard, but he had been expelled from there about the same time I had joined the battalion.

"Well then," the Greek sighed, "according to him, you two had been good friends for years, but I guess that only confirms everything all the more. That's why we didn't want you to know about Leo's demise, by the way. We all thought that, you two apparently being friends, you would tell him all about it."

The Greek sat down on one of the small chairs in our room, starting to prepare his equipment for the coming mission.

"You know, when I first arrived in Ukraine to join Azov, I was only in Kyiv for one day before I was sent to the front line at Marinka. It was a bit more chaotic back then, not like the bullshit when you came here later in the summer with all the waiting and other gay shit."

The Greek had only recently been able to trade his AKM in for an AK-74 and was emptying his magazines of tracer training rounds, replacing them with standard ball and armor-piercing bullets.

"Needless to say, I had also heard these stories and read about the so-called 'Swedish sniper' in the battalion, but I immediately realized something wasn't right the very first day. Our group lay next to an open field, covering some large buildings on the other side of it—five hundred, maybe seven hundred meters or something away. Fuck, who cares, right? Long apartment buildings, like those at the Polygon, you know? Several stories, many windows, that kind of shit. So, anyway, some random Ukrainian comes over to our position and starts babbling some nonsense, right? Eventually, we start to make some sense of it all, and it turns out there's some other guys being surrounded and pinned down nearby. Like, they're in a really fucked situation. Whatever exact kind of shit they had gotten themselves into, they needed us to come help bail their asses out."

The Greek found it difficult to talk while filling his magazines, all of them marked with white tape with the Cyrillic letters for "GREK" scribbled onto them. He put everything aside and began to properly gesture with his hands as he continued to convey the story.

"The Ukrainian, whatever his name was, has barely finished his speech as Mike, out from fucking nowhere, opens fire on the buildings, shouting 'contact!', okay? He had one of those civilian hunting rifles at the time, a Remington bolt-action with a scope on it. So, he starts shooting toward those buildings, saying they're full of separatists. We hadn't been taking any fire from there previously, and we had been laying there for quite a while, but he apparently could see all those Russians very clearly through his scope, so he opened fire, rapidly blowing through ammo in the enemy's general direction."

The Greek held his hands up as to point an imaginary gun aiming through the balcony windows, showing the reloading operation of a bolt-action rifle "Bang! Bang! Bang! Just like that, firing blindly, round after round, very quickly."

He kept the imaginary rifle at his shoulder, looking to his right at me over where the buttstock would have still been positioned. "I didn't really get what was going on at that very moment, but it did seem a strange way for a sniper to operate his rifle—very fast rate of fire, seemingly without aiming."

He stowed the invisible sniper rifle aside and proceeded to fill up his AK magazines.

"It was only a while later when I realized that all those separatists in the building—they were obviously just fantasies he invented to avoid moving into a real fight. And we didn't, as evidently this imaginary enemy force threatened our flank. Those guys we were supposed to help out—what do you think happened to them? Think they made it out in one piece? I'll tell you what: they didn't."

Greek's story gradually pieced together the mental puzzle I had been struggling with for a long time regarding our former commander. I cursed the fact that I had believed anything Mike had ever said, and all the time and effort I had lost because of it.

"However, he got even worse after the Battle of Shyrokyne in September, when we were shelled by the Russians," Greek continued. "You remember that story, don't you? Both Tor and I have told it several times over. Anyway, our group was the last to leave Mayak, the large brick building on top of the hill. When we were retreating, Mike was the first one out and took the only functional car for himself. There was something completely manic about him, like something was totally broken inside, you know? He tried pulling some bullshit later on, trying to convince us he needed to save some important general, or whatever the fuck his story was. Obviously, that was all his usual lies. We saw him: he was fucked, his empty eyes just staring into the void, retreating in full panic in the only car that actually worked well. That's why the rest of our retreat went as disastrously as it did. He made us squeeze together in cars with fucking busted gear boxes."

The Greek stood up and went to fetch some more small paper boxes of 5.45mm ammunition.

"You know, before the artillery had hit us properly, he was still perfectly okay. Sure, he made up excuses and all that, but he wasn't that bad. After Shyrokyne, though, he didn't ever want to leave the base. When we were ordered back to the front, he made up some bullshit about Tor's foot being messed up—while Tor was still sitting inside the same car! Obviously, there was no foot injury: Mike's morale was the only thing injured. And he wasn't alone in that. Many people had their spirits broken by the artillery that day. Even I, a real soldier, was in a pretty somber mood after that. I'm not joking, you understand? Grad, the rocket artillery, is fucking scary shit."

The Greek ended his monologue, looking at me for a moment.

"Well, that's all in the past, regardless. All irrelevant bullshit, really. We're heading out tonight. Shouldn't you be tending to your own equipment instead of listening to my nonsense stories, anyway?"

I decided to go with an AKS-74 no one else was using for my armament. Ammunition and magazines were aplenty, unlike my uneventful visit to the front last year. As the recon team consisted of something like twenty men in total, there was no need for me to carry the RPK to boost the squad's available firepower. The problem was that all my personal equipment remained specialized and hand-picked with the PKM in mind, not an AK or RPK. My large pouches were good for cans of belted-up 7.62x54R, but I had only a single pouch fitted for regular 30-round magazines.

I dug around in the mystery box of mixed equipment, scavenging for anything I could possibly use. It was mostly a blissful mix of different multi-purpose pouches and general crap, which I was forced to figure out how to make work as best I could. Nothing in the box truly fit my role; instead, I simply had to appropriate this or that assorted item and find a way to make it work for what I wanted to do.

After some experimentation, I looked at my sorry plate carrier and battle belt, thinking about how I would never have accepted being issued something this awkward even in peacetime. Being as this was in wartime, with no one to complain to, I realized there was no recourse other than to try to make it work, however bad and jerry-rigged it all was. It would at least hold a total of six AK and RPK magazines, in thirty and 45-round capacities, respectively. I complemented these with extra ammunition on readied stripper clips and in paper boxes.

I replaced my front armor plate with one of the properly shaped ones Mike had brought from Sweden, at least making the sad thing a bit less uncomfortable. The only thing there was plenty of in the pile was hand grenade pouches, and I managed to equip myself with three of them. Two RGD-5s for throwing, and an F-1 in case I was captured. The torture of prisoners was a well-known practice among the Russians, and a foreigner from Azov would get it even worse than normal. Because of the unreliability of the F-1, I also stuck a few spare 5.45 rounds in between the plate carrier and its MOLLE panels, just to make sure.

We had a scale in the room which I hung my loadout onto. Without ammo, rifle, and helmet, everything came to thirteen kilos, of which the armor plates alone were almost ten themselves. Once everything else added up, it all added up to almost twenty kilos in total. If I had been using the Polish vest we were issued, with its soft Kevlar panels, I would probably have been pushing above thirty kilograms in total. With that in mind, I felt I got off easy.

The weather outside wasn't exactly freezing, but staying still in a prone position would be a cold endeavor, which is why I tried to dress warm. I outfitted myself in a mesh undershirt and a warm sweater underneath my brown, windproof smock—light, so as to not sweat during a foot march through no man's land, but warm enough to hopefully avoid freezing to death from staying put for long periods.

Our cars were ready at the driveway outside of our quarters. A total of four Mitsubishi pickups would bring the large recon party to the front and then over it.

"Get in," said Långström, showing me the back seat of the leading vehicle. He then got into the passenger seat, and a huge Serb, who would be leading the adventure, took the driver's seat. The Greek soon squeezed in to my right side, pushing me into the mid-section of the seat, and Cuix, smaller and agile, took the left seat. I hadn't received any detailed information about the mission, other than that we were heading out to look for enemies around some village. Not knowing anything made me a bit uneasy.

Despite the cloudy skies, the air was still warm outside as the column of light-skinned 4x4 vehicles began moving toward Mariupol. The FM radio played music from some local station. The Serb seemed relaxed and nervous at the same time, keeping mostly quiet during the trip. Långström next to him was very talkative, seemingly to the Serb's vexation. Happy and excited like a child on his way to

Disney World, he made jokes and laughed, the Serb only meeting his enthusiasm when forced to with dull and gloom mumblings. As we approached Mariupol, the sky cleared, and I could see hundreds of small, bright, white orbs of light flashing in the sky far ahead. Soon, I also saw the long strings of red-colored tracers traveling like long snakes from ground level up toward the skies.

"Oh!" I shouted in excitement, leaning forward and pointing toward them. "Is that the Russian anti-aircraft artillery firing?"

"Yes," Långström replied, smirking, "they've begun firing more and more lately, as our recon drones have started fucking with their minds! Ha!"

The Greek, mostly silent up until now, piped in with his own explanation: "That's what they do, just firing blindly into the skies, expecting a different result. They never seem to hit anything, but it still makes them feel better for trying, fooling themselves into thinking they're the masters of their own destinies."

A Ukrainian military checkpoint began coming into view in the distance. We had reached the front line. In front of us, outside the windshield, lay no man's land and the enemy. I'd been nervously awaiting this final step for a long time, crossing this line in the sand. It had finally come.

Adrenaline slowly began to rush through my body, and the nervousness passed. We were going up against the Russians, finally. I had been waiting for this moment for months, years even, and I was finally here. I was ready.

We waited at the block post until the sun had set completely and the outside was pitch black. I met the Serb's gaze through the rearview mirror.

"If you smoke, you smoke last cigarette now," he said in his monotone voice, lighting two cigarettes in one go, handing the second one over to Långström. He took a long puff of smoke and continued in his simple vocabulary, "If I die here tonight, you find my girlfriend and tell, I am died in war, and not go with other woman."

Långström swallowed the smoke and coughed, looked at him, and laughed loudly and heartedly.

"Ha! What the fuck was that? You're a fucking regular at the brothel! You even fuck the ugly ones!"

The large Serb filled his lungs with another inhale of smoke, looked down on the floor and answered, again with the same monotone, albeit with a tinge of embarrassment: "Yes, but she don't know now. And you know, you don't tell this to woman, or she get so sad and cry."

The soldiers at the block post lifted the barbed wire obstacles in front of our cars, clearing the way for us to roll forward into the darkness with headlights turned off. The FM radio was off, and the inside of the car was devoutly quiet, with only the noise of the engine and the occasional bump over potholes making any noise whatsoever. I let my hand stroke down the side of my AKS in front of me. My fingertips felt different, as did the metal I was touching them over. The only source of light was that from the engine display, reflecting Långström's death-defying, infernal wide grin in the windshield in front of him. He was as alive as he would ever be—and so was I, for the very first time in my life.

After a few more mechanized kilometers, we arrived. From this point on, we would leave the vehicles to continue onwards on foot. I disembarked, double-checking my equipment as the large Serb approached me.

"You. New guy. Can you drive car?"

I immediately understood what he was asking. Guards would stay with the cars, and he wanted me to be one of them—provided I could actually man the vehicles.

"No, I don't have a driver's license," I told him, which was technically true, though I did actually know how to drive despite not having a license.

The Serb sighed with disappointment and walked over to Cuix, who immediately answered yes in his usual excited tone of voice. The Serb mumbled something, and I saw Cuix's facial expression turn sour, confirming my foreboding had been correct. Cuix would not follow the rest of the unit, but rather stay behind to guard the vehicles.

The group gathered into a single, long column and began moving forward through the open terrain. Most of the Ukrainians were unknown to me, and I only recognized one of them: Bonn, the big and rough-hewn man I had encountered the previous year when I unsuccessfully tried to retrieve Tor's AKM from Båtsman's house.

We began moving like a long snake along the tree lines between the large fields. The ground was mostly flat and should have been simple to cross, but what initially seemed like an easy stroll through the countryside soon proved treacherous. As the snow from the previous month had melted, its remains had sunk into the ground and created tremendous expanses of mud—mud like nothing I'd ever encountered back home. Even the grassy ground, which looked otherwise untarnished, couldn't be traversed without sinking into it. The mud and clay stuck under our boots, creating a new, heavy, slippery extra sole under them. Our boots stuck into the ground, and once you finally tore them free, they were covered in extra kilograms of mud. With each new step, another layer of mud would stick to the first one like drying candle wax, and then another layer would stick to that. Soon my feet were so heavy from the glue-like soil that each step became a physical challenge to perform.

The clay not only stuck underneath our boots, but also on their sides, the girth also expanding with each step. Soon it was like walking with snow shoes, and as they rubbed the insides of our calves, they soon began leaving bruises on them. After only a few meters, every man was forced to halt to try and get rid of the invasive mud stuck to him, but the tedious effort only helped for a single step. Scraping the mud off one boot with the other proved useless, as it would just stick on the other boot instead. Using a thin rifle butt to knock it off proved better, but soon it, too, was covered with sticky clay. Desperately trying to get the mud off the rifle, of course, only ended up with our hands and arms covered in mud, too.

As warm air blew inland this night, I tried rolling up my sleeves as far as possible. I removed my cap and opened up both my coat and sweater, but nothing helped. Sweat ran down my forehead, just as it did with everyone else around me. I saw Bonn in front of me desperately trying to rid his mud by kicking into a thick bush beside him, little good it did, as he, just like me, was covered in it again soon afterwards.

I thought back to all the stories of German soldiers experiencing the Eastern

European mud fields around these parts in 1941. I had never really taken those stories so seriously. Mud is mud, I had thought to myself. I've seen a lot of mud, too. How bad could it really have been? I had done many foot marches before, as well—the longest upwards of eighty kilometers in a single day. That march had gone through swamps and thick forest, and I had managed to pull through—the last few kilometers with a fractured foot, no less. Here, on the other hand, in the middle of the muddy Ukrainian steppe, we had only come a few kilometers, and it was already the heaviest walk I had ever endured. I thought I'd had all kinds of abrasions before, but I had never before experienced them in the inside of my calves until now, from where the thick mud continually scratched against them. Every step we took forward was a fight, and I tried my best to forget that each one meant one more on the return journey, as well.

I was just starting to regret not just saying I could drive a car so I could stay behind to guard the vehicles instead, when suddenly loud sounds of a firefight traveled across the empty plains. Occasional rifle fire and rumbles of artillery had gone off in the distance, but this was a lot more intense—and closer. Fierce automatic fire from several AKs came from the direction of our parked vehicles.

"Fuck," I mumbled bitterly to myself. "That's Cuix and the others dead, and our cars destroyed. Now we will have to walk all the way back to Mariupol."

The enemy were carrying out recon missions in the area just as we did. It wasn't unlikely one of their units would have stumbled onto our cars by accident and taken them all out. I knew Cuix was a good rifleman with quick wits about him, and I could only hope he was faster than the Russians.

"Nothing to do about that," Långström replied, wiping a flood of sweat from his forehead, but seeming otherwise unbothered. "The Russians have jammed all radio communications throughout this area, anyway, so not much we can do to try and contact them. We continue forward as planned. Then, when we come back, we'll just have to sneak up and see if they're still alive."

Cracks began forming in the thick clouds blocking the night sky, and above us the stars shined brightly through them. I had forgotten everything about how to navigate through them, but having forgotten my compass at home, I at least decided to memorize the locations of the stars, as long as I still knew somewhat which direction was east and west. Just in case we also ended up getting ambushed, split up, and disorientated.

We soon began to approach the area of interest, from where the group would conduct reconnaissance. The Serb halted me and two Ukrainians.

"You three stay, take positions here. Protect our back, while we go last bit forward."

The surrounding terrain was almost completely open and unused grassland. The only standout in that flat steppe was a tree by a road in front of us and two huge, dried-out, artificial concrete riverbeds of which we lay in the middle. I moved up to take position at the tallest point to be able to look ahead, while the other two kept themselves a bit further down on the slope. The rest of the group continued onward while we began our long wait.

The weather changed quickly. The wind's direction had turned and was blowing cold air from the seas again, and with it a mist which penetrated through my windproof smock and married with the already cold sweat against my bare skin beneath it. The wet moisture on the grass in and around us soon turned into ice crystals, giving a clear indication that the temperature had dropped. It wasn't long until my teeth began chattering. The others weren't meant to be away for more than a few hours, but every minute felt like an eternity.

I had never been a big fan of reconnaissance in general, having a difficult time staying in one position for too long, and the cold made my first real mission an increasingly unpleasant experience. I rolled out the sleeping mat I had brought with me, which helped a little bit. Having gotten rid of at least the cold ground, I then tried to crawl into the tightest fetal position possible, trying to retain whatever body heat I could.

"Don't move!" hissed one of the Ukrainians behind me—a small, young man with a thick, dark beard they called "Matros"—as he tried to adjust my position. "They will see you, fucking idiot!"

We were far off from any enemy position, and the terrain was flat and open. They seemed overly nervous to me. I bit my tongue instead of hissing back and curled back up, but despite the best of my efforts, nothing seemed to help. The cold winds continued blowing in, and I was tempted to crawl back down the slope in the hope the winds weren't as bad. If I did, however, nobody would be keeping a lookout. I tried distracting myself from the cold by finding something to look at, focusing my mind on that instead. The moon was bright, but despite its white shine, it had little to illuminate. The lonely tree next to me was about the only thing around.

The minutes passed, turning into hours. My teeth chattered on so hard it felt as if they would be heard from kilometers away in the quiet night. Suddenly, after what seemed like an eternity, I saw movement over on the left side: a group of men heading straight toward us on the road on the man-made hilltop. Seeing this made me forget I was freezing, and my jaw stopped shaking.

According to the time, the unknown distant men's arrival corresponded with about when the group had expected to be back, but there was no way to be sure who they were. I tried to hiss toward my other pair of Ukrainian mates to make them aware of the situation to no success. Instead, I grabbed my AKS, gently pushing the selector down to automatic fire and pulled the charging handle slightly back to make sure once more it was loaded.

Seeing the green steel case in the chamber, I gently moved the mechanism forward and set the magazine into the ground for support, letting the iron sights trail onto the men approaching us. They walked in a column with short distances, quietly, and at a leisurely pace. Even though they gave a faint glow, it was difficult to distinguish sharp contours in the darkness. They were already at about seventy meters distance before I could start making out their numbers: about fifteen men. It was a bittersweet observation, as it meant the size matched the rest of the group. At the same time, however, if they were the enemy, they outnumbered us significantly. Especially as it seemed as if I was the only one who had taken notice of their arrival. My heart was pounding as I put all my focus on holding the rifle steady, following the lead man in my sight with my finger slowly squeezing the trigger.

It was only when they were within twenty-five or thirty meters that I realized the man in front towered over most of the others. There were tall Russians around, sure, but if they were, they were usually very slender, as well. This man was tall as well as broad shouldered—obviously the large Serb. I put my rifle back on safe and exhaled a thick fog of mist from my lungs, indicating that my body temperature had increased significantly the past few minutes.

We reunited with the main party and began heading back through mud just as terrible as before. Some collapsed from exhaustion and needed help getting back up. As for me, I found comfort in at least knowing that from now, the road became shorter with each step. However physically demanding, it was also still welcome being allowed to move once more. An hour more of prone position in the cold winds, and I would have shaken my teeth to pieces.

Artillery and mortars had begun firing around us between the two front lines—no heavy bombardments, just each side harassing the other with a few shells here and there. From time to time, a heavy rumble, not unlike a distant thunderbolt, would travel across the steppe. Soon afterward, a thick white orb of light would throw itself from ground level up toward the skies from the other side, followed by the sound of another thundering echo a second later. It was not only a great distraction from the physical strain of the next step through the mud, but it also proved a great chance to observe and learn the difference in sounds of incoming and outgoing artillery fire. Had I not been completely exhausted, I might even have enjoyed the slow light show.

We soon arrived within sight of where we left the vehicles. The group paused a while in the shelter of a shrubbery to catch our breath and gather our forces, in the event that an ambush awaited us. I opened my water bottle, which was about to run out, and took a few sips, which caught Bonn's attention. He looked terrible and completely fatigued.

"You have water? Give me water, please."

It was obvious that his upper body strength and the extra weight his gear added hadn't helped him during the march. I handed him the bottle.

"Just don't drink all of it. I'm almost out," I told him. The broad-shouldered Ukrainian looked like he would have been able to empty a small lake, but he did in fact stop before having emptied everything.

"Thank you," he said, clearly disciplining himself to hand the bottle and its remaining contents back.

A few men from the group snuck up toward the cars to check the situation. They soon stood up, waving to the rest of us. Both the cars and the men left to guard them remained in one piece. We could all safely regroup at the vehicles. I soon saw Cuix among the cars, with a big smile on his face.

"Fuck, man," I told him, "we heard a lot of shooting in this direction. It sounded exactly as if you gotten ambushed and killed. What happened exactly?"

Cuix looked at me and answered in his terribly French accent, laughing all the while: "It's okay, my friend! Shooting came from your direction. I hear and am like 'what the fuck,' and think you all killed!"

"That must mean some other recon group got fucked up really badly somewhere in between us!" Långström pointed out, after which he also proceeded laughing.

Everything had worked out. Everyone was alive and well, and we didn't even have to foot it all the way back to Mariupol. We were just about to get into our vehicles as someone shouted "machina" (i.e., machine, or vehicle). Usually, it would be the order to embark the cars, but this time it was a warning sound. An unknown vehicle was approaching us, fast.

Our pickup trucks were all parked next to one of the main roads to Mariupol, between a couple of small hillocks and elevations in the terrain. To the east, a pair of spotlights appeared, coming from the enemy lines, getting bigger by the second. One of our pickup trucks was armed with a welded-on AGS-17 30mm automatic grenade launcher; one Ukrainian climbed up on the flatbed to man it, pointing it toward the road as the rest of us took up firing positions around the small hill slopes. Cuix, Långström, and I took the right side, aiming our rifles at the approaching vehicle.

"Fire on command only!" Långström shouted. "Nobody shoots without my command!"

The vehicle came closer. Everyone was at the tip of their toes, all sights locked onto it. It came closer and closer. Soon, we could identify it as a single vehicle, a car. More details became evident as it neared our hornets' nest. The headlights were close to each other, making it a small car. Very small and crappy, much narrower and lighter than our 4x4s.

"Fuck it," Långström muttered in anticlimax, "another fucking steel plant worker..."

Despite the ongoing war, civilians in the area were still forced to keep on working their jobs like usual. The salaries were bad, but they needed in. Mariupol's huge steelworks, the biggest employer in the region by far, still had workers not only in the city or on the Ukrainian side of the border, some workers remained displaced in what was now separatist-controlled parts of the country. The clock was about five or six in the morning and here was another one of them, an ordinary working-class hero driving through no man's land and artillery twice a day in order to get on and off his shift melting down beams of steel.

Suddenly, the crack of a rifle round going off broke the silence.

"Cease fire!" roared Långström, and the rest of us sang the same chorus.

Someone had apparently been a bit too nervous and pulled the trigger, but nobody knew who it was exactly. The bullet impact was nowhere to be found.

The car continued as before, albeit a little bit slower. At the wheel sat a middle-aged man, who now in his spotlights could see a large number of armed men, many still pointing their rifles straight at him. With a terrified look on his face, he met none of our gazes. He just stared straight ahead, keeping a spasmodic grip on the steering wheel as his rusty little Soviet car slowly coughed forward and past us. He was in great luck, as not only had the negligent discharge not hit him, but above all else that it hadn't caused all others to open fire, too, turning both the steelworker and his little Lada into dust within a second.

"Why the hell won't they just shut everything down?" I asked. "You can't have a fucking war with civilians going through the lines every day for work. What the hell is wrong with this country?"

Långström's gaze followed the car as its red tail lights soon disappeared in the distance west of us.

"It's all about money. It always is. Not even war is more important than money. That is all."

CHAPTER 21: FUNERALS

The reconnaissance missions continued every other night. Långström led all of them but rotated the personnel under his command. It was difficult to fit everyone into a single car. Besides, the group had greater numerical strength than was usually needed, anyway. It also gave each soldier at least some time for rest every now and then. This time, Chris and I had remained at the base overnight. Early in the morning, the Greek entered through the door, obviously excited.

"Fuck, guys! You missed something this time! We captured two of the bums as prisoners! Real fucking bandits, I'm telling you. You should have seen them!"

The sheer level of the Greek's satisfaction made me jealous that I hadn't been part of anything so exciting. Still, being able to rest a full night while the others were out freezing their asses off was obviously nice, especially since most of our recon operations remained uneventful, anyway. We halted our cars, got out, and marched to some predetermined position, hoping not to step on any land mines while doing so. Then we lay there, often without seeing anything. Enemy activity was usually low and we had never before even gotten close to actually capturing any of his men.

The Greek began to eagerly explain the course of events which had unfolded during the night: "Truth be told, this was actually a matter of coincidence! A simple fuckup, if you will. It could have ended up in many different ways. Anyway, we had been scouting around Shyrokyne, from one of the vantage points nearby. Then, as we start getting ready to pull back, all of a sudden, from nowhere—voices in the darkness! We threw ourselves in the ditch next to the road for cover. And so, then, there they came, passing right by us on the road, the bandits—both of them drunk, needless to say."

The Greek used the term "bandit" to distinguish between real Russian soldiers, be they volunteer fighters or from regular formations, and the common local separatist fighter.

"And the two of these, they were serious bandits. Fake leather jackets and equally fake Adidas pants, you know. That kind of shit. Only one of them came armed, and you could tell he was the leader, judging from that kind of oblong flat cap which higher-ranking gopniks commonly carry. You know they do this, right? It's like a badge of honor among them. It was the complete picture! Anyway, as we understood the two of them were alone, we simply seized the moment of opportunity and jumped out from our hiding place. The elder bandit dropped his rifle immediately! You should have seen their faces, when we confronted them in English! They fucking shit themselves, realizing the might of all NATO nations had suddenly fallen down upon them! Fucking laughable, really. Goddamn bums."

Russian propaganda from the very beginning had made a big deal out of the fact that there were Western volunteers in Azov. The picture which they painted, however, had made our numbers out to be much larger than they really were. The purpose, of course, was to turn the Russian-speaking population, both in Ukraine and Russia proper, hostile against us—to make Azov in particular, but also the Ukrainian army in general, seem like the foreign invading force, and thus illegitimate. A foreign army from the West, not the least different from the German one which had invaded the Soviet Union back in 1941. In order to complete everything, the Russian-language media and propaganda spread lots of stories regarding how the horde of monstrous Western mercenaries terrorized the population. Fake videos spread showing how we hanged pregnant women and crucified small children. There was no end to it, and whatever they had claimed the Germans to be doing seventy-five years prior, we were doing the exact same thing today. A large part of the Russian-speaking population readily believed this, however laughably ludicrous it was, and hated us because of it.

Though their propaganda had worked in that sense, it also came with side effects the Russians seemed not to have foreseen. Carefully and loudly painting the Ukrainian army as Nazis, murderers, and cannibals—Azov being the worst of the bunch—also gave us a strong intimidation factor. Hating us as they very well did, they were also terrified of us. When the two "bandits," as the Greek called them, were confronted by armed English-speaking men, they became so frightened that they were unable to act and surrendered immediately.

"The bandits just cried, praying for their lives like women!" the Greek continued on. "Utterly pathetic to witness! Adult men. The Slav has this grand image of himself as a brave giant or whatever—mostly bullshit, regardless, but these bums were definitely a disgrace to their race. We just hauled them back with us to the cars, and they kept crying all the way. Fucking children! Fuck!"

The Greek's previously enthusiastic tone increasingly shifted that of irritation. He continued, waving his hands around in front of him with increasing energy.

"Anyway, some loser from recon, Båtsman's old group, checked them before throwing them in the back seat, one in each car, seeing to it that they *positively* didn't carry any more weapons, right? So then, once we're all inside the cars, we start going back. I'm in the back seat with the lower-ranking gopnik, and the fucking pussy just keeps crying, right? Like, he starts hyperventilating like a

167

fucking adult five-year-old who just got fucked by his drunk uncle or something. Fuck! The annoyance I was feeling could hardly be described. Snot and tears everywhere! Jesus!"

His hand gestures were increasing, showing the momentum of the story was closing in quickly.

"Then, finally, I start to realize the bandit is trying to tell me something, right? He kept pointing to his pants pockets, like this," he said, pointing to his own side pockets. "So, what to do, right? There was clearly something in there making him uncomfortable. Maybe addressing this problem of his would stop him from crying all over the goddamn car, I thought, so I rushed to help him out. I tucked my hand into his filthy pocket, and guess what? The object was round and hard—but it wasn't his cock! Doubtful there ever was one in his pants, anyway, but fuck it, whatever. Guess what it was?"

The Greek laughed for a moment before clapping his hands together, signaling the finale.

"The bum had a fucking hand grenade in his pocket the whole damn time! And the *goddamn losers* who had checked them for weapons *swore they had done so properly!*"

The Greek always spoke loud and fast, but emphasized the most important parts of his speeches with a slower, but louder tone.

"Obviously, the bum could have blown us all up right there, had it not been for the fact he was far too busy being a crying pussy. Now he was just terrified that he would get a beating for trying to hide the grenade instead, doing everything to get out of it. But how fucking difficult is it to check a man's pockets properly! Jesus! The fucking fuckups these white niggers do sometimes, it's beyond me. I'm telling you—do you know why they didn't find the grenade? It's because they seriously believe that if you touch another man's cock by mistake, you become a homosexual! I'm serious, they believe that!"

The Greek ventilated some air out of his lungs and calmed down somewhat. "That's not the way it works, anyway. You only become gay if you let yourself be dominated by another man. You don't even become gay by fucking another man—no, not so long as you dominate him in the process. Quite the opposite— as long as you dominate another man, it makes you even *more* masculine. That's the truth! It's actually called 'Kolobaras.' It's Greek. You don't believe me? Look it up! It's real."

Although the front line outside Mariupol was arguably still calm, we had an increasing number of funerals to attend to. The two men who had previously borrowed my PKM, finding their demise on top of an anti-tank mine, already rested below ground. A commander from Recon & Sabotage, a middle-aged Crimean native with a black beard, was on his way to join them after falling victim to one of the many traps riddled across no man's land.

The battalions of the regiment gathered in Urzuf for the funerals. They were referred to as such, but in reality, they were merely ceremonies honoring our dead, as the bodies were afterward sent back to the families for proper traditional burials.

At each funeral, soldiers lined up in the same manner as during the normal morning counts, as we had done many times while stationed there. The coffin lay on a table in front with a small honor guard next to it. Battalion commanders gave some short speeches in honor of the fallen comrade, after which the soldiers formed a long, snake-like column, passing by the open coffin for a final farewell. The Greek gesticulated the sign of a cross as he passed by. The recon commander's face had been cleaned up, but it was impossible to cover all the injuries. His face was covered in holes from the many balls the Claymore-type mine sent out as shrapnel—injuries which seemed as if they caused sudden death.

It was also soon time for Leo's funeral. It should have been settled a long time ago, but issues with authorities and insurance in Sweden had made it a tedious process. Nobody wanted to cough up the cash to have his body transferred back to his native country.

"We should burn him on a pyre, like the Vikings did!" Tjeck suggested. "Then we can send an urn to his family, and pay it ourselves."

I thought that there had to be some other way, but the others had taken a liking to the idea.

"Leo would have liked that. I know he would have," said Långström, and the Greek agreed.

"I know you never liked Leo," the Greek said to me. "You saw him as an incompetent, bipedal disaster, and rightly so—not to speak ill of the dead. But he got a lot better during his last two months. Really."

"The guy was completely fearless! He was always the first volunteer for new missions, no matter what they were!" Långström filled in, clearly enthusiastic about this personality trait. "He was completely crazy! Ha!"

The Greek laughed, as well, taking a sip from the bottle of Lagavulin I brought for everyone to share during our discussions of the dead.

"Yes, it was a damn shame he would die so soon," said the Greek. "Had he only gotten a few more months, I think he could have become a pretty good soldier, really."

"Well," I said, having a sip of whiskey for myself, "in that case, I guess you can say that at least he ended his life while on top. A fortune bestowed upon far from many."

The Greek nodded and replied, "I agree. He really did enjoy himself here, actually having a purpose. I think this was as happy as he had ever been in his miserable bum life."

The Swedish Ministry of Foreign Affairs still refused to aid in the matter, so Azov themselves decided to pay for the transportation of the now weeks-old corpse. It was a lot of money, which the poor Ukrainian defenders probably could have found better use for.

"They still want to do the right thing, and that's commendable," said the Greek, "but one might think that maybe it would have been better if the rich Swedish state could have taken this one-time cost instead? Oh well, I guess they'll have more money to spend on immigrants instead. And after all, that's the most important thing of all. Nothing ever must be allowed to hamper progress."

While preparing for the funeral procession, part of the group was tasked with heading to the mortuary.

"Jesus, what stench!" Chris said, holding his hand over his nose.

The facilities were overcrowded, the corridors littered with bodies—some in a fresh state, others still open after autopsy. Overwhelmed by the war situation, the old Soviet-era cooling systems had become over-encumbered and broken down, increasing the rate of decay of the corpses. And Leo had already been lying there for weeks; his coffin was luckily not an Orthodox open casket, but a thick transport crate which was nailed shut.

Mike returned from the front again, probably among other reasons, to participate in the funeral. Långström grabbed him the moment he saw him, asking if he was the one who had talked to the media about Leo's death—something Mike still denied before leaving.

"I still don't think it was Mike. I'm absolutely sure of it," I told Långström.

"Whatever, who the fuck cares? At least he won't be returning to the group anymore," he responded. "That's all taken care of now."

It was a foggy, wet, cold morning on the day of Leo's funeral. The ground asphalt was covered in wet snow and puddles of water as the occasional light rain came down from the skies. The ceremony proceeded much like the ones before and the ones to come, the only difference being the closed casket. Chris, Bear, Fabien, and Cuix carried the large wooden box from the gate while I, Tjeck, and a few others walked as flag bearers in front of them. Next to the coffin was a black and white photograph of Leo. Speeches were given and flags saluted. Live rifle tracer fire was fired toward the skies. By the end of the ceremony, one of the soldiers from the long snake had left a knife on top of the coffin of the dead man.

Back at base, Långström took on the responsibility of contacting Leo's family, informing them that the regiment had solved the transport issues and their son was finally coming home. He stood in the corridor, looking through a stack of papers where arrival times and other information was printed down. He spoke in a very different tone than I was used to hearing from him.

"I need your bank account information here, as well," he told Leo's mother on the other line. "They do pay a kind of life insurance here for families who have lost members in the war. Ukraine is, as I'm sure you know, a very poor country, so it's only a couple thousand Swedish crowns. I understand it must feel like a joke, a mockery even, and it is, really. But I think you should accept it, anyway. At least it might help cover some funeral costs. I'll leave it up to you. There's nothing you need to worry about signing. I will solve all the major paperwork for you from here."

Långström put the stack of paper up on a large white board in the corridor normally used to plan missions and began to take down the account details.

"There is one more thing, and this one is very important, so you have to listen to me closely now, understand? Do not open the coffin when you pick him up at the airport. Just don't do it. As you know, all of this has taken a long time, and the condition of Leo's body reflects that. However much you want to see him, you can't. Seeing him now will only poison your memory of him. I'm sorry that I have to say this, but believe me."

Chris came out of the room looking like he had lost something. "Hey, mate, have you seen the second sniper rifle?" he asked.

170

Långström rolled up the papers and put his phone back into his pocket before responding, "Well, I guess Mike took it with him back to Kyiv. Bet it was very important."

The Greek overheard as he sat on the stairs to the upper floor, trying to get a good signal on his phone. "W-what the fuck? What the fuck are you saying? He took one of the guns back with him? To Kyiv? What the hell would he need that for? To post fucking selfies with it on Twitter or something?"

"Oh, whatever, fuck him now," said Långström. "We won't be forced to endure his bullshit over here any longer, and we still have one rifle remaining. It's fine."

Bear opened the door from the second room into the corridor, phone in hand. "Hey, guys! I just talk to Inkasator. He says we must come get him from hospital now."

"Has he been discharged? Is he well?" asked the Greek.

Bear shook his head. "No, he is still to be there because of his leg, but he says he is having fucking boring time, so he needs us to go kidnap him, and bring him back here."

Långström laughed heartedly. "So! It seems like we have a new mission for tonight, as well! Any volunteers?"

CHAPTER 22: ROCKETS OVER MARIUPOL

January 23rd, 2015

The reconnaissance missions continued, increasing in frequency. During our first trips across no man's land, the front line had been rather sleepy during the nights, with only the occasional cracks of gunfire and thunder of artillery every now and then, here and there. With each passing night however, activities increased. The gentle sound of small arms fire remained, but it was more and more regularly drowned out by heavy machine guns, mortars, and medium artillery.

We were now often stationed directly in Mariupol, at the third of Azov's bases in the general southern front. The battalion we stayed with was housed in the eastern part of the city in the Livoberezhnyi District, in buildings which had previously served as a local school. The main housing for soldiers came in the form of two large, worn down, old beige brick buildings, the northernmost of which is where we stayed. The accommodations offered were simple—or more accurately, nothing at all except walls with a ceiling above them. We slept on the floor in one of the deserted school rooms on the second of three floors. With the exception of electricity sufficient to power a light and charge radio equipment, there were no amenities, not even running water. Being based there did, however, save us around two hours of travel back and forth from Yurivka every time we were going out into the wildlands across the front, making the Spartan lifestyle more worth the effort.

As per usual, we had a recon mission planned for the night, but just as we were preparing to head out, we were informed that higher-ups canceled it. There would be no nightly hike across no man's land, and it was uncommon for missions to be canceled on such short notice. Especially without any explanation as to why.

After force-feeding myself whatever unspeakable horrors the kitchen babushkas had thrown onto my plate, I headed back toward our cold quarters from the dining hall down by the yard. It was dark outside with the exception of a few street lights. The ground was uneven from deep tracks made by heavy wheeled vehicles in the now frozen mud. The outside was quiet and empty, with the only life being a couple of Ukrainians having a smoke outside the large blue steel door leading into the stairwell of our school building. On the second floor, in the room next to ours, a couple of furry little hunter rats rested on some vacant beds, most certainly waiting for their night shift to begin. In the hallway between the classrooms, a dozen or so newcomers were taught basic weapon handling. The instructor was a sinewy young man in his thirties named Kozak.

To me, most Ukrainians went below the radar, really, but Kozak was one of the exceptions. He adorned his upper lip with a very large, dark brown mustache and his scalp sported a traditional Cossack haircut: a cleanly shaved hairline, front and sides in contrast with a decimeter-long piece of hair hanging down from the highest point of the skull. Under and below his large mustache, the long facial hairs also highlighted a strong and properly split chinbone. A distinctive look, for sure. Yet even without these features being visible—if covered by a helmet and balaclava, for example—his narrow eyes were just as eye-catching, giving an unusually confident, but also very calm and kind impression. In the large gray mass of soldiers, most of which all looked about the same, he stood out as if surrounded by some kind of unexplained energy of sorts. Observing him for a moment, I had a strange feeling that in the continuing story of Ukraine and this war, he would have an important role to play.

I opened the old wooden door to our quarters and found my other friends lying on the floor or sitting on some chairs pulled from the rubble in the corner of the school room. The current discussion was regarding new plans for the night. It was still possible to head back to Yurivka to get a good night's sleep in our warm and comfortable beds. At the same time, though, we rarely found ourselves in Mariupol with a free schedule. With everything the base and our rooms lacked, the city itself offered something special which the small village of Yurivka was painfully lacking—taverns and bars with alcoholic brews and spirits on demand. It being a Saturday, after all, we decided to defy the still strict regimental ban on alcohol and drunken behavior. Leaving our combat equipment behind except for a few spare magazines in our pockets, we grabbed our rifles and light machine guns and headed outside straight into the nightlife of a city under siege, eagerly searching for a suitable and dignified establishment to get shit-faced in.

As we exited the main gates, we turned west, following the main street until we arrived at an esplanade-like square. Taking a left turn, we soon passed what seemed like a small and unassuming little cafe of sorts. Långström, who had previously visited the place, said it was better on the inside than the covers showed and led us inside. The bar counter was very small, and the few tables next to it, most of which were already occupied, offered little space for seating. Consequently, the hostess, a blonde middle-aged woman, referred our company into an empty larger adjoining room offering both longer tables and chairs to accompany them.

The aura of the venue was as simplistic as it could have been, near identical to what you would find at a really cheap pizzeria in a depopulated countryside in

173

northern Sweden. Walls in dull colors adorned with similarly generic paintings of simple nature motifs, tables covered with brown and white wax covers and boring library-style chairs lined up around them—it was far from a five-star restaurant, but the room was secluded, and the beer was cheap. Best of all, the hostess informed us through Bear's interpreting that even though the sale of alcohol was banned after 9 p.m., there was no law forbidding serving it. We could buy as much beer as we wished as long we put in an order and paid before 8:59. The lady of the establishment would then come replace our empty glasses with full ones, allowing us to drink to our hearts' content, so long as we were all done by the time she would close up shop at midnight.

It was settled. We put our guns to rest under the table, hung our coats over our chairs, sat down, and began ordering beer and plates of cheese. As the clock began approaching 9 p.m., we reviewed our own value in combat, as well as our general state, using careful deliberations and advanced mathematics, dividing measurements of beer into minutes. We soon concluded that we would be able to down four more tankards of beer per person before having to head back to base, hopefully without earning a beating at the guard hut for being noticeably drunk. The nail was hammered down, the order placed, and the fog began to thicken.

Sitting in the cafe in Mariupol.

January 24th, 2015

The next morning, we were suddenly awakened by the entire school building shaking. An earthquake out of nowhere made the glass panels wobble as if they were about to fall out of their seats, or crack altogether. We all sat straight up from our sleeping mats. Nudging the drunkenness out of our eyes, wiping the hangover drool from our mouths, looking around trying to find answers in the man next to us—we all found nothing. Within a minute, the quake ceased as quickly as it had started.

"What the fuck?" Cuix asked. "Earthquake? Here?"

"No, bro, there ain't no tectonic plate edges here," Långström replied, already having begun arranging his equipment.

"Fuck, mate, that was artillery of some sort," Chris said.

The Greek followed up: "That was Grad, rocket artillery. Just as in Shyrokyne. I remember it vividly. That's Grad."

"Fucking war," I mumbled to myself, starting to pack up and ready my own mess of equipment. It just had to start up again, right now of all possible times, when I'm fucking hungover.

After the chaos erupted inside the base and turned it into a startled beehive, we were soon able to decipher somewhat what had happened. Much like the Greek said, the enemy had basically wrecked the better part of an entire block on the outskirts of the city using his powerful rocket artillery. No exact number of dead or injured could be given, and it was clearly mostly civilians. Långström quickly took charge. To begin with, we were to begin gathering up any surplus and non-essential personal medical equipment we carried.

We only had access to a single car as it was. Unable to fit everyone in it, three men would be forced to wait behind at the base. Richter was already gone, having joined forces with Inkasator. Långström decided he would bring Chris and Cuix with him as explosives experts, Bear as a translator, and Tjeck for support. Fabien, Greek, and I pulled the short straws and were forced to stay back at base instead of squeezing into the car.

Disappointed about missing out, I still knew there was no time to act like a dissatisfied, spoiled child about it. We emptied all of our superfluous medical equipment into a sleeping bag storage sack. Outside the windows, the driveway between the school buildings and the diner looked like a freshly poked anthill. Inkasator and Richter were on their way into a Ural truck they'd commandeered in the yard while Kozak directed the other drivers. Soldiers ran back and forth everywhere, loading themselves, weapons, and cluttered medical equipment onto the truck beds. Trucks and cars ran a gauntlet trying to get out of the gates, but soon enough, the yard outside was empty again, quiet as the grave.

The enemy rocket barrage had struck only ten minutes by car to the east in what was properly called the Eastern District, one of Mariupol's many Khrushchyovkas—large, simple, multi-family apartment buildings built in the 1960s named after then-party chairman Nikita Khrushchev. The buildings in the affected area were indeed large, boasting ten-stories; while they still stood upright, there weren't many glass panes remaining intact on them. A rocked had even found its way into the area preschool, blowing it into pieces. As it was a Sunday, though, only crayons and toys had become victims to the heat and shrapnel.

Burning cars and buildings covered the gray-colored urban area with dark smoke. Långström and the others, despite being quick on the spot, got there when most of the injured outdoors had already been taken care of. The only injured they came upon was an elderly man lying on the ground, bleeding badly and crying out for help. Unfortunately for him, though, as the group was closing in on him, their focus shifted away. A young woman, also lying on the ground, stole the scene as everyone instinctively hurried over to help her instead of the old man. The fact she was lying on her back, with the entire rear part of her skull missing, changed little. "Women and children first" still remained as the general instinctive rule of conduct, while the old man still bled and cried.

Cuix and Bear went up the stairwell in one of the buildings hit hardest, partly to see if there were any injured residents inside the wrecked apartments, partly to get a nice view of the destruction from the rooftop.

Still at the base, the Greek and I had come to the conclusion that the only healthy thing in the situation was to pass the time in the diner. We brought our equipment, leaving Fabien in the room in a supine position to focus on his own personal hangover. At the diner, after having looked at the daily ration of buckwheat which seemed to be yesterday's leftovers, we tried to figure out what was up with the text messages people had begun receiving.

Even before the rest of the group had traveled to the scene of the crime, a lot of people had started getting strange texts on their phones. As we had entered the diner, waiting outside a moment for me to finish my cigarette, we were met by a pair of Ukrainians looking through their phones for the same reason. The translations we got from them were simple but perfectly sufficient. The texts were made to seem as if they came from other Ukrainians, talking about how the separatist ground forces had successfully attacked Mariupol. The enemy was, according to the texts, in the very process of capturing the entire city, killing everyone in their path. Azov had obviously, according to these texts, fled immediately upon the sight of the superior force, and all message recipients were advised to do the same.

We were still in the eastern parts of town, well inside what was supposed to be enemy-controlled territory at the time. Apart from the previous earthquake, though, we had not heard or noticed any noteworthy commotion or gunfire in the area. With this in mind, it became rather evident that not only were these texts spreading false information—a form of psychological warfare we had not seen before, designed to spread panic and confusion among the city population and defenders—it was also a receipt that the shelling was indeed coming from the Russians, and that it was the start of an assault on Mariupol. Or at least, it was supposed to have been. Fascinated by being able to study these events in real time, I stood up and went to get another cup of coffee. Upon completing my mission and returning, I had only just sat back down at the table as the ground once more began to tremble.

The diner was mostly empty, with the exception of the kitchen babushkas behind the counter and the two Ukrainians we had met outside, sitting at the table just next to the entrance. I looked at the Greek and saw he was as surprised by the new incoming rocket barrage as I was. I felt my skin burning as I still held my mug of coffee, the hot brew spilling all over my hand with each tremor. The windows shook as violently as before, maybe even harder. The babushkas quickly disappeared behind the brick counter which separated the dining hall from the kitchen. The two Ukrainians at the table further down the diner sat up and started running toward the same cover. The first one threw himself over the counter like a bull in a china shop, landing so hard on the other side the loud crash was heard even through the rumbling noise of vibrating glass windows and stiff brick around us. The other Ukrainian did a far smoother pass, looking like a professional Olympic hurdler in passing the obstacle.

I looked at the Greek again, still holding the coffee cup so that it wouldn't tip over. My reflex to pull my hand away from the hot liquid running down my fingers didn't kick in, as my mind was too busy taking in the surrounding situation. As the quake began to subside, we felt confident enough that everything was fine, after all. We stood up from our benches, seeking contact with the Ukrainians who had previously taken cover together with the babushkas.

"It's okay, it's okay!" the Greek cried out to them. "You can come out. They're not attacking around here! All of it was far off!"

Moments earlier, the other part of the group out by the previously shelled area had concluded that there wasn't much purpose in staying on-site. All the wounded had been taken care of, and whatever unexploded ordinance remained was far too big for Chris to do anything about at the moment, anyway. The group piled back into the car, ready to return back to base as the second barrage came down on top of them.

The rockets had a very distinct sound as they ripped through the air above. It wasn't the same howling, screeching noise that tube artillery produced; rather, it was more like a loud hissing and cracking sound. I do not know if the latter comes from the rocket engines, still igniting left over fumes into sparks, but that's my best guess. Anyone who has been in a city with electric trams or similar knows the sound they make when their wires are having connection problems with their main lines of power. That sparking noise is eerily similar to what the Soviet-era rocket artillery would sing like, that very moment before impacting near you.

With the car still moving, everyone threw open their doors and ejected themselves out as the rockets—not only the 122mm Grad but also the even heavier 220mm Uragan—came crashing down around them. Långström tried to do the same from the driver's seat but ended up only halfway out the door before his upper body got tangled and stuck somehow. He tried to no avail to push the brake in with his hands while being dragged after the car, but it continued forward through the barrage. Långström still tethered to it, the car only stopped after hitting a light pole.

All of us had long suspected that Tjeck was the kind of young guy whose nerves might not have been quite cut out for battle, but at that moment, he showed that despite his nerves being a bit fragile, his heart made up for it. As two-meter-long rockets with twenty kilogram warheads of explosives were striking down all around him, only Tjeck saw Långström still stuck to the car as everyone else sought cover. In a moment of absolute bravery and self-sacrifice, he instinctively ran toward his commander and threw himself on top of him to cover him with his own life.

This flagrant example of heroism and selflessness was something which

Långström, however, had little admiration left over for.

"Get the fuck off of me, you fucking faggot!" he shouted, still stuck while feeding the heroic Tjeck several kidney punches in response.

As the barrage ended, the foreigners regrouped on the spot as a Ukrainian news team also in the area followed them. As Chris hurriedly passed by them through the smoke, the reporter saw it fit to try and get a quick interview, pushing her microphone up to him. Chris, not speaking the tongue and otherwise having more important business to attend to, covered his face from the camera and just said, "Outta my face! Outta my face, please!"

The news team tried to follow him but gave up as the distance quickly increased and the cameraman's attention was instead focused on a large piece of hot shrapnel on the ground. Chris' knee-jerk reaction from having a camera showed in his face didn't seem to be a big deal at the time but would soon prove to have enormous consequences.

Back at the Mariupol base, my cell phone rang. It was Långström. He was on his way back to swap people. As he dropped the others off to rest at base, Greek and I would accompany him back once more. Soon the car skidded up on the driveway next to the diner, and we got on board. En route to the eastern district, Långström gave a short summary on the events so far as I lit a cigarette listening to it. Our current directive was to prepare for a possible major assault on the city. The text messages intended to create panic still continued coming, and they matched propaganda videos recently posted on YouTube and social media. Videos showed tanks and armored personnel carriers with slogans like "To Mariupol" painted on their side armor side skirts. The rocket barrages were the last preparations for a major assault on the city.

We headed toward a gas station on the outskirts of the district—not the kind where you fill your car, but a building which supplied the residential areas with gas for heating via a myriad of steel pipes. We stopped a hundred meters away or so, and as soon as we got out of the car, I felt the stench of gas fumes traveling up my nose, the sensation getting stronger the closer we moved in on foot. Feeling how my lungs became tired from dealing with the fouled-up air, I asked the Greek and Långström if it was just me imagining this ominous stench. Their wrinkled noses were enough of an answer. Some shrapnel from the barrages must have cut the pipes or something, and I firmly put my cigarette out on my tongue.

On the map, the building seemed good to mount a defense from, with good visibility in all directions the enemy could conceivably attack from. In reality, it was a leaking fume death trap. A single spark, and everything would go up in flames. We looked at each other, deliberating on what to do. Suddenly, we were met by another pair of Ukrainian defenders coming out from around the corner, both with lit cigarettes in their mouths. Concluding that the position was already properly defended by professionals, which we wanted to keep even further away from than the gas leak itself, we got back into our car and left.

It had now been several hours since the first shelling of the city, and in spite of the text messages, no ground attack had so far come to pass. As to why, we decided to seek answers in enemy propaganda. Their media had in the past proven much more reliable than any other part of our own chain of information, provided the consumer could read it properly, of course.

The timeline was open and easy to access. During the morning hours, important information on current events had been posted all over social networks such as Facebook, Twitter—and in particular, VK. According to this information, the Donetsk People's Army had indeed launched a successful offensive to liberate Mariupol from the fascists. In fact, these stories confirmed what we had read in the text messages regarding how we, the cowardly Azov fighters, had fled in an unregulated rout at the very sight of our glorious adversaries. It was difficult to fathom, really, but according to social media, the separatist state banner was already flying over the city center.

This much we had obviously figured out already. The initial artillery strike had been aimed at fortified military positions outside the city in order to wipe them out, opening up a gap in the defenses for the enemy to exploit. Only problem was the artillerists had missed their mark by about five hundred meters or more, killing basically only civilians instead. Fucking up completely as they had, the wheels were, unfortunately for them, already set in motion, and there was no easy way for them to turn back from it. The enemy had been forced to improvise and begin the time-consuming effort of reloading their Grad and Uragan launchers and try again with a second volley. The second volley fucked up just as badly as the first one, again hitting only civilian areas; afterward, the Russians understood that their plans had failed. The prerequisites for an assault had changed completely. The element of surprise was gone, as the regular Ukrainian army, Azov, and all other volunteer units were all on full alert and, with the exception of a single military casualty, the body count of thirty dead and over one hundred injured were all civilians.

We scrolled back up to the most recent posts on social media. As time passed, information about the great separatist offensive and victory at the so-called Second Battle of Mariupol began thinning out. It was instead replaced with news about how the Ukrainian fascist regime had destroyed an entire residential area with heavy artillery. This was just another piece of irrefutable evidence pointing to the ongoing ethnic cleansing the Ukrainians were up to, because of fascism and stuff.

"Would you believe that?" the Greek said, beginning one of his many sarcastic monologues. "Fucking fascists, right? Ukrainian Nazis, killing poor civilians like that. Who would do such a thing! Horrible! Despicable! I'm calling Amnesty on this one, I'm telling you! We will let the UN sort this out. I, for one, will not stand idle against this fascism!"

"Fuck, mate, look at that shit!" Chris broke in. "They even show photos of the craters in their bloody propaganda. Fuck, man, you can see that the impacts originate from the east. How fucking incompetent are they?"

"They don't have to be competent," the Greek replied, having switched back to his serious mode once more. "The mob that wants to believe it will do so, even so if it said 'Made in the Russian Federation' in big lettering on those rockets which didn't explode. And that's all that matters."

The sudden shift in the enemy's propaganda channels made everything clear. Seeing as the barrages—both of them—had utterly failed in their purpose, the enemy now simply tried to blame everything on Ukraine instead of going forward with their planned ground attacks. We were unlikely to be issued orders to head over the front line for the time being. Not seeing any purpose in staying in Mariupol since there wouldn't be any fighting in the near future, we simply decided to return home to our comfortable life in Yurivka for the time being. The artillery strikes had arguably been among the biggest war crimes during the conflict, but to us at that very moment, such things didn't really mean much. Little did we know what bigger events had just begun unfolding around us.

Aftermath of the rocket attack.

Aftermath of the rocket attack.

CHAPTER 23: "OUTTA MY FACE, PLEASE!"

We arrived back at Yurivka just in time for dinner. In addition to the usual dishes, the local kitchen babushkas on this day offered freshly baked pirozhkis. They could come in different shapes and sizes, but this time they were served thin and oblong, similar to an average croissant in form. Still warm, straight from the oven, the soft and sweet wheat breads topped with a sparse cover of freshly hacked pieces of parsley lay waiting on the counter, free for the taking. As lovely as they smelled, I had learned from experience to be cautious with pirozhkis. The safest way to approach them was to break them open to see what was inside before burying your teeth into them. Sometimes the lightly sugary buns contained nothing at all, being just fine as they were. On a really good day, they would contain either strawberry jam or, even better, thinly shredded and fried cabbage, both pairing great with the soft bread around them.

I broke the bread, and immediately a foul and disgusting stench indicated that this was not such a day. Inside the appealing exterior hid a dark, brown-colored reek of boiled liver in sauce. The babushkas often decided to defile their beautiful creations with this concoction for no apparent reason other than that they must just hate life or something. Full of disappointment, I avoided filling my tray further with these counterfeit desserts.

I sat down at the table with the rest of the group, who were continuing their spontaneous review of the day that started in the car. The Greek stuttered and gasped for air as he struggled to describe the Ukrainians jumping across the counter in the diner, taking cover from artillery several kilometers away. This was accompanied by Tjeck's nervous laughter, still probably hurting from Långström's powerful jabs to his side. Cuix also tried to participate and describe something, but his terrible English prevented him from getting anywhere, much to the Greek's annoyance.

"Fuck, man, just say it in French already. Fuck!"

Chris sat down and mostly poked at his food, focused on the news flow lighting up from his phone in front of him.

After finishing up at the diner for the day, we headed out toward the small village shop to load up on the usual dried sausages, cheese, Lay's potato chips, and beer. The beer was usually offered in many varieties, including that Russian-brewed Polar brand that the Finn used to prefer. Considering the risk of a beating far beyond the usual for not only consuming alcohol, but Russian import at that, I gave it some extra thought but came to the same conclusion as the Finn. It was the best beer for the price tag in the store, and thus I picked up a few bottles and headed toward the counter.

After getting back, the group split in two and retreated to our respective rooms. I took a shower while the others opened up their laptops one by one. The Greek was eagerly following the current elections back at home which would conclude the following day. Långström played *Civilization,* and Chris sunk down further into the news.

It had been a long day, and I decided to take a long shower, increasing the water temperature to the point where my skin really burned. The small bathroom soon became as close to a sauna as possible. After cleansing myself of days of filth with the steam and hot water, I was ready to head back out into the real world and open up my own laptop. It was a thick and heavy Lenovo running Windows Vista, already six or seven years old or so. This laptop was rife with problems, and I had bought it before realizing what a disaster it was, but it somehow still ran—but only if the user knew exactly how to baby it during the booting process.

After going through all the steps to bring it to life, I, too, sunk down into the usual news from my own home country. There wasn't much interesting going on, really, just the usual bullshit. The only exception was a Swedish alternative news site I followed, which at the time held a very pro-Russian stance, who had written a short article about the events in Mariupol.

"Hey, man!" I told Långström, holding my laptop up toward his top bunk. "Here, take a look! Swedish media confirms the Russians captured Mariupol today. Now we know it's true!"

There was a knock on the door, and a Ukrainian opened it from the corridor.

"Hey, Swampy!" he said, looking at Chris. "You are famous!"

Chris stared at him confused as the Ukrainian waved him to come out.

"Come! Come see, you on TV!"

We all followed outside into the corridor where a bunch of people were gathered in front of the small TV hanging on the wall. The Ukrainian news team present in Mariupol during the shelling had made their report, and there was Chris covering his face in front of the camera.

"Outta my face! Outta my face, please!"

We all applauded our new TV star, who himself answered with an embarrassed smile while giving off a cautious and light laughter.

"Nice work, bro! Very professional evasive maneuver!" Långström said as we came back into our room, in a tone which made it difficult to tell if his cheers were genuine or sarcastic.

The Greek jumped up back onto his top bunk, spreading his arms out.

"The communists will win this election! This is the turning point, white people! Ha! Tomorrow, you will give...I mean, you will *loan* us even more money. Needless to say, we will obviously pay you back. We might just have to borrow some more money to do so. But you can trust us! Our government will be communist, after all, so you know you can trust us!"

The news story Chris cameoed in immediately began spreading rapidly throughout the Internet, mainly on YouTube. A few people—of varying intellectual capacities—gave their scathing analyses regarding the origin of the English-speaking man in Ukrainian uniform. Entertaining as the conspiracy engineers were, it wasn't long before things escalated further. Soon, RT (Russia Today, the main English-language propaganda tool of the Kremlin) picked up on the story.

In their studio, the host alongside a so-called foreign affairs "expert" they had invited onto the show speculated who this mystery "Outta my face"-man was. The show layout followed the usual propaganda script. The host, in order to appear as if he maintained a semblance of neutrality, asked leading questions. The chubby, somewhat Boxer dog-faced "expert" then answered in a way that in practice confirmed the Kremlin narrative. Was it possible that the "Outta my face"-man was a Western man fighting on the side of Ukrainian forces?

"Absolutely, I think the clips prove it beyond reasonable doubt," the fat man answered, as that was all quite obvious. However, he continued: "In fact, to me, the accent of the 'Get outta my face'-man seemed very much military-American. People have said maybe Canadian, maybe British. I don't think it's British. It could theoretically be Canadian, but it sounds to me like American military."

"Fucking wanker," Chris sighed.

"Have any of us ever met an American—military-American on top of that— who ever told anyone to fuck off and then politely ended it with 'please'?" I asked, to which no one in the room nodded their heads.

The fat expert kept on going, checking off the vital buzzwords such as "the Kiev Junta," "false allegations about a Russian invasion," "NATO intervention" and, of course, "Nazi sympathies." Soon, with the words "not proven but not disproven, either," the fat man confirmed that Chris, as well as the rest of us, were, in fact, probably all Americans from the infamous private military company, Blackwater. The actual rocket barrage on the city was barely mentioned except when the expert insinuated that we were somehow behind that, as well.

It was a good thing that we had bought snacks and beer, because the shitstorm unfolding on our screens in front of us was a pleasure to follow and study. We had all seen the propaganda before, but had never really seen it unfold like this in real time, let alone actually be caught up in the middle of it. Chris, who had initially been rather moderately amused by being called an American on Russian state TV, even began loosening up after a while. Sure, having the spotlights on him was cause for concern, but there was no denying the entire situation was also very comedic.

In the comments section underneath the report on RT, wild speculations were unfolding. Most of the people commenting were convinced that the "Outta my face"-man was indeed American. After all, the fat expert had said so, too. Others, however, weren't sure, trying to figure it out by themselves. Many thought he was Australian, others were sure there was a hint of some Brazilian accent in there somewhere. Almost nobody said he was British, probably for the same reason most people were sure he was American. Chris, increasingly agitated by this, threw his hat into the comment section himself.

"I am British, you fucking twats!" he pressed send to post in vain. The comments flooded in and drowned his out.

"Fuck me, mate, I bet this article will have over a thousand comments in an hour," Chris mumbled.

The Greek was less impressed by it. "No, bro, people these days have way too short an attention span. It will die out at seven hundred comments or something."

I saw how quickly the comments were pouring in. "It won't stop. There will be over two thousand comments within an hour," I said as the Greek, smelling an easy cash grab within reach of his paws, looked up at me.

"Are you retarded? Two thousand? Are you willing to put your money where your mouth is? I bet you five hryvnia, you bum. Two thousand comments before the clock strikes 10 p.m.!"

"Sure," I said. "Make sure you have the money, because you will lose."

"The bet is five hryvnia!" the Greek confirmed, and tension rose. The comment section increased rapidly as time went on, and the Greek soon begun to sweat.

"I want to change my bet. I say there will be twenty-five hundred comments. Can we agree to that, to a change?"

"Change it all you want," I said. "You'll still lose. There will be two thousand comments at 10 p.m."

"Ha! You shouldn't have done that you retarded Swede, Finn, or whatever the fuck you are," laughed the Greek. "You can see for sure that there will be more comments than that! Ha! Your money is as good as mine!"

The influx of comments began to slow down somewhat. The clock approached 10 p.m., and the comment section was closing in on two thousand. The Greek frantically tried to spam comments himself to push it up toward his goal of two thousand five hundred. The minutes ticked down.

"If it's 2250 or more, I will win!" the Greek gasped while typing comments.

Why do we even have to quarrel about this? Can we not just be friends? Send! *Let us build a multicultural melting pot instead of fighting with each other!* Send! *Would you fuck your own mother?* Send!

The Greek laughed so hard he had trouble breathing.

The clock was 22:00. The video had two thousand one hundred fifty comments. The Greek, drowned with massive disappointment, reluctantly opened his wallet and handed me an unfolded five-hryvnia bill.

"Fuck, mate, I'm getting a lot of warnings that someone is trying to gain access to my laptop," Chris said.

With the aid of a computer whiz friend back in Britain, he was soon able to find that most of the persistent intrusion attempts came from one and only one IP address. Looking it up yielded the name of a pizzeria in San Francisco.

"What the hell would a goddamn pizza baker want with you?" the Greek asked. "That's some bullshit right there. You're just making stuff up."

"Try searching for the physical address of that pizza place, not its IP," I told him.

Chris keyed it into the search field and up came the Central Intelligence Agency. The CIA was the first hit to appear.

"Well, fuck me, mate, the goddamn Americans, too? What the hell did I ever do to them?" Chris sighed.

"Think about it now," I said. "The Russians just pumped out propaganda stories about how one of their guys was caught on camera doing some shady shit in Ukraine. Like, 'Hey, Frank! Do you have any guys in Mariupol?', 'No, do you?' Ha, obviously they're trying to figure what the fuck kind of character you are now!"

"Well, they're not getting into this computer, anyway. It's as locked up as can be," replied a still somewhat grumpy Chris.

"Think of it this way," I continued. "Yesterday, we decided we would just go out and have a couple of beers and whatnot, no big deal. Today, an entire city block is demolished, a fuck ton of people are dead and maimed, and the fucking CIA is looking for us. Isn't that a good bar story, at least?"

Chris laughed slightly and agreed: "Yeah, mate, you're right about that."

Due to the now even further "deteriorating situation" in Ukraine, so-called crisis meetings were held in Russia. A seemingly worried Vladimir Putin took questions from his journalists. He complained about how the Ukrainians, without mentioning Mariupol with one word, were now escalating what he called "the tragedy" that was the "civil war." Clad in an obviously fake facial expression signing concern, it was obvious he was now using the situation with Chris to Russia's advantage.

"Who is fighting over there?" he asked rhetorically before continuing: "Yes, of course, there are some official forces, but most of the them are the so-called 'nationalist volunteer battalions.'" Putin put out his hands in front of him, as to show he was about to explain something important for the viewers. "These are not even an army; they are a foreign legion. They are a NATO legion."

"Wait a minute," I said. "Did Putin just call our little group a 'NATO legion'?"

Both Chris and the Greek looked at the video with large question marks above their heads. Långström just sat on top of his bunk, laughing and smiling.

"Fucking nice, brah!"

CHAPTER 24: NERVOUS OPPONENTS

A lone pickup truck stopped in the parking lot outside our quarters in Yurivka, bringing with it a flatbed full of equipment for the Recon & Sabotage platoon. As soon as my PKM had been destroyed, we requested a replacement. Hoping it had now finally arrived, Långström and I headed outside to search the great pile of guns and ammo. The contents included an AGS-17 automatic grenade launcher, two RPG-7 launchers, and plenty of RPG-7 rockets and boosters, as well as lots of small arms ammo boxes. Beneath all of this lay a large, green-colored wooden crate.

"This one for foreigners, what you order," the driver said, pointing toward it.

The box could easily fit a PKM and all of its accessories, but it seemed unnecessarily big for the task.

"I wouldn't be surprised if they gave us a fucking DShK instead," Långström laughed while helping me pull the heavy crate off the truck, and I agreed.

"Fuck me, man, sure weighs closer to a 50-cal than a PKM, at least."

I flipped open the side-locking tabs and lifted the lid. As previously suspected, there was no PKM inside, not even a DShK. I stood up next to Långström, and for a moment, we just remained there in silence, looking down at the contents inside: a complete 82mm light mortar, complete with legs, optics, floor plate, everything—much of it still wrapped in its brown factory paper.

"We need a machine gun," I told the driver. "What the hell use are we supposed to have for this...thing?"

The driver pointed toward the mortar. "You say, need big gun—is big gun! It is what is, was...eh...available." He shrugged his shoulders and continued, "If not want, I can bring gun back if—"

"No, you won't! We'll keep it!" Långström interrupted him before looking back at me, continuing in a lower tone of voice: "We can always try to trade it for something. Surely there are guys who want a light mortar out there somewhere."

The green-colored tube was stamped with the year 1945, yet not even the edges had a single flake of paint scratched off of them. The optics didn't have a single stain of dirt on the lenses. This was probably the first time the weapon saw daylight since it left the factory at the end of World War II.

"You know how to use these? Maybe we should bring it with us on recon, anyway?" I joked. "Imagine the enemy confusion, if they're suddenly hit by mortar fire coming from behind their own lines!"

As we were helping to load everything into our platoon storage room, a number of large vehicles came rolling into the park and courtyard next to us. We were not the only ones who noticed the eight-wheeled, heavily armed armored cars, which soon attracted spectators from all over the base. They were brand-new production BTR-3s. Although the chassis and main layout in many ways resembled their older predecessors, the Soviet-era BTR-70 and 80 APCs, the Ukrainian-produced BTR-3s were easily distinguished by their much larger gun turrets. The already powerful 14.5mm KPVT heavy machine gun was replaced by an even bigger 30mm auto cannon. To add even extra firepower, the new turret side also sported a number of powerful and modern Barrier-type ATGMs, more potent than any other anti-tank weapons the regiment previously mustered.

The interpreter, Richter, was among those who had come outside to watch. His eyes focused on the fighting vehicles slowly rolling down the narrow park road, he began talking in his usual low tone of voice.

"Not many of those have been manufactured. Azov are among the first to receive them."

The Greek had showed up, as well, sliding into the conversation: "There's some strange bullshit going on here. The government would never give Azov those kinds of things otherwise."

"Yes," the young but ever calculating and stiff Ukrainian replied, "something will probably happen around here, and soon."

I had already started to plan for my return trip back home, making the current atmosphere both unexpected and, frankly, even a bit unwelcome. Instead of reading into the surrounding signs like everyone else, I simply ignored them and focused on going home soon.

Nothing out of the ordinary going on here, I told myself. *Just a few more recon missions, then I'll head back home and sort my life out.*

The diner served up some unusual type of meatballs for lunch. They seemed a bit lighter in color than what you would expect, but still fine at first glance. As it would soon turn out, though, not only were their looks off, but their taste was, too. Still, they didn't exactly taste bad. In fact, they were great, only very different from anything I had ever eaten before. Above all, the minced meat offered an unusually rubbery consistency. Unable to place it, I recalled how, over the past few days, I had seen one less character present in the base's local dog pack. There had been six of them initially, all seeming to be siblings. Friendly mutts all of them, in both size and looks, very similar to Golden Retrievers. Recently, though, there had only been five of them running around. I looked down upon the mysterious meatballs, pondering a moment, then continuing to eat. Whatever it was, it was still better than the disgusting liver dishes they usually passed off as meat in the restaurant.

Back at the room, with nothing else to do than stare at the wall, I decided to pass the time by making my gun more tactical and cool-looking. There were spray cans in different colors out by our balcony, which some of us used to camouflage equipment, mainly the rifles. I had begun using the RPK as my standard armament, and even though it wasn't as eye-catching as a PKM, I still wanted to try and hide the fact I was a machine gunner somewhat. From Radik's PKM, I had learned that painting the long barrel would result in sickening fumes once it got hot. I therefore only striped it in a light beige color, focusing mainly on painting the thickest parts which would take more time to heat up, while still breaking up the contours well. With my old Weaver scope also attached, I painted this together with the rest of the rifle in a dazzle scheme of different colors, so as to not be mistaken for a sniper instead. The end result was fair, and you would have to get pretty close for the weapon to be identified as a light machine gun with a scope on it. To an enemy, I would hopefully be perceived as just another regular rifleman.

Following the rocket barrages on Mariupol, activity in the area had increased markedly and continued to do so. The pot had begun to boil underneath the lid, so to speak. Our schedule, which previously contained a lot of firing drills and free time, began to fill up almost completely with missions into enemy territory.

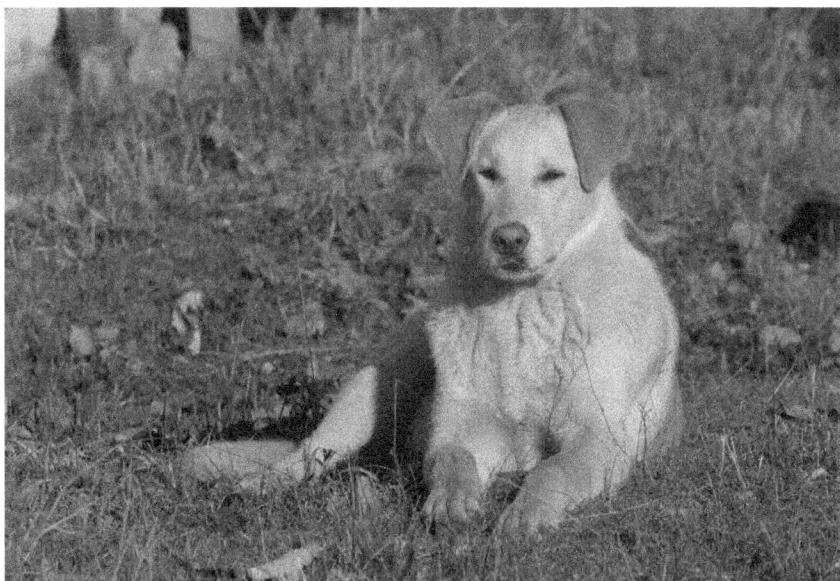

One of the dogs in Yurivka we may have eaten.

From the still Ukrainian-controlled large port city, three roads led east toward enemy lines. In a northeasterly direction, a narrow and winding road passed through several suburban areas along the Kalmius River. Outside the city, it led to the enemy-controlled villages of Pyshchevyk, Orlivske, and Pavlopil.

The main road leading east from Mariupol split in two just outside the last block post. The northern one went in a straight line through the open fields for about ten kilometers until reaching the small, enemy-controlled village of Kominternove. The southern one in turn led toward the coast, toward the most strategically important of the villages in the area—Shyrokyne. Also to the south, another small road led toward the same village in a less direct path, first meandering through the many small and interconnected coastal communities by the sea.

In between the northern road toward Shyrokyne and the one leading to Kominternove, there were two more enemy-controlled villages. The first and largest of the two, closest to Mariupol, was called Lebedynske; northeast, closer to Kominternove was the second, much smaller Vodiane.

This night, we were preparing to conduct reconnaissance in the middle of enemy territory at Kominternove. The route there would have us pass in behind the enemy front line, as we would have to sneak in east of both Lebedynske and Vodiane. This time, the entire group would be participating. Inkasator looked satisfied, the adventurous old man finally being restored well enough to come with us.

After driving a bit along the main road, we parked the cars to proceed the last couple of kilometers on foot. The terrain was completely open and left us vulnerable to ambush, but still less so compared to being seated in our soft-skinned vehicles. The moon shone through the thin cloud cover, and the temperature felt comfortable outside, at least so long as we kept in motion. The group began moving on the long, straight road in a column, longer distance between soldiers than usual. Bear walked about twenty meters in front of me, his tall and narrow stature illuminated time and time again by the light from the shell blasts in the distance. From having been calm just a few weeks ago, the front line now erupted like a volcano every night with fire and artillery going in full force.

First came the rumble and beams of light from the howitzers' muzzle brakes on both sides of the front. The shells howled as they passed above, then thundering as they landed in all directions in the distance around us. The blasts created perfectly spherical globes of white light, throwing themselves up toward the sky. As they then hit the very low clouds of mist above, they then bounced back toward the ground again, like electrified rings on water. They only appeared for a fraction of a second, but as each violent explosion was followed by two or three more, it created an ongoing, indescribably beautiful flutter of chalk-white light over the pitch-black landscape. An enormous, wonderful industrial thunderstorm.

In front of us in the distance, more and more of the enemy twin-barrel anti-aircraft guns hummed their long series of tracer rounds high into the skies above. With their long and snake-like brushstrokes, they covered their canvases in fruitless, never-ending attempts to down the Ukrainian reconnaissance drones. Thousands of red streaks of burning magnesium-filled explosive shells rode across the sky until first their fiery tails went out, only moments later self-destructing in small, white globes thousands of meters above us.

Just as in the skies, another lightshow was going on at ground level around us. Large and slowly falling signal flares of varying sizes illuminated no man's land in different colors. Glowing red tracers from light and heavy machine guns intermixed with bright green ones from assault rifles. The guns cracked and shuttered, sending their leaded stingers back and forth across the line on our side, initially straight as nails, then bouncing around where they ricocheted around after hitting treetops or the ground below them. Sometimes they would be accompanied by the huge, blinding white lights from RPGs or ATGMs, riding like shooting stars across the battlefield before exploding in bright fireballs, changing from light yellow to dark orange, before disappearing again moments later. Walking through this inferno was indescribably tense in an absolutely wonderful way. We were a few good men, all in the open and without any cover, all alone in the eye of the storm. We were heading toward the fire when the only reasonable course of action would have been to turn away from it all, yet we did not. In fact, we wanted nothing else than to come closer to it. I found myself enjoying life more than ever before.

We came to a tree line on the right side of the road, where the group rested for a short moment. A few hundred meters in front of us, the road we had previously followed headed up a small hill, then disappeared down behind it. From the top which sheltered us from view, only a few hundred meters more remained to the enemy positions in Kominternove.

"We'll split up here," Långström said. "Half the team comes with me, following the trail south, down past Vodiane, coming up on Kominternove from the south. Chris will bring the other half, heading across the road to the other side, scouting the village from the north."

"Right," Chris responded, "when do we meet back up?"

The Russians had signal jammers deployed in the area, and even though our radio communications still worked where we stood, we couldn't count on being able to use them closer to the enemy lines. Thus, we had to plan ahead before splitting up.

"Give it five hours before we meet back here. If something goes to complete shit, we'll meet up where we parked our cars instead." Långström waved his hand, indicating the short break had ended.

Chris gathered up his half of the team and hurried across the main road to the north, while Långström's half, to which I belonged, began making its way southeast along a pathway next to the tree line.

"If something happen, do not go down there," Cuix whispered in his lousy English, pointing toward some empty trenches in the tree line. "We stay above them, go down only if we really have to, need cover. Like they shoot at us, with artillery."

"Why not?" I asked.

"If I was a Russian, and I not use them, maybe I put mine in them and leave them for night while sleep," he explained while lifting his head, pulling his index finger over his neck. "Also, remember look at road, see where you put your foot. I like this road—very good place for mines here. Maybe he do, too."

We proceeded gently with a light step about a mile along the path before approaching a T-junction. In front of us, the path led up into an old, broken, and narrow paved road.

"Right, here's the path connecting the enemy-controlled villages," Långström whispered, bringing out the thermal binoculars from his pack. "Kominternove lies about fifteen hundred meters to the northeast, Vodiane only a few hundred meters in the other direction," he continued before taking Cuix with him to check the situation.

As the Frenchman swept for mines and tripwires, Långström surveyed the greater area with the thermals. He then waved toward the rest of the group to move forward. The lights were green.

"Let's do a comms check," he said, pressing the send button and whispering into his radio. Bear, the other man in the group carrying a radio, shook his head. Långström picked up his cell phone to check the bars. Nothing, no reception.

"Well, looks like we're alone from here on then. If something happens, we try to pull back as best we can, same way we came. If we get anyone injured or dead, each man does his best to bring him back, as well. Understood?"

Everyone nodded quietly. Långström again waved his hand, and the team moved forward once more.

The fiercest battles along the front line had died down during our foot march. The open fields were mostly quiet again, with the most common sounds being not thunder and machine gun fire, but gusts of wind swirling through the leafless branches next to the road. Even our cautious steps on the pavement were audible enough once more. It was so quiet that had I not used electrical tape around the ragged piece of shit bipod on my RPK prior to heading out, its light metal rattling would have been heard from miles away.

The terrain allowed us once again to shorten the distance between soldiers in the column back to the usual five meters. In intervals, we made short halts during the last stage toward our final position. Långström again made a quick sweep of the surroundings with the thermals, making sure there were no unpleasant surprises waiting for us. Once clear, we again continued forward.

Through the thick brush on our left, the enemy village soon began to appear. Although most of it was extinguished, some electrical lights remained alive, which penetrated the darkness just as easily as the maze of twigs between us. By the end of the tree line lay a small depression in the ground. Not knowing if it was simple sand dugout or a large crater from the war eighty years earlier, it nonetheless remained an excellent observation point. It was large enough to keep most of the group covered from wind, weather, as well as enemy eyes, while still offering the two men observing some visual protection themselves, thanks to the bushes and trees growing around it. Most of all, it provided excellent visibility over the entire southern section of the village.

On the north side, Chris' group had a shorter way to go than we did and had already reached their positions some time ago. There, however, they had a much more difficult time finding a good position from which to spot. Much of the village being hidden by another tree line and vegetation, there wasn't much to look at, and Chris was beginning to get impatient.

At the south side, we took turns observing in pairs, and soon enough it was mine and Långström's turn. I was looking through the thermals while Långström used ordinary night vision. The latter gave a crisp image but lacked magnification to write home about. The thermals, on the other hand, could use digital zoom; however, this was at the expense of the image being pixelated and difficult to make out. Both tools had their own advantages and disadvantages and had to be used to complement one another. The thermals were much more effective at picking up targets in the distance, but it happened more than once that what we thought to be enemy infantry at low magnification in the thermals turned out to be a simple pack of wild boars once we zoomed in.

We also soon began to suffer the same boredom as Chris, north of us. No matter how good the position was, it aided little when there weren't any enemies to observe. Every once in a while, some Russian would go outside for a smoke, but that was about it. The hour was late, and the enemy was sleeping indoors. The occasional Russian heading outside for a smoke gave us a little company, but far more often the cold sea winds coming from the south were our only visitors.

We had received a large amount of foot and hand warmers—small, disposable one-use bags which after being shaken would produce heat for a couple of hours. I used these not only in my gloves and boots, but tucked them into my pants and coat, as well. Not that it helped much. The cold winds cut through every layer of textile, stealing with it all body heat before continuing further inland.

"Fuck, it's cold tonight," I tried to whisper without hacking my teeth so hard the enemy would hear it. "My cock is so fucking small right now."

Långström tried to stifle a laugh and replied, "My cock is always big, but yeah, I think it's smaller than on average right now, too."

From down the pit, we could hear the Greek, cautious in tone, trying to seek contact with us, as he had his own thoughts on the subject matter.

"Surely, though, body temperature increases if you jerk off, right? I would do so myself, obviously, but I forgot to bring those hand warmers with me, and my hands are cold. Maybe someone else around here…has his hands warm and…would like to…give a helping hand? For the record, this is science. Nothing more."

Inkasator stood up in the middle of everything and took a few steps away to take a piss.

"Look, look!" the Greek whispered. "Look at that! It fucking does glow a little bit, don't it?"

On the north side, Chris' restlessness began to overpower his discipline, as well. Telling the others to wait back, he began sneaking up against the tree line closest to them. At ground level, there was too much vegetation obscuring the sightlines, so he stood up to try and getter better view instead.

Suddenly, our laughter was interrupted by a roaring noise. The air vibrated as a very loud but muffled series of extremely quick banging sounds crashed across the steppe, like the hooves of a thousand horses at full speed.

"What the hell was that?" I whispered.

I used to be able to distinguish the sounds of different weapons and artillery rather well, but this sounded like nothing I'd heard before. Whatever it was, I was unable to pin it down.

"It wasn't here, anyway," said the Greek. "It came from the north, from Chris' side."

Långström looked up as a second wave of blasts erupted. Flashing white lights and red tracers spewed over the ground through the trees where Chris' part of the group was supposed to hold up.

"Anti-aircraft guns," Långström mumbled, scratching at his again recently shaven mustache. It didn't take long before small arms and light machine guns also began opening fire, aiming their tracers toward the now smoking tree line and bushes.

"Okay, so, that went smoothly," the Greek said. "They're obviously fucking up Chris and the others. What do we do now?"

Långström, who had finished scratching his itching upper lip, looked around. "Not much we can do from here, anyway. If Chris put himself in that shit, it's his shit to deal with. Let's head back to the rally point and see if they come back. If not, we'll just have to head north and look for them."

I kept watching the enemy tracers shimmering across the field. "Wait a moment," I said to Långström. "There doesn't seem to be anyone firing back at the Russians. All the tracer fire and muzzle flashes are coming from the enemy side."

I pointed to where the enemy fired from and continued: "Look over there: red and green tracers. Listen to the cracks, see how the reds are coming from three different positions. There are at least three PKMs there, in addition to that anti-aircraft gun and all the AKs. Judging from that, if they've got one MG per squad, you've got about a platoon covering that side of the village."

"Okay, but what the hell are they shooting at, then?" the Greek asked in an annoyed tone of voice.

Långström grinned. "Chris must have made them jumpy or something, and now they're just firing for the sake of firing. To calm their fragile little nerves."

The Greek, one of the few men in the group who always wore a helmet, took it off and scratched his bald head.

"Maybe you're right. They've have always seemed scared of the dark, come to think of it."

Suddenly, there was an explosion in the enemy positions, and just a moment later, a second one.

"What the hell was that?" I asked.

Långström, who was also watching events unfold, pondered for a moment before answering. "I think one trench just fired an RPG into the other."

"What the hell are we even doing here?" I asked Långström, both of us struggling not to laugh.

Cold, bored out our minds, and tired, we soon realized there was no use in trying to be professional any longer. As the Russians began killing each other in front of us, we were all way too busy laughing to be concerned about the enemy presence in the area. It was worth the risk, as it made me forget how cold I was. Deep inside, part of me was actually hoping we would attract enemy attention—anything for an excuse to move around. Shooting ourselves out of the situation was fine by me.

"Look at it! This is *Star Wars*. Stormtroopers shooting wildly!" Långström laughed. "Anyway, let's get our shit together and hike back before these mongoloids gather up the courage to send out patrols. Come on now, column formation, just as we came in. I'll take the lead."

We started moving back in a calm and relaxed pace, Långström still halting every now and then to survey the area with the thermals. We soon arrived at the rallying point, meeting Chris and the others.

"Fuckin' 'ell, they started fucking shit up," Chris laughed. "I went up to the tree line to get a better view but didn't see anything, so I stepped back down again. Then, like fucking fifteen minutes later, they light everything up! Ha! Fucking took 'em long enough."

"Yeah, well, that's what they're like. Nervous and moronic," Långström replied.

"Not arguing against that, mate. It was welcome, though. Made it a lot more interesting to lie down there, watching them shoot all around. Fucking boring as hell before that."

We began to understand just how easily frightened the enemy was at night. It was well-known that the Russian soldier was heavily reliant on his commanders to make decisions for him. At night especially, all alone, we found he was very easy to provoke into opening fire. Getting him to *Star Wars*, as we began calling it—firing at random, for no other purpose than the mere act of firing—became an excellent way make him reveal his entire position and strength.

CHAPTER 25: THE LAST OFFENSIVE

Långström was telling the story calmly, but still with a lively affinity when his eyes suddenly widened, indicating his tale was about to take an unexpected turn.

"So then, just as I assume I've checked everything off, this Frenchwoman next to me by the bar just says, 'Sorry, I can't.' Why, I ask, of course, to which she replies, 'I'm actually working right now.' And that's when I finally understand what that wheelchair retard next to her has been doing there all evening. It was *her* fucking retard!"

We had just returned from the diner, sitting in our room in Yurivka. After being out on recon missions four or five nights in a row, we were quite tired and looking forward to finally having a decent night of proper rest. Långström jumped down from his top bunk, got a new beer from the fridge, and unscrewed the cap while continuing his story.

"Here I've been sitting, offering drinks, playing the entire fucking game to perfection. Like a fucking artist! I've won her. She is mine. And it turns out she is babysitting some fucking retard in a wheelchair! So, what the fuck to do? And mind you, this wasn't just some regular Downie—this was a real fucking mess. Fucking distorted face, limbs, and everything. Zero brain activity. That sort of thing, right? All my time and money wasted, unless, of course, I could make a final brave move. So, I obviously invite her to my hotel room, anyway. 'Just leave it downstairs. We won't be long,' I tell her, but still she says no, like the retard would ever know the difference. Fuck it, I say to myself. I grab her arm, saying 'come with me' and took her and her retard to the elevator. What do you think comes next? Well, fucking elevator is broken, that's what. I'm not giving up now, I think to myself, so I lift the retard out of the wheelchair, throw it up on top of my shoulder, and with the Frenchwoman in my other hand, I begin dragging both of them up the many stairs to my floor."

Långström had been in several conflicts and wars before, but his stories remarkably seldom contained details of them. "That's over three hundred years ago, in another time and life," he used to say, calling it ancient and boring history. Instead, his stories usually centered around women, drugs, robbing guys who were selling drugs while under the influence of drugs, and other kinds of nonsense.

"Then, just then, with only a few steps left, the fucking retard on my shoulder begins moaning. Loudly. You know, like retards do. And what the hell do you think happens then? The fucking retard starts pissing itself! On me! Right there in the goddamn stairwell. Warm fucking retard piss running down my back, I finally get my hotel room door open. I throw the retard—still moaning and pissing everywhere, by the way—in the shower and turn the water on and run water all over it."

Långström twisted his hands in front of him and moaned to give an accurate picture of the shower situation.

"Fuck, mate," Chris sighed while the Greek was loving every second of the story.

"So, the retard stays in the shower while I bring the Frenchwoman to the bed. She moans, obviously liking it, but I have a fucking hard time performing, as the goddamn retard is still moaning in the shower next door. Kind of throws you off, you know? Like, I mean, what if it drowns or something? What do I do then?"

There was a knock on the door and we proceeded to quickly hide our alcoholic contraband.

"Come in!" Långström said with a loud voice, and the door opened. It was Abdullah, our former aspiring interpreter.

"Come outside, we give information in corridor soon. Five minutes."

We all looked at each other, shrugging our shoulders. To actually head out into the corridor? For orders? I hadn't done something like it since my proper military service. We headed out of our rooms, lining up next to everyone else on one side while Abdullah waited for everyone at the end of the corridor. It wasn't only us foreigners, but everyone else, as well.

"Does everyone here understand English?" Abdullah asked, to which everyone nodded. "Good, good. Then I only need say all this one time." He took a deep breath, looking as if this was to be more like a speech, rather than a normal disclosure of information.

"Soldiers! Our commander, Andrei Biletsky, has determined that because the enemy is still in distance of artillery fire on Mariupol, we must push him out from there, to keep the city safe. We, Regiment Azov, must lead this attack, expelling the enemies from our country. We will take back all the enemy-controlled areas, from Pavlopil in north to Shyrokyne in south. Then, we will continue east to Novoazovsk, and push the invaders back into Russia!"

‘An offensive against the enemy, not settling to merely repel them from Mariupol, was a sound plan. The Ukrainians had finally lost the battle of Donetsk Airport only a few weeks earlier. The airport itself had been completely destroyed in the fighting long ago; consequently, there was no strategic importance left in it, but the moral loss was big. A Pyrrhic victory for the Russians, by all means, but a defeat for the Ukrainians all the same.

Now a new battle was going on north of Donetsk. At the town of Debaltseve, a pocket of Ukrainian forces was still holding the main railway line between the enemy capitals, Donetsk and Luhansk—continuing to deny the Russians the use of this important artery. With the airport finally captured, the enemy now focused everything on this actually strategic location, trying to capture the Ukrainians in the pot just like they had done at Ilovaisk in 2014. It was obvious the forces inside Debaltseve had no chance of success, but it was possible that a Ukrainian attack in the south would give them some room to breathe. If we, the Azov volunteers grinding our teeth like monsters, could push eastwards toward the Russian border, it might force the Russians to allocate units and resources to hit at us instead, thereby giving the soldiers at Debaltseve a better chance of escaping with their lives.

"What is our task in this?" Chris asked. "Are we going as usual infantry with the rest?"

"Recon units will work as infantry. It is beginning stage, then we will see what happens." Abdullah kept running through some other general points before reaching the end: "Remember. We are all volunteers here. This is not a game now. This is a real war now, and it will be hard. If anyone no longer wants to, you get off now. We attack tomorrow at sunrise. Whoever is still here then can no longer leave."

Everyone looked at each other. For my part, I had already overstayed my initial plans. I had other important matters to take care of back home and was planning to return any day now. Suddenly, everything had changed. I looked around at the others in the group. Bear was translating to Inkasator, who had previously nodded to the question if he understood English only because he didn't understand it, simply following everyone else's lead. Chris looked deeply concerned, as did Tjeck, while Richter was just as emotionless as ever. Långström had been excited ever since Abdullah uttered the phrase "real war," while Cuix and Fabien were speaking French to each other, seemingly trying to figure out what the speech had been about. The Greek was busy trying to ask Abdullah about some details he found important. Greek, full of excitement, spoke rapidly, however, so Abdullah had issues understanding everything he was asking.

I had not known any of these men for more than a few months, and I didn't really know in what light I should view them. Somehow, they seemed like as old of friends as any I'd ever had, like people I had known all my life. As I considered how much more I had lived in this short period of time than my entire life before, maybe I had actually known them for most of my life. In a way, they seemed like even more than friends—more like family, albeit a terribly dysfunctional one. Had I been a few months younger than my present self, I would have stayed for the excitement of battle. Now, I decided to stay for the sake of my very awkward and maladjusted friends. If they were going to shove their erect cocks straight into the Russian hornets' nest, so would I.

"We know we have people among us who work for the enemy," Abdullah said, holding out a black garbage bag. "Put your cell phones in here."

The Greek seemed unimpressed. "Are you taking our phones? So we can't tell the Russians about the great plan?" he asked.

"Yes!" Abdullah replied, holding the bag in front of him to encourage him to surrender it.

"So, what about our laptops, and all other means of communications around here? If we retain those, then what use is taking our phones?"

Abdullah was silent for a moment, pondering the question presented to him.

"Eh...we have your phones. That's it. And then, you can't write something about this on Facebook to anyone. You don't write your girlfriend or your mother, and say what you're about to do. Okay? If someone ask, you say everything is normal."

"So let me just get this straight. You're taking my phone to prevent me from ratting us out to the Russians. Then I also have to promise not to tell anyone on Facebook using my laptop, too, is that it?"

Abdullah shook the bag in front of the Greek. "Yes! Put your phone in the bag. Now!"

The Greek reluctantly complied with the order while mumbling, "This is retarded. Fucking retarded."

The rest of us threw our phones in the bag, as well, and returned to our rooms to ready our equipment. Chris loaded up on explosives and ammo. Bear was constantly knocking on the door, asking Långström what radio equipment he should bring, until the commander's patience met its end.

"Just fucking bring everything, you fucking idiot! Stop fucking bothering everyone with your fucking questions and fucking decide something on your fucking own for once! Fuck!"

Abdullah claimed the operation would only take a few days. I was hesitant about it, but against better judgment, I ended up leaving my sleeping bag. My pack was small and useless, and with the sleeping bag there was no space left for anything else. I discarded the sleeping bag and stuffed packs of cigarettes into every space available. A helmet cover I had previously gotten from Tjeck, equipped with a large pouch on the back meant for night vision battery packs, was also perfect for the purpose of storing smokes.

"Fuck," the Greek said annoyingly. "It's impossible to sleep while you people are packing your shit. I'm going for a walk."

"I'm all done. I'll come with you," I said. "Need a cigarette, anyway."

Free of clouds, the night was cold with a wide array of stars visible. The Greek headed down toward the children's playground, where he alternated between chin-ups and monologues while I puffed my cigarette, enjoying the calm lights above.

"Why are we doing this, anyway?" the Greek asked me while hanging freely on the iron bar. I had no solid answer to give. He landed back on the ground whereupon his loud thoughts continued.

"You know, sometimes I think I was just born to be a soldier. A warrior. I struggle to find real joy in anything else. Not that I haven't tried, I really have, but it never gives anything back, you know? A gay middle-class life—what's even the purpose of that? It's not that I'm a particularly good soldier or anything. In the Legion, I was always at the bottom tier of physical tests and shit like that. But there's something else about it. It really means something. To be a soldier. To be at war. To make war."

"Yeah, I was never the top of my class in the Coastal Jaegers, either," I said. "Rather the opposite, truth be told."

I had few other things to say. More than that, I resonated with what he was saying. There was something special about the war, but it was not as I had initially believed it to be. It wasn't about some special cause. What was so special about it, the main purpose everything else revolved around, was, in fact, the act of war itself.

"I love this," the Greek said, "though sometimes I wonder what it would feel like to actually love something else. I sometimes tell people I love other things, but I don't think I really do. I don't think I really can. That's just bullshit people want you to say. They want to hear it, so I say it. I just have a really difficult time understanding how to love—like, *really* love—something other than war."

The clock was approaching midnight. We were just about done with everything, still having around six hours of sleep ahead of us, which wasn't much, as we were already tired, but every hour seemed like a blessing. Before retreating to my bed, I decided to use a moment of my remaining time to enjoy one final shower of scalding hot water, enjoying it as if it were the last shower I would ever take in this life.

CHAPTER 26: ATTACK ON PAVLOPIL

Before even having had the time to properly fall asleep, we were woken up by a knock on the door.

"What the fuck?" the Greek mumbled, agitated. "It's morning already? Is it war already?"

I checked the time. It was still the middle of the night, still several hours left until we were supposed to be ready for our offensive.

"New plans," said the strongly accented voice which had just opened the door. It was Richter. "We're getting ready now. The battalion is starting, now."

Långström sat up in his top bunk like nothing, stretching out his upper body as was his routine. "Ah, Finland! Here we go killing!"

He jumped down, feet banging the floor like usual. He looked at me and asked, "You have any snus left? This would be a great time for one of those whiskey ones!"

"No," I answered regretfully. "I'm afraid I'm all out. I consumed the last piece of dried dust the day before yesterday."

Långström's smile fell apart for a moment, only to widen again into an even more wicked grin than before.

"Well, then! That's the Russians' problem now! No more snus. Right here, the monster gets released!"

Those hours of sleep we had been hoping for before the attack had suddenly disappeared, gone like a rug pulled from beneath our feet. My socks from last night hadn't even had time to dry properly, but they could luckily be switched out for others. My boots, however, were a different story: all the Gore-Tex padding was still damp. I did have replacements for these, as well, but as I was short on packing space, I had no way to bring more footwear than whatever pair I chose to wear.

I had the option between heavy rubber boots, my still wet Finnish high leather boots, and a pair of still dry low boots. The rubber ones were the most suited to guarantee staying warm and dry, but I couldn't see myself running away from a Russian tank wearing them. The low boots were recent issue we had gotten from the regiment itself. They were great in most aspects, really—lightweight, water-resistant, and very comfortable to use overall. The problem, however, which nobody seemed to have foreseen, was that the rubber soles on the Italian-made shoes had absolutely zero traction to grip onto snow and ice. At the moment, the ground was mostly free of snow, but as the temperature generally kept itself below freezing, a single day of snowfall could easily change that. The only choice was the wet Finnish leather boots, the same which I had used since I first arrived in Ukraine, my Finnish Jalas Phantoms. It's only going to be a few days, I thought to myself. I had managed a few days in wet boots before.

In order to avoid friendly fire, only MultiCam or Flecktarn uniforms would be allowed—uniforms which I lacked complete setups in. My only pants were A-TACS pattern, and I had to make do with it. It was only about minus fifteen degrees Celsius outside, but I had learned that the cold winds made it seem a lot colder. With two pairs of long johns and my old training pants, still in service from NCO school in Finland, I finally pulled the pants on top of it all.

The Greek looked at me pulling them up, highly concerned. "Bro, I've got extra Multi-Cams. You can borrow a pair of mine."

I told him it was fine. Even if the pattern itself was very different, the colors were the same. At a distance, my A-TACS would look like MultiCam to any observer, whether friend or foe. Besides, my A-TACS pants had great pockets for extra cigarettes on the legs, giving me an extra four packs in total. As for dressing the upper part of my body, the only thing available was my combat shirt. Its warm weather design combined with it being a few sizes too large for me made it easy to pull over other layers of clothing. I then packed my long, windproof, brown-colored smock into the slim space available in my backpack, hoping clothing regulations would loosen up as the offensive went on.

I brought out my helmet with a black skull and crossbones painted on the cover. The symbol itself had originally been white, and belonged to the Finnish fourth light squadron, the "Company of Death," a cavalry unit made legendary by their death's head helmet art. I had always had a difficult time motivating myself to carry a helmet; they're heavy, uncomfortable, and fucked with my hearing. However, as artillery remained a big concern, it may prove useful. The art was to motivate myself to actually carry it into combat. I had painted it to act like a totem for me. We were going up against the Russians, an enemy much stronger than us. We would need any help we could get, and seeking this in symbols was based on an old and proven warrior technique. My helmet and death's head would bestow power and protect me from my enemies, just like other symbols had protected warriors for thousands of years before.

"Everyone put yellow bands on their arms!" Bear cried out, arriving carrying two thick rolls of tape.

"Seriously?" Långström asked. "What was the point of wearing the same uniforms if we're supposed to mark ourselves up with this shit, anyway?"

"You will want to use those," the Greek said. "These savages will fire at everything that moves, friend or foe."

We assisted each other in taping up our arms.

"Here," Fabien said, pointing to his arm. "Make two for me, one under and one over the Azov patch."

It was pitch black outside. The base street lights illuminated the panoply of Azov's notorious and alluring blue-yellow flags waving and tearing violently from the strong winds. Inkasator would function as a battalion truck driver again while the rest of us loaded up a Nissan Navara pickup. Not everyone could fit inside, forcing a couple of men to travel in the cold flatbed. I was lucky enough to get a seat inside, and soon enough we rolled out to find our place in the ever-growing vehicle column at the base. Headlights dazzled and illuminated the surroundings as cars, trucks, and armored personnel carriers reversed and swung around to fit into the long caravan. The entire base was up and rolling. Trucks filled up with infantry just as the pickups. Our new infantry fighting vehicles, the BTR-3s rolled past us, their long-barreled autocannons proudly pointing toward the skies. The modern war machines contrasted heavily with their picturesque counterpart from the previous year's campaign, the battalion's single garbage truck converted into a self-propelled anti-aircraft gun. The tarpaulins covering its back were lifted and rolled up. The gunners carried forward and pushed the heavy, freshly cleaned and greased gun barrels into the rear-mounted weapons' receivers until they locked in place. Large metal boxes full of belted 23mm ammunition were laid on the loading racks on each side and then slid into each gun to feed it. The crew members each mounted their respective gun and cranked it to make sure everything functioned. The carriage swung around quickly from side to side as the fearsome and loaded twin autocannons angled up and down. The gunner waved to the driver. Everything was ready, start the engines!

The great column began rolling eastward. I was squeezed in the middle of the backseat with my long RPK in front of me. Långström was driving, and Richter sat in the passenger seat.

"So what are we doing? What's the plan? Are we taking Shyrokyne?" I asked with high hopes. I had heard many stories about the previous year's battles in the village, and I really wanted to see it for myself.

"We're going north," said Långström. "We're going to push the Russians out of Pavlopil and its surrounding villages."

Richter probably had the best overview of the situation, thanks to being native in the language. Ever observant of his surroundings, he filled in with further information.

"Kurt's company from the Mariupol battalion will take Shyrokyne. We will stay in the north. At least that's the plan."

I was a bit disappointed I wouldn't take part in the attack to recapture Shyrokyne. I would have loved to tell Tor about the experience, given that he was part of the battle when the Ukrainians had lost the village and its surrounding areas back in September the previous year.

Richter opened one of his sleeve pockets, picking out a cell phone. "Do you want to listen to some music? I have different MP3s on this one. It's mostly death metal and ISIS propaganda music."

Surprised, I asked how he had gotten away with not handing over his phone like the rest of us. Richter let off a rare gentle smile.

"I gave them my other phone. I always carry two of them, because I am not retarded." He then played some metal music unfamiliar to me, probably American from the early 2000s, and continued: "That thing about the phones and Russian spies inside the regiment is bullshit, anyway. As you know, they could still send messages without their phones, using their laptops or, unless they're retarded, through their second phone like this one. And they probably have done so. The enemy knows we are coming, but they still think they have a few more hours to prepare. They don't expect us to come now."

The sound quality from the old phone was terrible, and Richter kept switching between songs to find something to his liking.

"That entire phone bullshit was really about the Russians having control of the networks. They can see where phones are. A single phone here or there, it's no big deal. But, if suddenly fifteen hundred phones start moving though in a column on some road? That's different. Right now, though, all they see are all our phones still remaining back at the base. They are expecting to see them move out in six hours. Then, obviously Abdullah was talking about spies and shit just in case any of them were present at the lineup. It would make them a little extra nervous."

Richter finally found a song to his liking and blasted it, filling the car with Arabic chants.

"I like this one. It's the one ISIS usually play in their propaganda videos, when they murder infidels."

Långström laughed briskly while avoiding potholes in the bad roads.

"Are you some kind of Islamist Tatar, or are you an even more complicated character?" I asked Richter, who answered in his usual calm demeanor.

"Me? No, I'm white and Ukrainian, as you can see. But I like ISIS. They don't give a fuck about what people think. They just do whatever they want and kill their enemies without thinking twice. They're hardcore."

The car steered off on the main road to Mariupol as the Greek broke into the discussion.

"Praise be to Allah! I am actually pro-ISIS myself. Seriously! Listen here, think about it! They get to fight the venerable Turk, the communist Kurd, *and* the Russian! We only get to fight one of them—ISIS gets all three at once! Okay, sure, they're getting gangbanged by everyone at the moment, but fuck! Think about it! Fighting all three at once! *That's* a great fight!"

"Well, true as that may be," I broke in, "what exactly are we doing in Pavlopil today?"

"The battalion will take Pavlopil and the area around it," Richter explained. "Most will focus on Pavlopil, which is biggest. We will secure Orlivske, the smaller village west of it."

"Orlivske?" I asked. "Isn't that quite big, too?"

"Yes," Richter replied as he tapped around to find the perfect ISIS song on his phone. "It's about two hundred houses or something. We will check each and every one of them."

"Two hundred houses? We are ten guys here. Do they even understand how long that will take? Even if there aren't any Russians there, we'll still be there all day."

The roads were becoming better. Långström cranked into a higher gear, and the pickup sped up.

"It is what it is. Let's just do it and get it over with," he offered.

"If we are lucky, we will find some Russian soldiers hiding in there," Richter said. "Then we can kill them."

The column made a quick stop in Mariupol to organize with the rest of the regiment before continuing north through the outskirts of the city until finally crossing the front line. We met no enemy contact and soon disembarked outside of Orlivske to continue into the village on foot. The sun had risen but gave only a dim light as a thick blanket of clouds had formed above us. There wasn't much talk, but I could see that Richter was disappointed with the lack of greeting from the enemy.

"The pussies have already left, instead of staying and fighting," he mumbled as his gaze slid across the wintery horizon.

The small village lay in front of us. The landscape was flat, and the open fields around the large number of small houses were covered with a thin layer of snow. The settlement itself consisted mainly of two streets with small family homes and gardens on each side. Some houses were outright tiny, simple, and quaint. There wasn't much damage from the war itself, but rather more ravages from the teeth of time and alcohol-fueled neglect. The dilapidated houses made it clear that if any civilians remained in the village, they were unlikely to be people of great wealth.

"Where are the tanks, anyway? They were supposed to support our attack. I still haven't seen or heard them," I asked, to which Richter responded.

"They never came. They said it was too dangerous."

"So we just attacked without them?"

"Yes."

"Well, not much use in staying here, anyway," said Långström. "We'll start on the left street and move forward from there. Half the group will cover while the other half clears a building, then we switch and continue. I'll take the first half, Bear as my translator. Chris takes the other half with Richter. Questions? Okay, let's begin!"

I went with Långström's half to clear the first house. It seemed uninhabited, and the door was firmly locked. Bear pulled out his tomahawk he'd bought especially for a purpose such as this. When I first saw it at base a while back, I immediately recognized it was more of a display item than an actual practical tool but opted not to criticize his purchase since he seemed to like it so much. He slammed the axe into the wooden door, then a second time and a third time, breaking the long plastic handle just below the head. With Bear duly disappointed, we found other means of entering and cleared the building. Slowly but surely, we began cleaning up the village, one hovel at a time.

Soon we arrived at a house looking a bit tidier than the others. Bear banged firmly on the door, prompting audible voices from within. Soon the lock rattled, and the door slowly opened up. Inside the house, we were met by a middle-aged couple, dressed in dirty and simple rags. Bear explained to them in Russian that we had to search their home, to which they didn't seem reluctant at all; surprisingly enough, they seemed genuinely happy we were there—not even in a superficial way rooted in fear. The woman held her hands together and said something to me as I followed Bear inside. Even though there were no cameras around, the recent "outta my face" debacle with Chris was still fresh in my mind, and so I wasn't keen on revealing I wasn't Ukrainian unless absolutely necessary. Consequently, I responded only in body language rather than with words—a friendly smile and a nod to greet the woman based on her cues, not verbally answering whatever she was saying, which I understood nothing of, anyway.

Their home was small and cramped. The narrow hall and even narrower doorways were difficult to pass in full combat gear, especially while keeping the long RPK barrel pointing forward. The wallpapers could have used replacing some thirty odd years ago had they been able to prioritize it, though arguably the lack of flooring was an even more glaring issue. It wasn't the first time I had seen a cottage with a dirt floor inside, but I had never come across one actually inhabited. They kept their goat indoors to help heat the inside, just like I had heard my grandparents say that their own grandparents had. It was a surreal experience, like traveling through time. Were it not for a plastic-shelled radio on the kitchen table next to a paper juice box with a screw-on cap, the year could just as well have been 1941.

Orlivske.

There were no Russian soldiers to be found hiding under the beds or anywhere else, just as with the previous buildings and the ones that followed. We continued the search, but with the exception of the middle-aged couple and a few more like them, Orlivske was a ghost village. Outside, the occasional gunfire and series of light mortar impacts could be heard exploding nearby.

"Sounds like someone is having a better time than us, at least," Långström said while daydreaming of better places as we slowly worked our way up the street.

The winds soon brough to us the smell of burned explosives, a wonderful aroma that at least brought the lovely aroma of the battle we couldn't see. Our fear of spending the entire day searching for ghosts in the empty village was, however, soon interrupted. Before we had even had time to clear the entire first street, Bear suddenly halted, pressing his radio headset harder against his ear to hear better. He soon pressed the send button and replied "Plus," the Ukrainian command for message received.

"Enemy contact in Pavlopil!" he shouted. "They want half of us to come there and support the infantry in the village!"

Me, me, me! I thought. *Pick me!*—a wish soon fulfilled as I heard Långström say my name. The other half of the group would remain searching Orlivske. Långström, Bear, and I would go on to hunt actually existing Russians in Pavlopil, about two and a half kilometers east on the other side of the lake and river Kalmius.

The company currently engaged in combat would be sending a car to pick us up, while Fabien would head back to where we had left our own ride before we began clearing the village. He would then bring it forward to the rest of the group so they could reconnect with us in Pavlopil quickly if needed. Soon, a dark blue-colored 4x4 station wagon with Azov markings on the sides skidded up next to us.

"You take the passenger seat!" Långström shouted. "Be ready with the RPK!"

I hurried toward our taxi and grabbed the passenger door, which to my surprise fell completely off as I yanked it open.

A pause for the comedic moment ensued. The driver and I looked at each other for a second in complete silence, then at the broken-off passenger door, then at each other once more, and finally again back at the door on the ground. I wondered if I should lift it and try to hook it back in place, as I felt as I had just ruined someone else's car. The driver, who didn't speak much English, just grimaced and waved at me to get in as to say, "leave that shit, time is precious." Following the universal language of hand gestures and facial expressions, I got in.

After a quick ride over the river, we arrived in Pavlopil outside a building where the company staff had created their command center on the large kitchen table. Just outside it, the firefights had torn a hole in the thick gas pipe crossing above the street. The volatile fumes had caught fire, and the pressure violently sprayed the flames across the road. Wounded men screamed as they were loaded into the ambulance and driven away. Långström hurried into the command building next to the pipeline flamethrower with quick steps, ripping open the door to learn where the enemy was located. Bear tried his best to translate as the commanders in turn were busy trying to figure out the situation on the maps in front of them. Trying to find out the who, when, and where of it all seemed to take a minute or two.

Having initially followed my small team inside, I soon turned around and headed outside again to relax with a cigarette instead of being part of the stress inside. Some locals, of which there seemed to be a lot more here than in Orlivske, had come out of their homes trying to fix the fire-spitting gas pipe. Their efforts seemed to yield mixed results, as the fire kept spraying around, melting the snow on the road below. The water rippled abundantly, creating large puddles of mud in the pothole-filled asphalt.

The command staff decided that Orlivske would be cleared by regular infantry instead, and the rest of the group reconnected with us. The hunt for remaining enemy forces in Pavlopil was given priority, and we were the ones to go chase them down. Within minutes, the rest of the team arrived outside with Fabien at the wheel of the pickup.

Pavlopil wasn't a huge village, but it was still about twice as large as Orlivske. The streets passed crosswise between the buildings, making everything a lot more difficult to cover and control than the two straight roads in Orlivske. No one knew exactly where the enemy had retreated to, nor even their total numbers.

"Fuck! They're getting away!" the increasingly impatient Långström yelled at the commanders in an attempt to hasten their cluttered decision-making. "We have to act now! In a hurry! Faster! Speed, speed, speed!"

Someone shouted something, which Bear quickly translated. "They are moving northeast! They're trying to retreat, out of the village!"

One of the commanders looked at Långström and urgently said something in Ukrainian. Bear translated again: "We must take car and shoot the enemy while he retreats."

Långström ignited. "Everyone in the car! Fabs, takes the wheel! I'll take the passenger seat. Chris, to the right! And Bear, left in the back seat. Windows down, guns out! The rest of you, take the flatbed. I want Carolus and the RPK on the roof facing forward, the others around him covering the sides. Quickly now! Go, go, go!"

We hurried toward our pickup and climbed aboard. Långström threw open the passenger door with a grin on his face as he embarked.

"This is rock and roll!"

I unfolded the leg rest and threw the light machine gun onto the roof. Fabien hit the gas, jerking the car forward, gathering speed. I held on to the edge of the roof while locking the RPK into my armpit. The roof edge offered little grip, and I squeezed my fingertips around so hard I was worried my nails would break off beneath my thin goatskin gloves. Gravel sprayed from the tires. Fabien drove like a madman, skidding in the curves between the small houses in order to catch up with the retreating enemy, hopefully able to catch them by surprise in a flanking maneuver. We would, if successful, perform a proper drive-by at them. It could work, as long as they weren't prepared for it. I wished I would have been carrying a PKM instead, with a hundred rounds at the ready. The RPK only offered forty-five rounds of fire before forcing me to reload, giving no hope of achieving fire superiority over anything but a completely unprepared enemy. With any amount of aimed fire directed back at us, our unarmored Nissan wouldn't offer us any protection. Luckily, I was far too busy trying to hold on for dear life to avoid flying off the platform during sharp turns, leaving little available focus to calculate the actual odds of life or death.

We drove around the village, hearts pounding heavily, selector switches set to full auto. Despite the violent ride and the risk it posed, I rested my finger directly on the trigger, ready to empty everything upon the enemy without a moment's notice in order to disorient him. If that failed, we wouldn't get a second chance.

We soon circled back to where we had departed, without meeting any enemies along the way. We had waited far too long, giving the enemy time to leave the village. The commanders explained that the enemy had last been seen in a tree line just east of the village, on the other side of an open field. With no further orders, we insisted on clearing out and taking possession of some buildings instead of sitting idle. If nothing else, we could use the cover in case the enemy would strike us with mortar or artillery fire. Just a few hundred meters away was a row of buildings suitable both for protection and defensive positions. Långström took a liking to one of the two-story buildings.

"There! That one on the other side of the road is ours! Follow me!"

Långström leading and the rest of us quickly following behind, Bear lost his Ukrainian Velcro patch from his arm in the quick dash. We halted at the crossroads to survey the situation. There shouldn't have been any enemies left in the town, but best to still be careful and observant. The roads seemed clear.

"Over there, next to the white brick house," Bear said, pointing at a lonely two-story building along another street, just over three hundred meters away from our position at the crossroads. Behind the building, the northernmost one in the village, a number of men hunkered down, seemingly as if to hide.

"I'll move up a bit to get a better look," I said.

I hurried across the road to an opening in some bushes seemingly offering better visibility. I sat down and lifted my RPK to have a look through the scope. The weather outside was still dim, and the unidentified men stood in the shadows. Even with the 3x magnification of the scope, it was impossible to make out what pattern of uniforms the men wore without the sun to illuminate them. I unfolded the leg rest and laid down on the ground to get a more stable picture. Långström and the others followed up, taking firing positions around to cover me.

"Well? Well? You see anything?" Långström asked, obviously eager to shoot them down.

I waited for someone to make a move out of the shadows, but so far the only thing I could be certain of was that they were indeed infantry, the silhouettes of their AKs clearly visible. Soon enough, one man ventured out of the shadows for a short moment. Wide yellow armbands adorned his arms, and I exhaled and lowered the rifle butt into the muddy ground beneath me.

"Yellow armbands," I said. "They're our guys."

"You see?" the Greek said to Långström. "It's a good thing we carry these armbands."

Richter was noticeably dissatisfied. "That's too bad. They were in a very good position. Very easy to kill from up here."

"Fuck those guys now," Långström reminded the group. "We're taking control of that building now, then we clear the neighboring ones, as well. Chris will take half the group and begin with the one on the other side of the street. The rest come with me. Let's go!"

After we secured the houses, more Ukrainians began clearing the rest of the street. Chris and Cuix made an extra sweep of our house's interior to double-check for hidden traps and explosives. Most of the others being busy, I decided it was time for another smoke break outside.

"Got any more of those?" Långström asked, following me outside. He usually stayed away from cigarettes, but in distressing times of empty snus cans, there wasn't much else on offer.

Occasional clatter of small arms fire could still be heard in the vicinity, but this noise had long since become nothing more than a common part of the facade. Suddenly, however, a much louder sound hit us—as loud as artillery and explosives, but much sharper than anything I had heard before. It cut into my eardrums like a knife. As I tried to determine the origin of the sound, a much more common sound followed it from another direction, the usual loud yet muffled thump from where something had exploded. Smoke rose up from the tree line where the enemy infantry had last been seen.

Seconds later, the same loud, razor-sharp sound struck through the air like the sound of lightning. Then again, the same dull sound of a large impact.

"Over there! On the hill!" Långström cried out, pointing to the other side of the river to our north.

I looked over in the direction of his hand, struggling to spot anything when suddenly a bright flash of light in the middle of the open field lit everything up. A second later, the sound of a lightning bolt hit us once more. A lone tank in the middle of the field was firing its main gun.

"Is it ours?" I asked.

"Don't know. I think so. It's coming from our side, and it looks to be firing on the enemy."

I lifted my RPK to look through the scope. The T-64BV main battle tank did indeed have the two white stripes on the front and sides of its hull—the markings of a Ukrainian vehicle.

"So they finally showed up, but only in one tank?" I said and sighed, thinking of our promised tank support that never materialized—until now.

Långström laughed. "That might very well be. They usually need a few hours of stiff drinks of vodka first, to boost their courage stats."

"Do you see any others?" I asked, sweeping the field with my scope. "There should have been at least a platoon, three or four tanks."

"Well, maybe there was only enough vodka around to boost the confidence of a single crew?" Långström laughed as the tank continued firing.

Another 125mm high explosive fragmentation shell soared across the field, striking the ground at least two hundred meters in front of the suspected enemy position.

"Ha! Would you look at that! You're right, they *are* fucking drunk!" I said.

Långström laughed as the tank fired again, this time striking even further away. The shell hit inside the village, in the roof of the building where we had previously observed our fellow Ukrainians in the shade. Now less than three hundred meters away from us, the sound of the impact increased readily, rocking both of us to take a step backwards.

"Fuck!" I said. "Well, now their aim is really fucking off."

Långström nodded and replied, "They *are* fucking drunk!"

The next tank shell also hit the house. I lifted my RPK back up and looked through the scope to see the men around it crawling for cover. One of them threw himself face-first down the basement stairs as the third shell exploded in the wall, throwing large pieces of white bricks around. Suddenly, I had a startling realization.

"Fuck, man! They're not missing their intended target. They're hitting exactly what they're aiming at. They're shooting at friendlies!"

Långström observed quietly for a moment and then nodded his head in quiet agreement with my thesis.

"Someone should try to inform them about this via the radio," I said as a fourth explosive shell hit the building, tearing out bricks and throwing floor panels around in the air.

"Yeah…" Långström mumbled, his gaze following the large pieces as they were thrown into the skies, slowing down in mid-air before journeying down to earth again.

One of our Spartan armored cars in the village was opening fire back at the tank. I lowered my gun and looked at Långström. He seemed completely out of reach, enchanted by the scenes of beautiful ultra-violence playing out in front of us.

"Dude. Come on. You could at least try to make an effort, you know?" I said, knocking him on the shoulder.

"Ah! Yeah, yes!" he exclaimed, snapping back into reality. With quick steps, he ran hurriedly back inside our occupied building.

"Bear! Fuck! The radio! Where the fuck are you? Bear, you fucking asshole!" he roared once inside before we found our surprised radioman in the kitchen.

"What's the problem?" Bear asked with a confused tone of voice.

Långström grabbed him by the shoulder and shouted in his face: "That fucking Ukrainian tank is smashing our guys down the street! Get on the radio and set it straight! Fast!"

Bear immediately pushed the transmit button on the radio, speaking frantically while hurrying outside. Soon after, the tank ceased fire, and the previously targeted soldiers began crawling out of the ground. Everyone seemed unharmed, though still a bit shaken—with ringing cars, surely. As loud as it was out by our position, it must have been much worse where they were.

With our building screened and secured from booby traps, I headed upstairs to find a good spot from which to cover the enemy direction. A small bedroom with large windows gave excellent vision to the northern edge of the village by the lake, the fields to the east, and the tree line to where the enemy had retreated. It was warm in the room, which was initially welcome, but soon became overbearing, causing me to sweat like a pig. I soon realized the radiator was the culprit, being improperly placed below the window. Hot water vapor pushed through it so hard that the entire thing vibrated noticeably. Turning the regulators on its side did nothing, and there didn't seem to be any way to turn it off. The room felt like a sauna, and not even opening the window made much difference. A mirage of heat sprayed outside through the opening, making it difficult to see— particularly through my rifle scope—as the vapors contrasted with the cold air outside.

In addition to the hot air, I realized that the position had yet another pressing issue. If even a single bullet punctured the radiator, it could start spraying scalding liquid at me. The pressure inside the room might even make it explode, for all I knew. Unsure of what to do, I looked for other suitable positions but found nothing. The only window on the upper floor with a view toward the enemy was the one with the death trap inside it. Lacking better ideas, I covered the radiator with mattresses and other pieces of junk in the hopes that it would save me from getting sprayed with hot water in case the enemy fired on my position.

There was no mitigating the heat itself, however, and as I didn't wish to sweat through my clothing in the otherwise cold climate, the only solution was to undress. I removed everything but my boots and pants, as they'd be more time-consuming to get back on should something happen. The Greek came into the room, applauding at the sight of my sweaty and moist naked chest.

"Very nice, very nice indeed! You know, should you get cold, I'm always here for you if you need some extra...heat."

It would turn out much later that—at least so I was told—the tank we thought was part of the original support plan was there for an entirely different reason. A long-range reconnaissance patrol (or LRRP) from the regular Ukrainian army, completely unaware of our offensive, had been passing through the area on their way back to friendly lines. Some Azov volunteers had spotted them coming in on foot from the direction they'd have expected Russians, and consequently mistook them for enemy soldiers and fired on them, injuring two of the LRRP soldiers. As the LRRP unit came under fire, they radioed a distress call about enemy contact from the village of Pavlopil, and a tank from a nearby block post was sent forward to give them fire support.

It's a good thing we have these yellow armbands, as the Greek had said.

CHAPTER 27: RECON OUTSIDE LEBEDYNSKE

Everything seemed to calm down; as such, I no longer considered it mission critical to continue monitoring the area from my firing position. After using an old sweater from a nearby laundry basket to wipe away my excess sweat, I dressed up again. Enemy activity had been nonexistent for some time, so I took the opportunity to explore the rest of the house.

It was large, with many rooms, and had nice furniture which the soldiers now rested on. Fabien and Tjeck had both fallen asleep, but Tjeck, snoring quietly on the living room sofa, still looked ready for combat with his helmet still on his head and his AK firmly gripped across his chest. This house was in a class of its own compared to the small dirt floor cottage we had searched through in the neighboring village. The family who had once lived here could afford to flee, either to Russia or other parts of Ukraine, unlike the poorest of people around these parts. Other desolate houses we had come across had often been looted, but this one seemed to have been spared from thieves and bandits so far. Most of the silver and glass remained in the cabinets, as did the living room's widescreen TV—all very desirable items for roaming gopniks. Inside the closets, some of which had the doors left open, plenty of clothes remained. The exception might have been one of the particular rooms which, judging by the mirror and boy band-themed posters on the walls, had probably belonged to a girl in her mid-teens. In there, the closets were mostly empty.

The front doors were left wide open, and as the sun began setting, cold winds rushed inside. My toes slowly began to ache from being trapped inside my still wet socks and boots. That's when it hit me: if the previous inhabitants of this place had left behind valuable items like nice china, silver, and their thick TV, then they must have been in such a hurry they left behind some dry socks, as well!

I pillaged from bedroom to bedroom, looking through all the drawers as my feet hurt more and more. I had already changed socks once, but the heat from my upstairs firing position had soaked the new ones in fresh sweat immediately. I still

213

had a pair of thick, unused, knee-length socks in my gear, but I was keeping these in reserve for emergencies. They were after all the most expensive socks I had ever bought at sixty euros a pair, and I would not break them open until I absolutely had to. Only when I had reached my breaking point, the absolute point where I wouldn't be able to go on any further, would I bring out these super-socks. Until then, my current socks would have to suffice, but scavenging for dry footwear was one way I could overcome this.

I searched for a long while but ended up finding nothing. Whoever used to live here had taken with them every goddamn pair of socks! All sorts of items of value were left behind, but not even as much as a dirty pair of worn-out cotton socks remained.

Just as I began to accept the demoralizing fact that my efforts of pillaging the east for riches was leaving me empty handed, new orders came in. The foreigners along with the rest of Recon & Sabotage would return to Mariupol, where we would hold up as reserves. I would have preferred to stay in Pavlopil. With the door closed, the house here was warm inside. There was everything from sofas to actual beds to rest on, much unlike the cold, hard floors of the Mariupol base. Orders were orders, though, and we proceeded to load ourselves back up in the pickup. It was my turn to ride in the flatbed, and together with Långström, Richter, Bear, and Tjeck, we crowded together on top of our personal equipment and ammunition. Lacking space, I still managed to find a comfortable position at the tail end, leaving one of my legs hanging out. Passing through one of the checkpoints on the road back to Mariupol, there must have been a news team in place taking pictures of our vehicle column. Several days later, long after the offensive, I would see myself half-lying down like a slob with my leg hanging out in a photo published by the BBC—in great focus, no less—in an article about the conflict in Ukraine.

Our wait as a reserve force in Mariupol was short-lived, and we soon re-embarked our pickup in various sloppy postures again. The Foreigner Group was to be moved up as reinforcements to Lebedynske, the larger of the two villages between Shyrokyne and Kominternove. The Ukrainians had captured Lebedynske, just like the other strategic areas of first-stage operations, but Russian recon units remained a threat just outside the town, harassing the Ukrainian positions. We were heading over to see if we could do something about it.

The sun had set, and darkness settled over the landscape once more. It was still cloudy outside to the point where even the sun's rays may have had difficulty reaching the ground below; certainly, the moonlight wasn't about to. Fabien sat at the wheel, all lights off so as not to attract any attention. The road was barely visible in the darkness, forcing us to keep a relatively low speed. Långström sat in the front seat with the thermal camera but was unable to use it, as the windscreen blocked out all heat signatures. I was in the back seat, having replaced my RPK with the group's AKS-74 with a night vision scope, watching our right side through the green lens.

Out of nowhere the car suddenly jerked violently, and the loud screeching noise of unlubed metal on metal knifed into our ears. The car screeched to a halt, not as if we had collided with a firm obstacle, but as if Fabien had simply slammed the brakes for whatever reason. We traded looks between each other, though no one had any answers to offer.

"Out! Fast!" Långström hissed. "Firing positions, three-sixty around the car!"

Opening our doors, we hurriedly got out and formed a circle around the inexplicably halted vehicle. Our pickup had stopped on the asphalt road, open fields, and narrow tree lines surrounding us, providing no protection whatsoever.

The mystery of why Fabien had slammed the brakes was quickly solved as we noticed a long obstacle of barbed wire stretched across the road. The steel cables had twisted themselves around the front tire axels and wheels. By sheer chance and good fortune, the wire hadn't punctured any of the tires. While Fabien went to work cutting the mess open, I kept my rifle and night vision scope pointed toward the tree lines. Above us, the cloud cover began to crack open. The bright rays from the moon stretched down toward the surrounding fields. The temperature was constantly changing, and most of the morning snow had already melted away. It was actually quite pleasant outside, as we were accompanied by more and more stars in the sky.

What a lovely war, I thought to myself as I looked for movement among the trees, all the while enjoying Fabien's beautiful muffled French cursing behind me.

His cursing clearly worked, because we resumed the journey toward Lebedynske without losing too much time. Once we arrived, we parked on a large driveway in front of one of the largest buildings in the village. Wide and long, cast in concrete two stories high, its location suggested it was some kind of town hall. Inside the wide entrance level between thick pillars, the sparse lighting shimmered against the large, white-colored doors and windowed walls on their sides. Behind us, on the other side of the driveway, the tallest building in the village towered over everything with its four stories of white bricks, its short side facing toward us. It seemed as if it had been an apartment building, and if so, at least two of the top floor apartments had been completely erased. Some huge explosion had torn away roof and walls alike. It remained unclear what kind of projectile had created such devastation, but whatever it was, it must have been sizable.

We disembarked, ascended the stone steps of the town hall, and entered what was now the local Azov command HQ. The entrance hall was large, open, and mostly empty. The occasional fighter moved in between the other rooms, often passing by a couple of tables situated on the left side of the great hall. These tables also caught our eyes, as they were equipped with a number of boxes of water bottles and various foodstuffs. No one in the group had gotten to enjoy a meal— not even a terrible one—since the previous evening. No one complained about their hunger, but the sight of food quickly aroused the basic instinct nobody had found the time to think about till now. Like ferocious birds of prey, we fell upon the defenseless and simplistic little provisions center.

The menu on offer was a simplistic one. There was the usual canned buckwheat porridge with pieces of unknown animals mixed into it, as well as an assortment of different kinds of mostly unsalted and tasteless biscuits. A number of already half-eaten loaves of bread in varying degrees of dryness also lay scattered across the table. The best and most-coveted provisions were the small packages of foil-wrapped soft cheese, few still remaining. While Långström was away trying to find someone in charge, the rest of us took the chance to gorge as much of this down our throats as possible. Best to make the most of the opportunity, as it was unclear when a similar selection would be displayed in front of us.

Långström soon regrouped with us next to the table, where he began explaining what he had learned about the situation.

"They are a strange bunch, these people. They say the area has mostly been calm, except right here behind this building and northeast of the village—enemies sneaking around since the sun went down, shit like that."

"So that's where we're going then?" Chris asked.

"No!" Långström shouted in a sarcastic tone of voice, spreading his arms for emphasis. "Of course not! That's impossible! They said we couldn't go there, because there were Russians in that area! It's far too dangerous! Instead, they suggested a lot of other places around the villages we could go look, where there aren't any enemies, as if there's any fucking point in doing that whatsoever."

The Greek looked disgusted, but didn't verbalize his opinion for once in his life.

"So, what then? We go somewhere and look at nothing the entire night? That's the plan?" I asked.

Långström bit off a large piece of bread and proceeded to press it into his mouth together with half a pack of soft cheese. He snorted contemptuously and continued, still with food in his mouth.

"Well fucking obviously, we go north where there are actual Russians. And the villagers here can stay here and continue sorting their lives out."

The town hall was located in the southeast corner of the village, and behind it lay mostly open fields with some large warehouses just over five hundred meters away. These could very well house Russians, but they were far too risky to approach due to the open terrain. On the main road leading to them, though, a smaller one turned north along a tree line. This thin section of forest was also a supposed area of Russian activity. The small road went straight north for about five hundred meters, evenly along the outskirts of the village. From there it continued five hundred meters more until it ended at some larger, unguarded buildings in the woods next to some more open fields. The last half of the road and the buildings were well out of sight from the town itself. Had we been the enemy probing and reconning the town, we might very well have chosen that path to approach our target, as well.

"We will head up the road along the tree line, toward these buildings," Långström said and pointed at a map on the table. "Fabien drives but stays a bit behind the rest of us while we cover the area on foot. We move forward slowly until we get contact. Any questions? Alright then, let's head back to the car. Move out!"

As we reached the junction where the small road exited from the main one, we were immediately greeted and exposed by a large dog. Tied to a small building on the other side of the road, its loud barking echoed over the otherwise quiet area.

We looked at each other, unsure if we should continue.

"Oh, fuck that dog, it doesn't matter! We keep going, forward!" Långström hissed, taking the right side of the small road.

I took the opposite side, still equipped with the AKS and night vision covering, forward and to our left. The rest of the group continued on both sides behind us with a couple of meters in between each man, with Fabien tailing in the car at crawl speed. We soon came up on a number of smaller buildings on my left side of the road. Closing in, it became evident that unless there were any Russians inside them, they had been uninhabited for years, as they were completely ruined with neither doors, windows, or roofs; only brick walls remained.

Långström slowed down his pace, allowing me to take point to check the dangerous angles. Aided by the moonlight, I used my naked left eye to watch my step for rubble or traps inside the ruins, keeping my right one looking through the green night vision optic in the dark corners. It was both therapeutic and a bit exciting to clear out the ruins slowly and methodically, all the while fresh and healthy little gusts of wind from time to time blew pleasantly between the walls. Not finding any Russians occupying the skeletonized houses, we then proceeded forward like before.

On our left, we finally came up on the point where the town ended. With five hundred meters still to go, we gathered up a short moment before continuing. With the pickup, we soon arrived at the large buildings where the road ended. After finding these also empty, the question arose weather we should return or try to press forward even more.

The road was gone, but the tree line continued for another four hundred meters along the edge of the field, where it turned around to encapsulate the area. It would surely have been fine to cross on a hot summer day; however, the melting snow had left the field both wet and sticky with thick mud. Crossing it on foot alone might be difficult, but trying it with the car carried even more risk.

"Let's try it," Långström whispered. "If we can get across that field on the other side of the tree line, there are more roads on the opposite side. If we get to them, we can cover a much greater area afterward."

We started walking, the pickup following behind us just like before. Everything looked promising. The field was not as wet as we had suspected, and the car pushed forward nicely, albeit a bit bumpily, over the uneven terrain. Problems, however, soon began to arise. The field sloped down slightly, the ground immediately getting weaker and wetter. Boots began to sink, and the tires began to slip. One by one, we slinged our rifles on our backs to help push the sinking vehicle forward, without much success. Sweat ran from our foreheads and our muscles ached, but it was no use. Halfway through the field, we found ourselves inexorably stuck. Moving forward was useless, as the ground only got worse and worse. Our only chance was to try the reverse gear and push the pickup back up to where we had come from.

If the engine been quiet before, during our slow journey to this point, the exit was a whole lot noisier. Fabien hit the gas in reverse, but the tires kept slipping without the car moving anywhere, except possibly deeper down into the thick mud.

"Spread out!" Långström ordered. "Keep a lookout on the tree lines around us!"

This was the second time tonight we were stuck in a very unfortunate situation. I first climbed the flatbed, then carefully onto the roof of the car as to avoid buckling the thin metal plate and creating even more noise. Not a safe position to be standing in, by any means, but it gave a great overview of the area together with my electronic optics.

Fabien continued trying to reverse as I noted the red tail lights reflecting into the ground below the car. They weren't particularly bright, but in the darkness, they were like beacons to any of our enemies trying to work out where the screeching engine noises were coming from.

"Quickly!" I hissed at the others. "Get something to cover up those fucking tail lights, fast!"

Luckily, we still had those wide rolls of yellow tape in the back seat. They weren't very thick, but applying several layers over them still helped dim the lights considerably.

While I kept looking through the night vision, the others kept pushing; with great effort, our pickup finally came unstuck again after an hour-long struggle. We took the same route back, still cautiously, but at a much-increased speed compared to previously. Upon leaving the small road, we were also once more greeted by the angry barking mutt, signaling to our surroundings we'd returned safely. It was getting late, and we knew nothing about what the next day would bring.

"Nothing for us to do around here. Fuck it, we'll head back to Mariupol instead and try to get some sleep," Långström decided.

CHAPTER 28: LÅNGSTRÖM'S RAID

There wasn't much sleep to be had in Mariupol. Not because we were woken up to perform some tasks or otherwise found ourselves disturbed, but because of the cold. The radiators were still dead in the empty school room. Lying flat on the floor, I placed my wet insoles and socks on my chest beneath my clothes in a faint attempt to dry them with my body heat. It worked moderately, but was of little use, as the boots themselves remained cold and wet throughout the night.

During the day we remained mostly idle on the floor, spending the time talking the usual bullshit.

"That's pretty rich coming from a second-class Turk, anyway, acting as if you're still entitled to an opinion," the Greek called out, using my Finnish heritage to insult me.

"Doesn't that make you a third class Turk, though?" I asked.

"No, no, no," the Greek shook his head. "Y-y-you've got it all wrong, you bum. That's the Kurd. The Kurd is the third class of Turk! Don't you know anything?"

"So then, that makes you what? A fourth class or something?"

The Greek looked straight into my eyes, pointing his finger at me. "Did you know Jesus was Greek?"

"What?"

"I will prove it to you!"

Långström had left us to recruit some officers for a reconnaissance trip. Though boring to remain at base, it didn't matter much to us. Getting some rest was nice and very welcome, although having it in Yurivka in a real bed and a place to dry my clothes had, of course, been far more pleasant.

The Greek ranted on about Jesus' supposed Hellenic heritage before walking off and leaving the room, saying he had to take a shit. I had gotten lost somewhere in the first ten minutes of it all and asked Chris if he could fill me up on the rest.

219

"No idea really, mate," he replied. "I mean, man, my English is good. It's my first language, but fuck me. I struggle to understand the Greek even on a regular basis. Not only is his vocabulary huge, but fuck, man, he doesn't know how to pronounce a single fucking word correctly. And his fast pace of talking! Shit, my brain can't handle it. I usually just check out halfway through his speeches."

It had already begun to darken outside when Långström returned, all hurried up with brisk paces.

"The Russians counterattacked!" he exclaimed in excitement. "They've already taken back a lot of what they lost yesterday! Get your shit together, fast! We will have work to do soon enough for sure! Ha!"

We hastily began preparing ourselves and our equipment without knowing what for. Långström was hurriedly heading out toward the command room in the next building, excited and hoping to grab a good mission.

"We have to be quick about this now! Otherwise, all the cool missions will be handed to others, and we will end up guarding some smelly block post or whatever the fuck leftovers remain," he said, quickly descending the stairs toward ground level.

Having finished my equipment preparation, I followed him. I was going out for a smoke, anyway, and found myself both worried and childishly curious to find out what was about to happen next.

Långström lifted a camouflage netting which formed the door to the command room and stepped inside. I glanced into the dimly lit room but decided to wait outside. There was no reason to join in on the chaos I could already see going on in there. Radio equipment cracked and sparked. Commanders and lieutenants raved on between each other—all of it in a language foreign to me, anyway. The only sound thing to do was to keep a polite distance.

Despite the noise, it was still possible to somewhat distinguish Långström's voice in English, but it was difficult to make out exactly what he was saying. Something about "Why in the hell did they retreat?" and what sounded like accusations about someone being a "fucking pussy" and a "coward." Most of the other words disappeared in the cacophony echoing between the concrete walls. Långström's tone, however, was much more easily distinguishable and gave clear indications on how negotiations were going inside.

At first, I could tell our leader was very determined, then clearly agitated. It sounded as if he once more was trying to lay out a plan of action nobody higher up would approve, something rather standard by this point. Soon enough, though, the volume and tone of his voice sank down, sounding more disappointed and resigned. It seemed as if he had given up. Maybe it was for the best, I thought, still worried about what insane ideas his mind had cooked up for us all to partake in this time. However, I couldn't help but feel the wet cloak of disappointment cover me up, as well. After all, it would have been fun to do something completely crazy. The odd thing, however—and it stayed with me like an annoying itch— was how easily he seemed to have given up. Långström used to nag and rave endlessly when trying to get his ideas approved, but this time he laid down his arms unusually quickly. If he really wanted something—and judging from the initial tone of his voice, he did—why had he given up so easily this time?

He soon lifted the netting again and came back outside, seemingly completely calm.

"What happened in there?" I asked while lighting two cigarettes, handing one of them over to him.

He gave a short summary, telling me how the enemy had regained control of Kominternove.

"The Russians apparently just fired a shitload of Grad over it and overwhelmed the place," Långström said, stretching out his arms in front of him. "And nobody saw the problem with that? Who the fuck retreats just because you get fired upon? That's fucking half of what makes this shit interesting! Fuck!"

"So what are we going to do then? What's the plan?"

"Well, as the Ukrainians withdrew, they left all their ammunition behind— even though it was still loaded up on the truck bed! Imagine that, huh? An entire fucking KrAZ! It's still there in the village, by the main square. So, of course, I offered to head in there now, under the cover of night, and blow it up before the Russians have time to empty it."

"You mean we're just going to walk in there, into the middle of all the Russians?" I asked, somewhat hoping he was joking. The plan was crazy on the same level as his previous idea about stealing one of the enemy tanks—possibly even more dangerous.

"Well," Långström said as his eyes fled away from meeting mine, "they said we weren't allowed to do that, so I just suggested we go over there and do some regular recon instead. See how many Russians are in the village. Shit like that. And they were fine with that, some slow and boring recon."

I breathed a silent sigh of relief. How in the hell he had imagined we would get into the village, surrounded by open terrain, without being detected was one thing. But then getting out again, especially after blasting a truckload of explosives and ammunition in the middle of it? Laying down in a tree line simply observing the Russians from a distance didn't sound very exciting, but compared to Långström's idea, I was completely fine with that.

We went back to our building and slowly climbed the stairs back up to the second floor. One of the tables up there had a ton of the group's equipment on it, namely ammunition and explosives. Långström saw Chris standing near it and immediately grabbed him firmly by the shoulder.

"Hey, Swampy! How many explosives do we have?"

Chris looked up at him, surprised. "Explosives? Mate, why do you ask? We have a lot, buddy."

"Great! Get some ready, because I'm going to need a big explosion!"

"A big explosion? How big?" Chris asked, sounding surprised.

"A really fucking big one. We're going to blow up an ammo truck!"

"Say nothing more, mate. I've got exactly what we need!" Chris replied and proceeded to dive down into boxes full of miscellaneous grenades, shells, and bombs he had been collecting during his now five-month stay in Ukraine.

I grabbed Långström. "What do we need explosives for? I thought we were just doing recon? What the fuck is going on?"

Långström took off his ugly balaclava, revealing his crazy brown hair pointing in all directions, defying any sort of military hairstyle regulations.

"Don't you get it? We need the explosives to blow up the ammo truck the Ukrainians left behind."

I was confused. "But they said we weren't allowed to do that, didn't they?"

Långström finally turned his face up at me again, meeting my confused-looking eyes with a shifty look in his own and a sly, ever-increasing grin below them. He looked like a child who had the perfect mischief planned, but a lot more disturbing.

"Yeah, sure, but if they don't know we're doing it, anyway, do you really think it bothers them? This is going to be awesome, man!"

He then hurried away only to secretly, outside of any commander's knowledge, arrange the practicalities. As soon as he was back, he gathered the group around and listed the names of the people who would be involved.

"Swampy will be responsible for the explosives and the blast itself," he said, looking over at Chris. "It's said to be two vehicles in total, a large KrAZ truck and one Geländewagen jeep. They should be in the middle of the square still, unless the Russians have moved them. What kind of explosives have you got?"

Chris pointed toward the tables where he had laid everything down and explained, "I've got two TM-62 mines and about a meter of sausage, mate. How's that sound?"

"Sausage?" I asked, already confused and now doubly so.

Chris stepped over to the table where he lifted a thick piece of sausage-shaped canvas not unlike a firehose, but filled not with water to put out fires, but rather TNT to start them.

Långström in his ugly balaclava, on our way back after capturing Pavlopil.

222

"It comes from one of those mine-clearing vehicles," he explained. "You know, those things that fire off very long cords of explosives to instantly blow paths through minefields, right? I...I found one of those vehicles a while ago and I...uh...I found myself in a position where I could...well, you know...I thought it would be useful at some point...yeah..."

"That's perfect!" shouted Långström. "This is going to be a fucking great explosion! How are you setting it off?"

Chris picked up a red block of TNT and a rolled-up piece of black powder fuse. "I have a starter with about ninety seconds of delay here."

Långström smiled with great content and pointed toward the others. "Richter! You're my radio guy. Cuix, Tjeck! You two empty your backpacks and load up these explosives in them."

Tjeck and Cuix nodded and began fulfilling the order. Richter replied "plus" in a monotone voice. Långström's gaze fell upon me.

"Carolus, you'll come with us as fire support. Do you want to bring your RPK, or a PKM? I arranged for us to borrow one for tonight, in case you want it."

The choice was easy. If we were going into the middle of an enemy-occupied village to stir up trouble and something went wrong, we would need a lot more firepower than 45-round stick mags to shoot ourselves out of it.

"Get me the PKM. How much ammo do we have for it?"

"Ammo?" Långström laughed. "Bro, we have more ammo than any of us can carry. How much do you want?"

"I can probably get by with three hundred rounds on me, but we could probably use at least a couple hundred more in case everything goes to shit."

Långström called upon Cuix again: "Hey! You and Tjeck make space for a hundred rounds of extra PKM ammo each. I'll bring an additional hundred rounds myself, as well."

As far as the plan went, it was retardedly simple. We would get a ride from our friends at the Recon & Sabotage platoon. They would drop us off at our usual starting point just in front of the small hill west of the village. The Ukrainians would then fall back but stay nearby to function as a quick reaction force in case something went wrong.

"Bear, you stay with the pickups so I can quickly get a hold of you, in case we need assistance," Långström ordered.

Bear replied that he had received the message and would double-check that all batteries and radio equipment were in order.

After our drop-off, we would continue much like our previous recon missions against Kominternove. We'd arrive at our usual locations for observation; as such, our goal this time was to simply continue further. The vehicles marked for destruction were located in the middle of the village, and should we manage to reach that far, we would destroy them. There was no assurance that radio contact would work, but if it didn't, the number of explosives going off would be more than signal enough for the QRF to come retrieve us, which they would do along the main road, just under the enemies' noses.

I zoned out somewhere in the middle of Långström's briefing, too busy thinking about beautiful women and their naked tits. The plan itself wasn't really anything I needed to understand, anyway. I knew roughly when, where, and how. The only thing I needed to focus on was following the others and being ready to rapidly lay down a fuck-ton of firepower in the proper direction, should it become necessary.

Långström came into the room carrying a PKM in its handle and handed it over to me.

"What do you think? Does it look good?" he asked.

I gave it a quick but thorough examination. The machine gun was in good condition—clearly not one of those dubious Ukrainian-made ones, but a proper piece of Soviet weaponry. The bipod sat firmly, or at least as well as it could with its design. No seriously rusted out areas. All screws and rivets were solid and in place. The bolt moved back and forth smoothly, and the firing pin wasn't missing or broken. Even the safety worked as it should. I dug around through the great pile of 7.62x54R rimmed ammunition and made sure to fill my belts up with as many armor-piercing incendiary rounds as was available, placing extra tracers at the end of the belt to signal the need to reload. I then loaded the belts into newer production canvas ammo cans, thus getting rid of the loud clutter of metal on metal the original steel ones produced. I hooked one underneath the gun itself and added two more hanging from each side of my waist.

Outside the schoolhouse, our battle taxis were ready. The Ukrainians had dressed warm with scarves and balaclavas over their faces. The engines idled, emitting thick diesel fumes. The headlights illuminated the frozen, cracked mud parking lot through the burning fossil fuel's thick mists. There was a total of five 4x4 pickups—Mitsubishi L200s and Nissan Navaras. Three of them would transport our group as well as soldiers from the QRF. The remaining two would function as fire support—one technical equipped with a 12.7mm NSV heavy machine gun, and the other with the powerful 30mm AGS automatic grenade launcher. The sound of metal rattled as the heavy cartridge belts were lifted into their feed trays and the covers slammed shut. Engines shouted as they fired up. We loaded ourselves and our equipment onto the flatbeds, and our transports began bringing us through the base gates. Mariupol's nightscape was lit up by incoming vehicle headlights, street lights, and glowing billboards along the road, everything flashing as we passed them. We turned off our own headlights as we arrived at the final block post. Shortly after, we continued eastward through an empty landscape, where the only light was that of the bright moon above.

CHAPTER 29: A BIG EXPLOSION

The night was bright and starry. The snow was gone, but the moonlight shone over the fields, lighting up the surroundings. No man's land was still controlled only by the ravens. As dark as the sky was, their even darker figures, numbering in the hundreds, contrasted in the skies above us, pitch-black wings and feathers reflecting the white moonlight at their edges.

The pickup wheels bounced over the bad roads as ice-cold winds swirled over the flatbed. The vehicles bounced violently over potholes, and we held on tight as to not lose equipment or fall overboard ourselves. The sound of diesel engines driven by subpar Eastern European fuel married with that of shaking metal. I enjoyed a sparse moment of music through my MP3 player—some good retrowave tunes to get in the mood.

I double-checked my machine gun to make sure everything was in order once more.

"Please help me now," I whispered, addressing my pleas to anyone willing to listen. "Help me do this. Help all of us now."

Our transport stopped next to the road by the path which would lead us into the southern approach to Kominternove. We disembarked and gathered up.

"Good luck! I will see you guys soon," Bear whispered from the top of his flatbed as the pickups and our QRF backed off. The rest of us continued forward the same way we had done so many times before, only now with very different intentions.

Upon arriving at the intersection where the path led into the small, paved road between Kominternove and Vodiane, we halted like usual to check the surroundings. Långström picked up his radio and whispered into it. Shortly after, a crackling response from Bear returned to sender. The Russian jammer equipment, which had previously killed off all forms of communications at this point in the terrain, had either been shut down or withdrawn further away. Radio contact with our Ukrainian friends waiting by the main road remained.

We continued forward on the paved road on a column, Långström in the lead. We soon passed by our usual observation post and continued forward, toward the ruined buildings next to the southern entrance to the village. Tjeck and Cuix took off their heavy backpacks to rest a moment as the rest of us covered around the ruined brick buildings.

Map of Kominternove.[1]

"Carolus!" Långström hissed and waved at me. "You bring Tjeck with you and check out the enemy foxholes on the road behind us. Make sure some asshole isn't sitting there, ready to shoot us in the back."

The asphalt road turned left and into the village, but there was also a gravel road leading in the other direction into a forested area south of us. We knew there were a number of small, entrenched positions along it. There were seldom if ever any Russians occupying these at night, but best to check to be safe. The recent fighting going on in the area should, quite seriously, have increased their motivation to stay awake at night. I very much disliked the idea of leading the adventure carrying the PKM, being a lot less agile than the average rifleman. I did, however, also recognize that it would be an exciting challenge in itself trying to sneak up on an enemy with a belt-fed machine gun. I fastened all my equipment properly and tightened the cartridge belt running from the ammo can which fed into the weapon, to move as quietly as possible. With everything secured, I then grabbed Tjeck, telling him to follow me at a good distance and keep a lookout to cover my six.

"We will check out the foxholes first. Then we continue down the road, two hundred fifty, three hundred meters or so. I go first—you stay ten meters behind me. Make sure nobody shows up and tries to fuck me in the ass or something. Be quick about it—shoot first, but make sure you don't hit me, okay?"

Tjeck nodded gently as his eyes fluttered a bit. He was clearly nervous, but

---

[1] Editor's note: Kominternove has been renamed to Pizuky since the events of the book.

remained collected. I switched the machine gun selector forward to automatic fire mode.

"It will be fine. I'm heading forward now, crossing the road. You'll begin following once I've reached the other side and have continued a few more meters along it."

I hastily but quietly began squatting my way across the road to the other side, where some tall, dense bushes gave cover, concealing me from the moonlight. Everything else in the terrain was open. I then proceeded forward, slowly, and crouched, footsteps light and soft as to not cause friction and noise from the gravel beneath them. I turned around quickly to see Tjeck following me before heading over to clear the first foxhole. There was a three-point pre-war wire fence in front of the holes, so I couldn't easily get in behind them. The only way to approach them was from the front, walking by the road. I soon spotted the dug-up dirt around the first manhole, and it seemed unoccupied. No coughing sounds, no white smoke from exhaled air or cigarettes came out of the ground, but I still proceeded just as cautiously, finger resting on the trigger. Only once I was right next to it did I make a quick move to rise and point the machine gun barrel straight down into it.

The foxhole was empty except for a pile of cigarette butts at the bottom—those at the top clearly being recent additions to the furnishing. The tobacco had not yet discolored their thin, white paper wrapping due to the wet snow or moisture, as they would have had they been discarded prior to the snow melting during the warm daylight hours.

We continued forward, checking all the foxholes in the same manner. All of them came up empty, outside of leftover, cheap, Russian-brand cigarettes. With the pits secured, we continued along the road down toward a lightly forested area. It was difficult to know what lay ahead, as the road reached a small peak in height prior to it, but upon reaching the top, the line of sight increased. I could easily see a couple more hundred meters down the road into the forest section. I remained in position for a moment, looking for some lights or movement. It was difficult to tell, though, as the moon didn't reach down through the trees, so I instead closed my eyes and tried to listen. It was quiet, the only sound being the occasional cold winds swirling across the road and through my clothes. I waved over to Tjeck that we were done around here and turned around to head back to the others.

Our rear was free and the journey up toward Kominternove continued.

"Follow me," Långström whispered. "Short distances, three meters."

"Really?" I whispered back. "That's the plan? We just walk straight into it?"

Långström smiled back at me. "Trust me, it will work!"

Between the ruins and the edge of the village, there was nothing but road and open ground for about five hundred meters—not so much as a decently sized bush, not even a ditch to take cover in. There was no good way to sneak into the village, so Långström decided to not even bother trying. We just walked along the road keeping close distances between the soldiers, as if we didn't have a care in the world. As if we belonged there.

If some Russian saw us—and someone must have—there was, of course, nothing suspicious about it. If the Ukrainians were about to attack them, they most certainly wouldn't do so with just a single squad. And, most definitely, that squad would not attack in a column at walking speed with rifles hanging on their chests, pointing downward. Who does that, really? Any Russian who saw us would naturally conclude we were a bunch of merry friends of his returning from a patrol or something similar. Sure, maybe he would find it strange how he couldn't remember hearing about a patrol due to arrive in the middle of night, but that wasn't out of the ordinary. Would he risk waking his aggressive lieutenant and ask about it? No, of course he wouldn't, because he liked to sleep indoors, instead of in those empty foxholes down by the road. Looking right at us, the Russian obviously wouldn't think anything of it. Obviously, he would go right back to sleep in his sweat-stained bed instead. The squad-sized group of men entering the village from the south wouldn't be any of his concern. Nothing for him to ask questions about. He'd mind his own business, just as he had been taught to do.

We were inside the village. The moonlight shone bright as a spotlight, and our profiles, glowed white under it. Our target should remain in the middle of the village, straight ahead. Instead of taking the shortest route there, though, Långström took us up on a smaller road leading up around the northern edge of the village. In doing so, the length of the path increased from seven hundred fifty to over a thousand meters in total, but by doing so, we would literally enter the square through the back door. Furthermore, by not heading in straight through the village, we kept our flank open as long as possible, avoiding trapping ourselves with enemies on both our sides before we were eventually forced to.

Everything seemed fine, but we didn't get far along the outer edge of the village until the first dog got wind of us and started barking. We couldn't tell whether these were wild mongrels or pets chained outdoors, but there were probably varieties of both. No matter what, as soon as the first one began barking, a second one joined in. Then a third. Then two more, and so on. We hadn't reached halfway before the entire village sounded like a large kennel with a cat thrown inside it.

Had the Russians been sleeping through our entry, they were guaranteed to have been awakened by now, most certainly able to smell trouble coming their way.

"Billiards!" Långström hissed at the man behind him, who repeated it backwards until the entire squad had gotten the memo.

We ditched our slow pace and started advancing quickly, each man covering until the man behind him moved up and hit him in turn, sending him forward, like balls on a billiard table. We continually picked up the pace until it could be called outright aggressive. We moved between the shadows where we—thanks to the stark contrast from the moonlight—effectively disappeared into the darkness. Nobody said anything. All communications went by hand signals and eye contact, pure feeling and simple routine. Seeing the fighters work their way up the street, I couldn't help smiling with great content. *This is beautiful, this is fucking beautiful!* We were a greased-up piece of machinery, fast and precise, efficient and flawless. All our marches and practice runs during the slow days in Urzuf and Yurivka had paid off. Everyone knew exactly what he was doing, and how he needed to act in relation to the soldiers around him. Everyone was attentive, quick, professional, and above all, part of a perfectly molded chain—a chain I was loving every moment of being a part of.

We covered the last five hundred meters very quickly, without being detected, and soon arrived by the square. Thick, large, old conifers rose in front of a worn, old, white building. Much like Orlivske compared to Pavlopil, the town hall of Kominternove was the old, small, and neglected counterpart to the fancy great hall of Lebedynske. We had come out from the villa's area, and the barking dogs slowly began to subside. No Russians appeared to be around, and the target vehicles remained in place where they'd been left. On one side stood the small Geländewagen, a decently sized jeep that managed to be completely dwarfed by the huge truck next to it. This wasn't some old Ural or weak UAZ like we had been riding the previous year, but a new production KrAZ-6322. The model was nicknamed "Soldier," and rightly so. At almost thirteen tons and able to carry its own weight in cargo, its V8 turbo-charged diesel engine alone was the size of a small car. We secured the area around the impressive truck while Cuix climbed up toward the rear of it to lift the cover and look underneath. He forced himself not to yell out in excitement, but instead hissed in his French accent.

"Ah! Ammunition is still here—it looks like everything still. Lazy, stupid Russians. They didn't even try to empty it," he whispered in an excited tone.

The size of our coming explosion had just increased by ten-fold. A few meters away from the truck lay a lone, wooden ammo box which had caught Tjeck's attention. Cuix had a gifted situational awareness, allowing him to perceive everything happening around him, and this was no exception.

"Shh!" he hissed loudly at Tjeck, quickly jumping down from the truck and heading toward him. "Don't touch that! You touch, maybe dead, okay?"

Tjeck stepped back as Cuix slowly started waving his hands around the box, looking for wires stemming from it. His concern around IEDs and similar boobytraps seemed to border on an obsessive-compulsive disorder, but then again, the young Frenchman had previously fought the Taliban and other Afghan warlords.

To each his own, I thought. *Better that than not being careful.*

Now, it was time for Chris to work his magic.

"Take position to cover the main road," Långström waved at me.

I did as he said. On the other side of the square and the worn pillars of the white town hall building, the main road passed straight through the village. To our left lay Mariupol, to the right deeper into enemy-controlled territory. I unfolded the bipod and got down on the ground, focusing to the east toward deeper Russia, but also with an unobstructed view in the other directions. We were now in the middle of the hornets' nest, and they could come from anywhere.

I wasn't positioned far away from the others by the truck, just over thirty meters. However, it was far enough that I couldn't contact them. In a way, Långström was a master at what he did, despite being obviously mentally disturbed. I had never had a reason to complain about him before, but this time he had left me in a position alone, something I disliked with great magnitude. I tried to wave to the others and seek contact, looking for a friend to pass the time with, unsuccessfully. The others were enamored by the explosives being fitted into the engine blocks, as if caught in a trance. I was close to them, but still all alone.

The main reason I wanted a buddy to stay with me wasn't because I was nervous. Sure, I was nervous insofar as nobody wants to be alone in enemy territory, and two pair of eyes are better than one. No, the main reason was that I wouldn't be able to warn the other in case of incoming danger. I could of course just open fire or yell, but both of those would expose us and might not be the wisest thing to do in every given possible scenario.

There haven't been any Russians around so far, I thought to myself as I lay down again. *Maybe they're still asleep.* It did not, however, take long until my gaze caught movements in the distance, coming toward us along the main road.

"Fittans skit sen då," I mumbled to myself, recognizing the silhouettes of two men carrying what appeared to be rifles. They moved slowly at a walking pace and were still far away at two hundred fifty meters or so, but they kept coming closer.

Shouting a warning out to the others would surely reveal my position, but I couldn't very well get up and go to them, as it was more important to keep an eye on the suspected enemies, to see what they were up to and where they were heading. Had Långström assigned a friend to the position with me, this wouldn't have been a problem, as I could have just sent my buddy over to warn the others we were about to have company.

I looked around for anything I could use to grab my friends' attention. The old pine trees had dropped cones all over the place; I gathered up the ones closest and tried to hurl them at my friends. The cones were light, though, and throwing them from a prone position was difficult. I couldn't throw them far enough to hit the backs of any of their heads. I needed something better. My eye caught a large piece of artillery shrapnel just an arm's length away. It was the size of the palm of my hand, and the heavy, twisted piece of cast iron would surely fly well through the air, hopefully even mildly hurting whomever it hit.

I took a deep breath and prepared for the perfect throw. My right arm swung away, and I tried to increase the momentum by also adding my legs into the effect. Obviously, as I always had terrible ball control, my throw was completely fucking off. The piece of shrapnel hit some of the hanging pine branches and fell straight down back into the ground, making a small thumping noise. The others remained obnoxiously unaware of the approaching Russians, now close enough that any more movement on my behalf would surely alert them.

Carefully, as to not scrape the metal bipod loudly against the ground, I moved my PKM into firing position. I lifted the stock and placed it firmly against my shoulder, resting the "shoulder thing that goes up" on top of it—the *actual* shoulder thing, not the barrel shroud. My front sight traveled from the side until it came to rest at chest level of the man on the left. They were now less than a hundred meters away. I decided that when they came within fifty, I would open fire and kill them both. A single burst, traveling from left to right. With about ten rounds, I expected I could hit both with at least one incendiary round each. They should be gone from the earth before falling and hitting it, and blown-out lungs cannot provide screams. The sound of gunfire wasn't uncommon at the front, and so long as nobody screamed out in pain, it was still possible it wouldn't attract that much immediate attention.

I needed to cut down both men in a single swipe. If one of them got away, the survivor would immediately run back to his friends and tell them exactly from where they were fired upon. The Russians would then be all over us in an instant. In that game of numbers, we wouldn't have a chance. I needed to kill them both, in one quick sweep, without either of them screaming.

I painted an invisible line in the terrain ahead. If they passed it, they would leave this earth. I would annihilate them. I counted their steps as they came ever closer to their demise, mumbling curses at them as the distance between me and them shortened.

"Take a fucking right turn now. Take that right turn now, you fuck. Head away, get into that fucking house instead now, motherfucker. Don't come any closer, you piss-filled fucking cunts."

They crossed the line where I told myself I'd gun them down at fifty meters, but I didn't open fire just yet.

"Okay, you fucks, you get ten more meters to live. Last chance, assholes."

I mumbled on as I focused my front sight and began squeezing the trigger. Their death sentence was close at hand—and likely ours, as well.

The Russians walked a few more steps before suddenly turning away from the road, searching refuge in one of the houses to the side. I exhaled all the air from my lungs. Maybe they were two soldiers sent out to examine what the dogs had been barking at and didn't find anything. In that case, I wondered if they, on their way back, ever realized how close they actually came to finding what they were looking for.

I heard something behind me. Turning my head, I saw Cuix hurrying toward me.

"We are done. We will have big boom soon now! Come on, we get out of here now!"

I followed him back to the trucks where everyone was gathered. On the way, Cuix lifted up the wooden box Tjeck had taken an interest in before. It contained several light mortar rounds, as it turned out.

"We make big explosion, anyway. Why not use?" he whispered with a big smile, carrying the box over to the truck.

We gathered in an arc around Chris, holding the fuse and a match in his hands.

"Right, mates," he spoke softly. "The moment I light this, we have sixty seconds, and we probably want to be out of here by then."

Långström got Bear on the radio, calling our taxis to get ready: "We're almost done here. Get ready."

The plan was to leave in more or less the same way we had made our entrance—perhaps in an even more shameless way. Once Chris lit the fuse, we would sprint on the main road westwards, straight through the enemy positions on the edge of the village. There, right under the Russians' noses, our friends from Recon & Sabotage would come in with their soft-skinned but quick steel diesel horses and pick us up. With luck, it would all be over before the Russians even knew what happened. It was a crazy plan, but crazy to the point it might just work. After all, crazy had worked so far.

"There will be a fucking lot of running now," Långström said and pointed to my PKM. "I can carry that thing, if you want."

Our leader was definitely the fittest of us all, while I was arguably one of the least—besides maybe Fabien and Inkasator, who were both twenty years older than me. I looked up toward the road. It looked exceptionally long and had a slight elevation to it. However, I knew it was less than a thousand meters.

"Dude, fuck no," I said. "I can run a thousand meters with a PKM and ammo. How useless do you really think I am?"

Långström shrugged his shoulders and said it was my choice. I wondered if this wasn't just the fact he was suspecting our exfiltration to fail, but that he expected us all to die, anyway, and that if such would be the case, he'd like to go out in a blaze of glory with a belt-fed machine gun in his hands.

Chris kneeled and put the match and fuse together in his hand.

"Fire in the hole!" he hissed loudly, pulling the head of the match against the matchbox's striker strip. The flame ignited brightly with a loud puff as white smoke and the smell of black powder filled the air.

Långström pressed the transmit button on his radio: "Sixty seconds! Pickup, now!"

We began running for all we were worth, out from the square and turning left onto the asphalt road, then straight ahead, westwards. The night was silent. Our boots hit the firm ground with loud bangs as heavy men with heavy equipment rushed forward. Rifles, ammo belts, and equipment rattled loudly through the cold winter night. We snorted like thrusting horses and sprayed thick white carbon dioxide smoke from our lungs, which was lit up even further by the moonlight. In front of us by the horizon, strong and blinding headlights soon appeared shining right at us. Our brave, death-defying Ukrainian riders were charging right into enemy lines to pick us up, Bear and the others standing firmly on the flatbeds with rifles and heavy machine guns at the ready.

The pitch-black and white, moonlit landscape around was then suddenly, for a moment, lit up in a powerful, fiery yellow color. Like that from a burning golden sun, the colors then changed to a warm orange and further into blood red. A wonderfully beautiful play of colors let itself flow across and cover all surroundings, but only for a short moment before then blackening out the world more than ever again. The ground then shook so hard as if it itself was made of pure thunder. I stopped running. As I turned around, a huge white pressure wave hit my face, as an incredibly loud bang vibrated throughout my entire body and organs. As the sound of the explosives died out, it was followed by the applause from shattering glass and windows all across the village. Behind us, a huge mushroom cloud of burning chemicals and oxygen rose up from the small square. Darker parts of red fire meandered through brighter and more lively flames of yellow. The enormous globe of fire might have only existed for a few seconds, but the heavenly and bewitchingly beautiful view of flame and destruction will always remain vivid in my mind and memory.

I now finally understood why we had brought so many explosives with us to blow up a 13-ton truck already loaded with more of the same. I wasn't the only one who had halted. We all stood there in the middle of the enemy line, Russians all around us. If all of us, despite being prepared for the explosion, still turned around to find ourselves bewitched by the majestic fireball, wouldn't the abruptly woken up Russians react the same way? Of course, they did, though a lot more distraught than us.

Our pickups skidded across the road, making sharp turns to point their rears toward us, whereupon we in turn started climbing the flatbeds. The Russian did nothing about neither the cars nor us. He was completely lost in the thunder and lightning, the magical dances of fires in the skies and the shock it left behind. It would only take a few seconds for him to get back to his senses, but those were precious seconds we were taking full advantage of.

I hurried up toward the pickup and grabbed my PKM to begin with, lifting it up on the flatbed. Then I felt that its carrying sling had fallen off my shoulder during our hasty run. The thick piece of canvas was now locked at my waist, stuck between my plate carrier and belt. I tried to lift myself up while still having the gun on my chest, but it was impossible. The gun stuck to the rear end of the car, and I couldn't get up. We were in a hurry, and I had already spent five, maybe even ten seconds trying without success to board the vehicle.

Two Ukrainians from Recon & Sabotage sat on the flatbed looking at me, probably wondering what my major intellectual issues were. They were both wearing balaclavas, so I could only see their eyes reflecting against the moonlight. I gave them a resigned look, telling them I was giving up on this. I had already wasted enough time, and there were a lot of men in danger right now because of it. Assuming they didn't understand English, I couldn't even tell them to just leave me behind. I was as ashamed as a dog. "I can't get on board, just leave me behind," I tried to tell them, not with words, but with disappointing looks and shrugging shoulders.

The Ukrainians looked at each other, nodded while seemingly sighing, and stretched forward toward me. They bent over the edge of the car rear, each grabbing my ass and belt. They then flexed backwards with a strong thrust, literally throwing me up onto the flatbed where I landed on my back atop the cold metal floor.

The car engine screamed, and the smell of burned rubber filled the cold air. I sat up, out of breath, looking back at how we quickly left the village. A large fire raged where our target had previously been located. Our armed technical pickups which had stopped by the road to provide fire support each reconnected to our column. The Russians fired red flares into the air, not only from Kominternove, but across the entire front line.

We had made it. We had infiltrated a position in the middle of enemy territory, blown our target up, and got out again. All of it without a single shot having been fired.

CHAPTER 30: FRONTLINE RAVEFEST

Thursday, February 12th, 2015

After our successful sabotage mission on Kominternove, we returned to Mariupol for a well-deserved rest. However, it didn't take many hours before we were woken up, again called back to the front line.

"I'm beginning to distinguish a pattern here," the Greek said in an ill-tempered way while packing up his personal equipment once more. "Go to Mariupol, rest, sleep, eat, shit—*no*! Go back to the front! Reinforce! Defend! Fight for glory! Fast! No! Go to Mariupol! Fuck!"

He tied up his backpack and sat down next to it.

"Azov is starting to resemble a real army more and more by the day. It hurts, you know. It really does pain your heart seeing your battalion—*regiment*, I mean—turn from a hardcore, rough riders-style fighting unit into a gay regular military force, in which fucking nothing works, and only the standardized military nonsense prevails. Sure, we are getting standard uniforms, actual weapons, and all that, but fuck! At what cost?"

Now accompanying Recon & Sabotage, we would once more be returning to Lebedynske as reinforcements. The sun had just risen as we disembarked our pickups outside the town hall. The air outside was very cold, but the weather was lovely. The sky was free of clouds, allowing the strong sun to send its bright, burning hot rays down upon us.

Our exact purpose as reinforcements remained unclear—not only to us, it seemed, but to our formal Ukrainian platoon commanders, as well. We would simply remain in the village and exist, waiting for something to possibly happen. Where and how exactly seemed to be up to us to decide.

Inside the town hall and current HQ, there wasn't much to do, hardly even enough chairs to plant our asses on while we waited for this "something" to occur.

Most of all, it was cold and drafty. The large group of newcomers headed outside to look for better accommodation elsewhere. Just a few hundred meters west along the street, behind the bombed-out, four-story apartment house, lay a large, oblong, two story white brick building on the right side of the road. It seemed stable and artillery-resistant enough to serve as shelter for both Recon & Sabotage and us few foreigners tailing along.

Upon entering, it became evident our new address was that of the town school. Unlike the one in Mariupol, it wasn't completely cleared out. Apart from broken windows, occasional bullet holes here and there, and the expected degree of vandalism, it was still as it would have been left on the average Friday afternoon. We gathered up in one of the larger rooms on the southeastern corner of the upper floor. The school building itself was like any other brick building: colder inside than outside. However, with the sun shining from the south side through the many large windows, the temperature inside actually got rather pleasant quickly. We were even forced to open a few windows after a while to let in some cold fresh air from the outside.

School desks and chairs were lined up inside the room as one would expect. In the back right corner next to the windows, there were also some old but very fair-looking armchairs covered in sun-bleached textiles. I sat down in one of them and looked out over the rest of the room. I then proceeded to immensely enjoy the most comfortable place for my ass to sit upon in a very long time.

The walls were covered in paintings of animals, plants, and the like, revealing that this used to be the biology classroom. On some tables to the sides remained a few burners and bottles, suggesting that chemistry was also taught in here. The chemistry equipment being short in numbers was not because of it being a poor school precinct, but rather a result of these items being coveted by the roaming packs of local gopniks who had gotten to the school before us. In a small connecting room in the left-hand corner, there was also the main chemistry lab itself. A myriad of small bottles and ampoules were lined up on top of the cramped shelves and cabinets in the small, dimly lit room. Chris immediately hurried to have a look, probably hoping there was something volatile or explosive in the mix which could be useful. Disappointed, he was quickly forced to give up, as the local ferals and survival artists had already had ample time to gobble up everything of interest and value.

The biology section and main classroom were more interesting to investigate, especially that which dealt in human biology and evolution. Obviously, I could not understand anything in its written form, but since the material on offer was tailored to high school students, much of it was designed to be taught in pictorial form, anyway. I decided to go for a walk around the rest of the building. Downstairs stood Richter, looking at a number of flags lying on the floor, all of them crumpled with footprints from wet and muddy boots.

"Do you know what flags these are?" he asked me.

I looked down at the dirty rags. They were dark blue with a small image of Ukraine in the middle and a yellow text on the side. I tried to read the Cyrillic lettering.

"Partija...Regionov..." I slowly read out loud. "Some political party, for this region?"

Richter had studied me as I tried to read and now looked down at the flags again.

"It was our former president, Viktor Yanukovych's political party. He was popular in this area. He still is. Or well, he was, until right now," he said. He kicked the dirty flag and walked over it with his muddy boots back upstairs.

The Greek had found a small storage room for books in one of the second-floor corridors and was digging through it. On a chair next to him, he had placed a couple of specific books he seemed intent on keeping. I picked them up to take a closer look. Both were thin; the top pink one had drawings for children on the cover.

"Seriously? Out of everything you can steal here, you're going to steal children's books?" I asked him. "You're sick. You know that, right? Get help."

"Bro, these are textbooks, to learn the basics of the Ukrainian language!" the Greek defended himself. "You know, you should actually try to learn the language yourself. It's very bad manners to stay in another country without doing so. You should at least try, even though I wager your simple, half-Finnish peasant brain would have a hard time learning anything other than pidgin."

I explained how I did not intend to stay long enough to learn the language. And even if I aimed to learn it, now wasn't the best time. I wasn't about to start reading children's books during an active offensive. That said, I had nothing better to do at the moment, so I stayed with the Greek, looking through what other kinds of literature were around. It wasn't a great use of time, but hardly worse than sitting down staring at a wall.

On the top shelves lay a number of thick, large, and black-covered books—simple in their design, with a bolded text on the back of them. "Biblia," I managed to translate from the Cyrillic text, stretching out to lift one of them down to ground level.

"Where did you find that?" asked the Greek vehemently. "I could use one, too. You know, I actually believe in God. Before I came here, I was actually considering fighting the unfaithful dogs in Syria."

"You mean joining ISIS, against those who do not submit to Allah?" I asked.

The Greek sighed and bit his lip a little. "No, no. I'm not bullshitting now. Seriously, I had found a small Christian monastery down there, outside Aleppo. They were setting up their own militia. I have always wanted to go to a monastery. I have long felt an urge for it. I'm serious. I was thinking of being their sniper. A sniper of God. A sniper monk."

Richter had sneaked up behind us and stuck his head in through the doorway.

"You're not stealing shit, are you?" he said in a serious tone. "We are not thieves; we do not steal shit. Niggers steal things, we don't. Do you understand?"

The Greek pointed to his children's books and explained that, yes, he did intend to steal books, but for good reason. Richter looked at the colorful booklets and thought for a while.

"Okay, you can steal those. But then you must learn Ukrainian also. I will test you on it in the future to see that you do. Do you understand?"

"I'm going to steal a Bible, too!" the Greek continued defiantly and pointed at me. "Carolus also intends to steal a Bible! In fact, he has already done so! We are both men of God now, and you cannot stop us from receiving Jesus into our hearts! You hear that, infidel?"

"Okay, whatever. Steal Bibles, too, if you have to. Be niggers," Richter muttered, leaving us to our pillaging run.

I had not intended to take the Bible at first, but now that the Greek had already branded me a simple thief, I began to more strongly consider it. The bulletproof plates we carried were both heavy and strong, but I had heard rumors of people getting injured despite wearing them. They could produce spalling on the inside when hit, and I lacked any sort of soft Kevlar beneath them to protect against it. The Bible was both large and thick. I opened up the vest and tucked the book behind the chest plate, and it turned out to fit very well.

The Greek looked at me. "Did you put the Bible under your plate? That's good, that's a good idea! I should do that, too."

Richter stood some distance away in the corridor and looked quietly through a window. Next to him was a chessboard and two chairs. Most of the pieces seemed to still be there, and those that were missing could easily be replaced through improvised means, like using cartridges.

"Shall we play a game?" I suggested to the quiet and austere young Ukrainian. "You play, don't you? I haven't played myself in many years, but we might as well use the opportunity, seeing as we have an actual board and everything."

Richter looked at the board and answered in his usual autistic tone, "Yes, I can play chess. We can play a game if you want."

"Best of three, then!" I confirmed, after which we put our weapons aside and sat down by the table.

Because he was so quiet and morbidly calm, Richter was difficult to read. A game of chess, on the other hand, could say a lot about a person—whether he was eager, daring, or calculating. In fact, it was mostly to try to better understand who Richter was that I had suggested a game. Of course, I also looked forward to playing in general, and to win. As I had played against my friends at a younger age, I had rarely lost a game, and the young Ukrainian would probably not be a problem, either.

A while later, after losing four times in a row against Richter, the charm of playing had worn off for me. I suggested we go and see what the others were doing instead. Richter smiled gently but triumphantly.

"We can do that, if you do not want to play another game."

The rest of the group stayed in the large biology room where we joined in. We entertained ourselves with the blackboard and the chalk crayons, where we tried to draw riddles and figures for the rest of the group to figure out. The answer was often one of our group members, albeit in a degrading situation and tone. A stick man who hunted two little stick men representing children, for example turned out to be, quite obviously, the Greek.

"Ha, it's the Greek!" Bear laughed out loud. "It's the pedophile Greek!"

We all proceeded to mock our Mediterranean friend. The Greek stood up for himself and defiantly defended against our laughter, saying it was his people who had actually given the rest of us both civilization and democracy.

"We never wanted your democracy," Cuix shouted at him. "France would have been better without it!"

The Greek explained whether we wanted it or not did not matter. "You took the offer of Demos. You used it—in your own way, I might add—and now you will pay the bill for it! It's about time you actual, white Europeans pay up! Greece needs money, you know, and you're in debt! Pay up, whiteys!"

We played hangman, charades, and just fooled around in general, with Bear documenting parts of it through his helmet cam. The atmosphere was just as we were a bunch of kids ourselves—a group of bored students with the classroom to ourselves, waiting for a teacher and lesson which never came.

The sun began to set, and it soon became cold in the room. We began closing the windows, but it only bought us minutes. Without the warm rays from our life-giving star coming through them, the large classroom quickly began cooling down. Unless we wished to freeze, we needed to find a new, smaller room where we could lock ourselves in to keep warm—especially as it looked increasingly likely if we would remain here overnight.

We found suitable accommodation at the other end of the building. Still upstairs, the room was much smaller, and the windows had already been barricaded. The building still offered a little electricity, and a lone lightbulb in the ceiling gave light. We unfolded our sleeping mats on the floor at the far end of the room and then sat down around a table with chairs at the other end, near the entrance.

Bear had been inspecting the basement of the building to prepare for the possibility of artillery forcing us below ground. Coming back upstairs, he brought a stereo system he had found in the catacombs, complete with proper speakers and everything. We eagerly connected the socket to the wall but without any sound coming from the large, electronic device.

"Well, obviously, it's broken," I mumbled disappointedly, something Bear did not see as a problem at all.

"Turn the light here so I can see," he said. "I can fix this, no problem."

The electronics were dead; the buttons gave no response.

"How are you planning on doing that?" I asked Bear, who had already begun to unscrew the silvery plastic casing with a knife tip as a chisel.

"This is simple electronics. I studied electricity at the University of Kyiv before the war began, remember? Electronics are my thing. This is baby's work."

It did not take long before the display came to life with players and speakers. The moment everything was plugged into the wall socket and began to suck power, though, the light in the ceiling died. There was not enough juice for both shine and music.

"I know!" cried Bear, as if with his own light bulb over his head. "Wait here, I come back soon!" he said and hurried out of the room, closing the door behind him so as not to let out any of what little heat remained.

The rest of us who were not already sitting at the table sat down to await our talented electrician's return. A new face, or at least new to me, came instead.

"Who are you?" I asked the young man. He looked to be around twenty years old with short blond hair. His youth alone was increased by his somewhat oversized Flecktarn uniform, giving him an almost childish appearance.

"You can call me Becks. You know, like that German beer," he replied in a thick accent yet surprisingly good English.

"Do you belong to the recon platoon, or anywhere else?" I asked.

Becks explained that he was a completely new recruit. "I was meant to be drone operator, but when I came here, they said there are no drones, so they just give me Kalashnikov instead and send me here," he said, pointing to his AK-74 leaning against his chair. "I have never even fired a gun before."

"How did you miss Becks?" the Greek asked, looking at me as if I were a tad mentally disabled. "He was with us in Lebedynske already."

It was true that I didn't pay much attention to most Ukrainians, whether in the battalion, the companies, or the platoons. I was never very social and had no interest in making new friends—especially not when any of them could die at any moment, anyway.

"They want to place me as normal infantry soldier, but I like to be with Foreigner Group, so I ask to be moved here," Becks explained as I studied him.

He was young and completely lacked military training, as he admitted himself. We didn't need any more recruits without military training. The fact was, however, that even from the brief conversation, it was clear that Becks was quick-witted and clever. Not just because he spoke good English, but from his way of being in general. He was also honest and didn't present a false image of himself. Another interpreter for the group wasn't a bad thing, and it was always better to have an honest and clever guy without formal education than a certified and schooled idiot.

"Well then," I said, extending my hand to the young drone operator. "Welcome to the Foreigner Group, I suppose."

Loud noises emanated from the stairs out by the corridor. It was Bear returning from the basement with a large, old, ugly, but very much functional diesel generator.

"Now! Finally! Soon, we will have music!"

And just as he had predicted, it wasn't long before the generator was running in the corridor outside with all wires and cables connected inside our room. The CDs that came with the sound system began spinning, and heavy electronic music soon echoed inside the room. It was loud techno, rave, and obviously hardbass. Cuix dug through his backpack to find a small glass bottle with some kind of green-colored liqueur in it.

"This is from my hometown," he said and handed the bottle over to me. "Try it! It's very good liqueur!"

French spirits had never been close to my heart in general, but I accepted the generous offer and took a small sip. I lied and told him it wasn't so bad, which it actually was. Booze being booze, though, we all shared the bottle between us as the party got livelier and livelier. It didn't take long before other Ukrainians from Recon came over to see what was going on. The music thundered to the point where the thin plywood boards covering the windows flexed from the thick bass. The entirety of Lebedynske could hear us—and probably the Russians we had met

239

in Kominternove last night, as well.

Soldiers danced wildly and ferally around the floor. The air became hot and humid from sweat like in a sauna. Thick winter coats came off for the first time in days and soon also the tank tops. I rarely indulged in dance, but for a feast like this, I was willing to make an exception. Roughly carved out and tattooed skinheads and militant football hooligans created a violent mosh pit in the middle of the room. The music finally killed the bulb in the ceiling, so the men with flashlights mounted on their rifles provided light instead. Chris started the strobe in his, letting the light from his AKS muzzle flash throughout the room. The night had everything a club would offer, but also that from a warzone. The only women we needed to dance with was Kali, Freja, and Athena, and they in turn danced with all of us with great pleasure. In what was an indescribably human experience, the goddesses of ultimate reality and war were tenderizing and priming our unknowing souls for what was about to come.

The violent war dance had finally exhausted us, though we were in the greatest of spirits. The clock had passed well over midnight, and it was time to close shop, at least if we aimed to get a somewhat proper night's sleep for once. Just as we were lying down, though, new orders came in. We would be returning back to Mariupol.

"Ha! I knew it! The pattern!" cried the Greek.

We geared up, quickly packed our belongings, and left the school as well as the warzone nightclub, everyone guaranteed to be a good memory richer, and me with a thick Bible under my breastplate.

Back in Mariupol, it was business as usual: same boring room with the same cold floor. We did not, however, have much time to rest on it before we again, as the Greek's pattern stated, were woken up once more.

It was now the early morning hours of Friday the thirteenth. We were going to Shyrokyne.

CHAPTER 31: SHYROKYNE

Friday, February 13th, 2015

"Finally!" Långström said happily. "I have participated in several wars. I have been shooting at other people, and other people have probably been shooting even more at me. I have tried all kinds of drugs imaginable, and I have gotten STDs while in the jungles of Africa no white man has had before. I've done most things, really, but I've never been bombed for real before. You know, getting bombed by a real, modern army. A real enemy. With really heavy fucking artillery! Now, maybe I'll finally get it! If I can get just that, being bombed for real, at least once, then I can retire afterwards. Then I've done everything."

It had still been dark outside when we left the Mariupol base, but the sun found time to rise during our short car ride toward Shyrokyne. The cloud cover was, however, quite thick, leaving the surroundings dark and gray, more dead than alive. The small villages along the coast seemed deserted outside of the hundreds of ravens still littering the air and fields around, thoroughly observing our column passing through their lands. The atmosphere was as gray as the world around us. I had once aspired to get to the famous Shyrokyne, but now it just felt like another small, useless village among many others. It was cold and I was freezing. My feet hurt. I was in no condition to be able to long for war and combat. I only dreamed of a hot shower and, most of all, dry socks.

We turned right from the main road, continuing up a small hill, which was unusual in the Ukrainian steppe. Once atop it, we halted outside a large, strong white brick building: "Mayak," as it was called, or the "Lighthouse." It was a three-story building with nothing special to it, but the third floor was slightly smaller and rose above the rest of the flat roof, so the analogy wasn't that far off. It was also close to the sea, its white color reflecting against the dark surroundings—with some imagination, it actually did resemble the beacon of a lighthouse somewhat. Clearly visible from atop the Mayak's roof, the mostly deserted village slumbered below.

Map of Shyrokyne.

Shyrokyne looked much like the others, with lots of small and medium-sized villas and gardens that spread through a network of small roads; the percentage of larger houses was possibly higher than in the other villages in the area. The main difference lay in its general size. Still being a village, Shyrokyne was by far the largest in the area. The community seemed—from Mayak's roof, at least—to be at least twice as large as Lebedynske.

"Will you please stop holding your finger on the trigger all the fucking time?" Chris growled indignantly at Tjeck.

"I do it out of old habit. It's okay, anyway, because it is on safe. I know what I do," Tjeck replied naively and insecurely at the same time, just as he did every other time someone pointed out this unwelcome behavior.

If it was cold outside, it was much worse inside Mayak. Even though the walls on the whole remained where they stood before the war, the shelling had destroyed every single window. The cold air inside had not been warmed by the sun's rays all winter. My feet hurt indescribably as soon as my boots hit the cold concrete floor. The rooms were dark, as blankets were hung over the windows in an attempt to trap some kind of heat inside.

A sharp, piercing sound suddenly echoed loudly in the room. Everyone backed away and looked around. The only one who stood completely still was Tjeck, near one of the doorways. Around him, a cloud of concrete dust slowly fell back to the ground. He had finally, by always holding his finger on the trigger, negligently discharged his weapon. Luckily, the bullet went straight down into the floor. Like a fifteen-year-old who had dropped his tray in front of everyone in the school cafeteria, he just stood there, huddled and ashamed.

"I fucking told you!" Chris said, pointing at him.

Tjeck laughed a bit cautiously and looked around. "Oh, sorry guys. I guess it wasn't on safe, after all. I'm a bit nervous, a bit twitchy…It won't happen again. I'm going to take my finger off the trigger now."

Ringing ears were annoying, but nobody was particularly angered. It was a good thing that Tjeck had finally let a round fly, because now he learned not keep his finger on the trigger, without anyone bleeding to death as a result. We sat down in a vacant, dirty sofa in one of the otherwise empty rooms, waiting for further orders. Fabien, Tjeck, and I soon got a lot more comfortable, as Tjeck unfolded his sleeping bag for us to use as a blanket. Even the pain in my wet, cold feet slowly began to subside.

Like many times before, our group was soon divided. Chris, Bear, and Inkasator—who had finally joined back in with us now that transports weren't going around as much—would go with Abdullah to the northeastern outskirts of the village. There they would climb a large tower from which they would scout in the enemy direction.

"I'm bringing the sniper rifle and night vision AK with me, but I doubt I'll be needing all these magazines," said Chris, opening his vest-mounted mag pouches, handing three 30-rounders over to me. "Here, take these. I'd bet you will have better use for them than I will."

He hastily got his equipment together, grabbing the large black sniper rifle hardcase bag in one hand and extended his other toward me.

"See you soon, buddy," he said, clamping my palm before heading out through the doorway into the light outside.

"See you, too. Take care, man."

The rest of us did not have to wait for long, either. A Cougar armored car rolled up on the cobbled courtyard outside. It would give us a lift down to Shyrokyne itself and the building in which we were to be accommodated. The vehicle was even nicer on the spacious inside than outside, even smelling of new car instead of sweat, old blood, and grease. It was only crewed by a driver, so as the others were loading up, I climbed directly up into the turret ring to man the NSV heavy machine gun.

"Do not shoot at anything in village," the driver shouted while waving to get my attention. "They are friends, yes? Don't shoot gun in village."

Did I look like a complete moron? After a moment of feeling offended, I came to the realization that from the driver's point of view, it was probably safer to make sure—lots of jumpy people around, no way to know what kind of retard might be climbing up into the turret.

The cover had cracked open, and the weather appeared to become just as good as the day before. Perfect for a nice buttoned-up ride through the Ukrainian village. The view from the Cougar's roof behind the heavy machine gun was great and offered a good overview of the area. The houses were, for the most part, mostly intact, just suffering from the usual neglect. The buildings looking the worst were actually just unfinished construction projects, as apparently the village had been expanding prior to the outbreak of war. It did not look half as ruined as I had expected, based on the stories of the previous year's battle. The Russians had shelled the place with heavy artillery back in 2014, but it appeared as if most of it must have landed outside of the village area, in the fields and around the Mayak.

The cool but fresh sea winds soon stopped blowing over my face as the armored car began slowing down. We had arrived near the southeastern outskirts of the village, not too far away from the beach. Next to us stood a tall, orange-painted metal wall.

"We're going in here," Richter said, lifting up his backpack and stepping through the open white gates in the middle.

The house inside the gates was a proper one. Nothing like the small, weak, thin wooden huts that looked like some bad weather would turn them over, but a sturdy fortress built from brick. The main part was made up from red ones with an entrance in the middle, two large metal garage doors on the left, as well as a slightly smaller top floor made from the same material atop the roof. The right side of the building was a slightly smaller extension, looking somewhat weaker in its construction but still made from bricks, this time painted in white. None of the walls were quite as strong as the Mayak's but were definitely still bulletproof. The windows were all still there, and it looked as it would be warm indoors. Outside the main entrance were a number of comfortable-looking sun loungers and even a nice hammock. On the lawn in front of the house, there was even a bathtub where a brave enough man could relax should he wish to do so. That was, of course, provided that said hero could arrange for water to actually fill it—none of which would be coming from the taps, at least, neither hot nor cold. All gas had also been cut off, but there was a small iron stove in a small room in the middle of the house, right next to where the white extension met the main red one.

Our house in Shyrokyne.

Another group from the Recon platoon had already settled in and made it cozy for themselves. In the kitchen, the first room to the right of the entrance hall, lots of food items had been stashed: soft cheese, hard cheese, bread, doctor's sausage, and all the paper-tasting biscuits one could ever want. We had also received new food shipments of barbecue flavor potato chips and soft drinks—Pepsi Lemon in fifty centiliter PET bottles. We could eat well, and there was no risk of starvation; however, needless to say, all of us would probably have agreed to swap all the potato chips in the world for a single actually warm, cooked meal.

Some of the Ukrainian recon soldiers entered the kitchen, bringing with them a man dressed in dirty civilian rags, whom they sat down on a stool in the middle of the room. One of the Ukrainians was carrying a civilian-looking backpack and a disassembled shotgun, a cheap 12-gauge over/under Baikal. The middle-aged man was terrified and stammered as they interrogated him. What was he doing there, sneaking around the area while armed? We stood back a bit as Richter quietly translated a shorter version of it to the rest of us.

"Old man claims he was just hunting, so he says."

A Georgian who was leading the questioning showed the old man what else he had been carrying: a GPS and two cell phones.

"Do you use these for hunting around here, as well?"

Tears began to run down the old man's cheeks. He was sobbing and dripping snot all over the place like a child, to the point he could barely speak any longer. It was, to be frank, a simply pathetic display.

"Ask him what he was hunting," I whispered to Richter, who turned over to the man, then back at me.

"Birds, he says. He says he was just hunting birds."

I waved over to the man carrying the backpack, signaling I wanted to take a look in it. Inside there was an old, worn leather belt for 12-gauge shotgun shells filled with plenty of ammunition, which I began inspecting.

He had quite a mixed assortment, and a few were indeed birdshot. However, most of them were of different variations: Brenneke slugs, 00 buck, and even some reloaded mixed pellet loads, only to name a few. The oddest of the bunch was a red plastic case which, when cut open, revealed three large lead balls following each other in a see-through plastic sabot. It was highly doubtful the things would even work, so it mattered very little.

"These aren't bird hunting shells," I said. "These are for hunting pigs. Or man."

Even though the interrogation never turned physical in any way, a never-ending stream of tears flowed down the old man's cheeks. It was obvious he was working for the enemy, and he knew very well that we knew. Being a village local, he was obviously not out there to kill anyone, carrying his shotgun only for self-defense. His exact mission remained unclear, as the bird hunting story was all he would admit to. His GPS, however, made it quite evident he was marking our positions and relaying them to the enemy.

With the spy still held up in the house, the Georgian asked if we wished to tag along to search his address. His home wasn't far away, just a bit further south, closer to the beach. It was more of shackle than an actual house, but it looked cozy enough. Some soldiers remained to guard the outside as we entered. It was a lot larger inside than the unpretentious outside made it seem, but it remained mostly void of anything of interest—no obvious military documents, no maps, no intel of value. We already knew the man was armed, but the already open gun cabinet held nothing more inside except for an ancient, rusted out, yet still functional old pellet gun. On the bench next to the cabinet were plenty of 12-gauge shells and tools to reload them. Lots of nice old full-length brass cases, gunpowder, and primers. The old man, crying like a baby, was obviously not fit to use such volatile chemicals, anyway, so I rounded everything flammable up into a bag and brought it back with me, thinking it would be a neat welcome back present for Chris once he reconnected to the rest of us.

One of the types of shotgun shells the old man in Skyrokyne was using to hunt birds, supposedly.

Outside of the gunpowder, shotgun ammo, and components, we returned empty-handed. It made little difference, though. Sneaking around with a gun and with a GPS tracker, it was plain as day what the old man was up to. We soon packed him into the back of a small Soviet relic of a police car, together with the evidence to be held against him, and drive him back to Mariupol. From there, the official long arm of the law would process him.

"He will be free in a week or so, anyway," Richter said as the police car exited through the gates.

"They took him in with a gun and a GPS. Of course, they'll jail him for espionage," I said, looking through my loot while relaxing in the hammock enjoying the warm sun.

"This shit was able to start because of the police. They did not take care of it back in 2014, they won't do it now. They are either lazy and retarded, or they work for the Russians." He paused a short moment to ponder before continuing, "Actually, usually, it's both."

There was no fighting in the area, except the occasional heavy mortar shells landing around. The impacts sounded as if they would have been rather close by. On the left short side of our home, a staircase led up to the small penthouse on top of the roof. I hurried up in order to see where the shells were landing. I was sure that it must have been inside the village, but despite the good top view, I couldn't find any clear signs of smoke rising. The shelling continued, just one or two at a time with fairly long intervals in between—just enough to likely be harassing fire, rather than sporadic, random shelling.

As darkness began to settle outside, we sought our way indoors to warm ourselves. The Georgian and the other Ukrainians had already made sleeping quarters in the main building, but there was plenty of room left over in the white brick extension to it. Our group settled down near the small iron stove, where we made a fire. It was the first time in several days we were in a heated room, a very pleasant experience. The forged iron stove door was opened to insert a couple sticks of wood and then closed back up again. The homely sound of the metal door being closed was followed by a noisy crackle as the dry wood began to combust into flame. I could finally take off my boots and socks to dry them. My great worry that my toes would be black or even stick to my socks as I pulled them off turned into a sigh of relief. Everything remained where it should be, even though my feet came out concerningly pale. There was no electricity on offer, but we made do with candles we found in the kitchen. The small flames gave a pleasant, soft, moving light flickering across the small room.

Tjeck in the kitchen of our Shyrokyne house.

"You know, it's been six months since I was last here," said the Greek. "It's strange, I have strange emotions about it. Last time, back then, I actually didn't think we would make it. I was quite sure I would die here. I even cried a single small tear after the battle. That's the truth. I didn't think we would make it out alive."

He went on to talk about the battle in 2014, when the Russians forced Azov and the other Ukrainians out of here.

"We were out in the field, west of the village. The Russians were already inside by then. I had that AK with the scope on it and saw one of those fuckers, standing on a rooftop inside the village. I was about to get him when I noticed he was standing there completely naked, so I hesitated to shoot. He was just standing there, looking at us with his binoculars, right? But he was completely naked. So, I got stuck thinking, like, what the fuck? I'm expecting to kill men in uniform but can't see any. I can't see who he is, so I hesitate for just a second. Only then, after a moment, it dawns on me: that's the fucking Russian directing the artillery on top of us, with his fucking binoculars. He had simply undressed so that we wouldn't shoot him, to fool us into thinking he might have been a civilian. Anyway, I lift the rifle and back up to take aim, to fucking kill him already. Fuck, man, I think the fucker is about to get it now! But then, I don't see jack shit through the scope. The fucking fires around have spread by that point. Grass all around us, all of it ablaze, creating a mirage blocking my vision. All of it a fucking blur! I shouldn't have hesitated, and just gunned him down at once. Fuck, I wish I had!"

Our main task still remained. We would simply stay in the area until further orders, and such a tedious activity had never been as welcome before now. By our warm stove in the cozy candlelight, simply existing and waiting was a mission as perfect as any, and we were completing it as professionals. Our group had only one other secondary objective. Together with the others from Recon, we were tasked with keeping a lookout by the cliffs overlooking the beach throughout the night, two men at any time. Långström and I had performed our first shift for one hour during the daylight hours. My next turn would be with Tjeck at midnight. There had been time for a few hours of sleep before, but the good company and very chill atmosphere was to cozy to say no. We decided to stay awake until our turn came.

It struck me that I hadn't been very attentive during my first walk over to the lookout position. I had simply let Långström lead me, following behind without making a map over the place in my mind at all. I asked him to go over the route for me once more, just to be sure.

"You follow the street toward the beach, into the warehouse area. Pass the trucks on the left. Find the narrow path along the cliff edge and follow it."

It didn't sound too difficult, and Långström gave simple-to-follow instructions, but I knew I had fucked up similar missions before in my life.

"Did you understand any of that?" he asked, annoyed, as he could clearly see my low levels of concentration left me kind of lost.

"Absolutely. I understand. The warehouses and then the path by the cliffs, 100 percent," I lied. It was more like 75–80.

Outside but still inside the yard, we were met by a large dog I had never seen

248

before. It somewhat resembled a German Shepherd in color, but much larger with thicker fur. Its face was also far from as slim and aesthetically pleasing as the pure breed, but rather more brachiocephalic with dropping cheeks full of saliva. There seemed to be something Newfoundland mixed in there, but being ugly as it was, the large drooling face was also a friendly one. An older Ukrainian from Recon, broad-shouldered with a short, thick gray beard, stood outside smoking. He looked at us and asked in simple English if we were heading out toward to the guard shift, then pointed toward the large hound.

"Do not worry! He follow you there. He lead you, help you find your way."

We laughed a bit at the joke and thanked him for providing us with this escort. It was obviously one of Shyrokyne's many wild dogs, only a whole lot bigger than the usual mutt. These wandered around everywhere, carrying out their own missions, but often gravitated toward soldiers if any were around. Just about all the wild dogs were incredibly friendly and fantastically loyal company, much more so than any kind of pet. Especially if offered some food, as they often were by soldiers.

We headed out, following the same route Långström had described. To our surprise, the hound actually did seem to follow us. Soon he even took lead, with the exception of a few short detours he made inside the warehouse area among the abandoned transport trucks. Judging from the dull growls and screaming beeps which followed, it was likely he was striking down smaller rivals which had made the grave mistake of wandering into his territory at night. Once he had asserted his dominance over the area, he then returned to us, leading the way over to the path through the high grass over by the cliffs. Below them, the now deserted beach and shoreline previously filled with tourists opened up next to the calm Sea of Azov. After a short walk on the narrow, sandy path past bushes and thickets, we arrived at the lookout position where Cuix and Greek were waiting for us.

The hound sat down next to the shallow foxhole, waiting for the new guard to replace the old one. As soon as Cuix and Greek had their equipment ready, he then proceeded to escort them back to our house again.

There were barely any clouds outside, and the skies were filled with bright, shining stars. The wind was almost completely still with no heavy gusts tearing through the grass and bushes to disturb the silence. In the silence of the dark night, the only noise was the buzzing of enemy reconnaissance drones flying around above us. Sometimes they flew high up, hovering slowly across the village. Other times they came in low, passing by close and quickly. They were clear signs of bad times ahead, something we had it far too good to give any thoughts about. With the stars and moon shining above us, the larger Russian cities were visible through their bright lights traveling upwards toward the sky, reflecting back over the calm waters of the sea to our right. Far away to the east along the shoreline lay the largest cone of light. It was probably not the certainly weakly lit Novoazovsk, but more likely the large Russian cities of Taganrog and Rostov. On the other side of the sea in the south, a number of slightly weaker beams shone on the other side, the largest of them Yeysk, only fifty kilometers away across the sea.

The clock began approaching one in the morning, and our shift was close to an end. We could hear cautious footsteps behind us as I readied my RPK. Contrasting the moonlight, a number of figures soon appeared coming through the bush, two taller ones and one much shorter. The short one came in first, sat down near the pit, and panted contentedly. It was the hound which I soon began to refer to as "the pig," due to its ugly face but nice personality. He had brought our replacements with him and now waited faithfully to lead us back home again.

Långström and I had requisitioned a sofa bed in the house extension, just next to the doorway to the room with the stove. The others spread out in other parts of the building on top of beds or in armchairs. The stove had heated up the house very well. I took off my boots which had not had time to dry completely prior to our guard shift, leaving them next to the stove for the night. I then opened my backpack and took out my ludicrously expensive, unused socks and hung them over the back of a chair next to the sofa bed. Life was good again, and it would be even better the next day when I would pull brand-new wool socks over my sore feet.

"Fuck it," I told Långström. "Let's sleep in tomorrow, whatever happens. We deserve it and could use a couple extra hours of actual sleep"—a proposal he willingly approved.

CHAPTER 32: THE SHELLING BEGINS

Saturday, February 14th, 2015—Valentine's Day

My deep sleep was slightly disturbed by detonations as the ground shook a bit, but there was nothing to stop me from falling asleep again. I noticed how some of the others got up from their beds and sofas in the room, arguing about something while getting dressed. This disturbed me way more than the dull thumps and vibrations coming from outside. I turned around and pulled the blanket over my face. Långström also remained while the thumps outside continued. Soon a clear voice was heard in the room, seemingly directed at the two of us. It was Richter. He was in full combat gear, standing at the doorway between us and the stove room.

"The Russians are firing with heavy mortars," he said.

Långström woke up and looked at him, clearly annoyed. "And?"

"They will hit us soon. We have gone to the shelter. You guys should come, too."

It was not uncommon for the Ukrainians, in particular, to take refuge as soon as they heard something even remotely resembling artillery. Statistically, this was just another one of these common instances.

"Go away, we're sleeping in this morning," I said, equally as annoyed as Långström. With my newly awakened gaze, I blurredly saw Richter shrug lightly.

"Well, I tried to warn you. You do whatever you want. I'm heading for shelter."

It had just started to get light outside. I lay down again and pulled the blanket over my eyes to get the surroundings darkened again.

"These fucking Ukrainians and their paranoia," I mumbled to Långström, who confirmed that they indeed were very annoying, always running for cover as soon as something happened.

It thumped outside again, louder than last time, and the ground shook more. We flew out of bed and looked at each other.

"Fuck, that one did seem pretty close," I said.

A moment later, another came down, even closer. We finally understood what was going on and quickly got up to put on our boots and vests. We hadn't had time to finish before a shell crashed right outside the house. Everything shook as the sound of shattered glass broke into the orchestra. The Russians had used a few rounds to zero in on us. Now that they had their angles calculated, they would soon commence firing at full force. The shelter was in the garage, and we found ourselves in a scramble to get to safety. It was really just a small staircase leading down to an ordinary, small, cold store basement, but it was the only place that could be considered safe. To get there, we needed to get through the small room with the fireplace and the kitchen. After that, only the small entrance hall remained, leading directly into the garage. I ran first with Långström in close pursuit, but we only made it a few meters.

Just before I reached past the large kitchen window, it was all blown inside by a violent explosion. I had to throw myself backward, landing on my back. Glass and shrapnel shattered through the room and smashed the cupboard doors, the porcelain inside them clinking and shattering. The thin curtains swirled around from the air pressure as the heat ignited the oxygen in the air. The white curtains danced around through the flames above me, like a woman in a bright summer dress dancing in the warm summer wind. It was indescribably beautiful.

The mortar rounds crashed all around and on top of the roof above us. We needed to find some other safe spot, back to where we came from. Långström was just on his way back into the bedroom, as the windows there were also blown in, just like the ones in the kitchen. Shortly after, another shell hit the ceiling, smashing the roof down into the room below. Glass and shards darted through the dust. This route was also blocked.

The small fireplace room was the safest stop remaining. There were no windows there, and in one of the corners, we were also protected from the window openings from nearby rooms. We crawled up together into the corner and pressed ourselves as hard as we could against the walls, finally finding a moment to tie our boots and fasten our vests.

Långström looked around, newly awake. I wasn't sure if he was scared. I had never seen him scared before and thus didn't know what it would look like. He definitely looked confused and anxious. It might have been fear. I thought that, if he really was scared, we were in a really bad situation. Surely everything was bad as it was, but as long as we had him, I was sure that we would make it out of any situation. At the same time, if Långström got scared and we lost his leadership, then we would be really fucked. I needed to try to lighten the mood somehow.

"At some point, I think we ought to try and listen to those Ukrainians," I said in between the loud explosions. "You know, for once, this time they were actually right about something!"

Långström looked at me. His slightly anxious expression quickly turned into a smile followed by a loud laugh as a thick beam in the room next to us was thrown to the floor by another shell impact. I picked up two cigarettes, lit them, and handed one over to my friend. He had seemed lost there for a few seconds or so, but now he was back in his usual excited state. The 120mm shells rumbled around us in an act of industrialized violence, one after the other. Sometimes, even though the main explosive force was directed on the other side of the roof, their heavy tails would punch through both it and the roofs. Some of them got stuck in the ceiling above us, some penetrated and struck the floor below. It was apparent that he loved every second of it. Finally, his dream of being bombed for real was fulfilled.

For my part, of course, I also thought it was very exciting. I loved the strong smell of burnt explosives which filled our lungs. We actually had a rather great time sitting there, joking about how absurd our situation had suddenly become, how we could never get a single fucking morning to sleep in, and so on. However, all the fun did not diminish the fact that I was, of course, quite worried. The walls seemed to hold, but the roof collapsed more and more with each explosion. The ceiling would not last forever, and it was impossible to know for how long the Russians would keep up their angry barrage. Fire and black smoke sprayed in through the now completely broken glass and windowpanes around us. I felt like a mouse hiding in its nest, but outside wasn't some hungry cat , but a snorting, wild, fire-breathing mad dragon, ferociously trying to break in.

The mortar fire ceased as suddenly as it had begun, without the roof above us giving way. We got up quickly, dusting off our now white robes before we hurried off to the shelter. There the Greek stuck up in his head, helmet on top of it, turning around to see us coming.

"Fuck! How fucking difficult does it have to be to just die? I bet money you peasants were dead, fifty hryvnias! Fuck! This is your fault. You owe me that money back now!"

He stepped out of the shelter, looking as if he would continue to talk about his great disappointment, but something else suddenly caught his eye outside the window.

"What the fuck? What in the actual fuck is that?" he asked himself while raising his voice. He turned around, down toward the basement stairs, stretched out his arm and shouted angrily, "Quickly! Hand me my AK!"

A hand came up from the hole holding his AK-74. The Greek grabbed it and hurried toward the window, where he took aim upon a dark figure hurrying away from the scene.

"Fuck! He got away! The fuck! The damn fucking old man! I'll kill him!"

"What the hell are you raving about?" Långström asked.

"It was that fucking old man again! That fucker with the GPS we took in yesterday! The fucking cunt was out there, with a fucking cell phone in his ear and probably the same fucking GPS in his hand! It was him who led the mortars on top of us, but he got away. Fuck!"

A shrieking noise outside was quickly followed by more explosions.

"Mortars!"

"Cover!" everyone shouted as we threw our asses down the shelter stairs.

The old man, who had apparently been released by the corrupt or incompetent police earlier than even Richter had guessed, appeared to have been there to see the effects of the shelling. He must have seen and heard the Greek's singing, as well as found that our building remained standing, so he had called the mortars to give us another run for our money.

The atmosphere of the shelter was dim. It was very small, just a few square meters where now up to twenty men crowded. I could not bear to try to squeeze in, but instead sat on the stairs. It was not as safe, but I reasoned enough so. In addition, I had both a seat and, unlike the others, plenty of room to stretch my legs, which still felt a bit stiff after my interrupted sleep. One of the Ukrainian Recon guys took the moment to enjoy some canned pickles left in the basement by the previous civilian owners.

"As soon as they stop firing, I'm going to find that old man and kill him," the Greek continued, an idea the others were clearly also floating.

"I'm going to stab him," muttered a bitter Långström while he continued to dust off his uniform. "I'm going to stab him in his stomach, and while he is bleeding to death, I'm going to throw him down the cliffs."

Richter tried to urge the two to reason: "That will never work. He is far away at this point. We must come back later and get him when he sleeps. We know where he lives. This way is more effective."

It was difficult to know what was going on. Our only contact with the outside world from below the dark basement was the small trachea-like hole in the wall. In addition to circulating oxygen, it could also gossip about what was happening on ground level. Of course, we heard and felt the shells that hit our house very well, but during the short breaks between these impacts, the pipes led down sounds from the more distant surroundings, as well. Other detonations around the village growled through the tin pipe and down to us, a clear message that it was not just us who were getting a good beating. It sounded as if the entire front line was being fired upon, some positions perhaps with even heavier artillery than us. There were also things that sounded lighter, like the light automatic grenade launchers. The quick knocks that gurgled through the pipe sounded really funny. As if someone were farting extremely loudly, repeatedly, at a very fast pace. Fart jokes being the most essential for any comedy act, needless to say, several of us had a difficult time keeping the laughter at bay.

The mortars again fell silent, and so we could once more leave our hole in the ground. The large steel doors that made up the garage door were perforated by large shrapnel holes which the morning light shone through. The pickup in the garage belonging to the recon team seemed to have survived the worst of it, though. The rear window was smashed to pieces, and there were some holes here and there in the body, but the tires had survived from puncture wounds, and there didn't seem to be any fluids leaking out of it. The Greek poked his AK outside the window, looking for an old man who, just as Richter had predicted, had not stayed for a second time. Everything seemed quiet. The only thing that was heard was the low crackling of Richter's radio.

"The Russians are attacking," he translated calmly. "We should move to the next street to support."

Långström took over immediately, much more hurried and excited than the quiet Ukrainian.

CAROLUS LÖFROOS

"Let's go! Forward, to the front line! Follow me!"

CHAPTER 33: THE BATTLE OF SHYROKYNE

South side

We pushed open the large garage doors and moved forward at full speed. The front line itself was the easternmost street in the village. We were a street behind this one, and the fastest way there was across a large muddy field, about two hundred meters. Långström was a fast distance runner, as well as in the sprint, and set off like a bullet, angling in the sharp turns by the street like a race motorcycle on his way out toward the battle. I followed right behind him but had no chance of keeping up. The rest of us could only try to follow behind to the best of our ability. The ground was difficult to pass over as the soft soil gave way. Boots sank down and sucked into the mud. Thigh muscles tightened and ached with each running step. Looking out over the field and the front line forward of it, I could see Långström was already far ahead.

Soon the snapping sounds of bullets whipped wildly through the air as they passed above the field—and our heads. They seemed to come from everywhere in the enemy's general direction. Between us and the enemy lay the easternmost street, two dense lines of small houses on each side of the road. The front line and Azov's main positions protected us from immediate danger during our advance. The bullets we could hear were the ones passing high, over the houses. I thought that if there were so many bullets going over the rooftops, then how many weren't hitting the walls on the other side?

The sounds of the bullets' fiery whiplash gave renewed strength to my legs. The pain caused by cramping muscles disappeared, and we soon arrived at the first row of houses, making our way through the courtyards out to the main street. We were met by complete chaos; the infantry in the first line had also taken artillery fire just like us. In the smoke from craters and shattered houses, riflemen and machine gunners poured fire against the enemy from between buildings and behind rubble. RPG gunners loaded their launchers and threw themselves into the hail of bullets where they halted, standing up straight, completely exposed while carefully aiming their tubes toward the Russians in front of them. The concussive blasts from their rockets then threw up large clouds of dust behind and around them. Pressure waves bounced between buildings down along the street. The very moment they released their anti-tank rockets toward the enemy, they then threw themselves back behind cover to reload and then repeat.

Our adrenaline-fueled leader was trying to find a Ukrainian counterpart who could give us a position to defend.

"Stay here! Stay in reserve! Until we see where you are needed" the Ukrainians told us.

We remained in a rearward position, reluctantly, impatiently, and eagerly awaiting to be released the last few meters forward and engage the enemy.

There was plenty of small arms ammunition, RPG rounds, and single-shot anti-tank rockets prepared, lined up along the house wall. The enemy needed to attack over a completely open field. Although their artillery had softened us all up well, it did not appear that we had taken too heavy losses up until now. Långström began looking around, counting the members, finding the numbers falling short.

"Has anyone seen Greek? And Tjeck? Where the hell are they?"

We became aware that the RPG gunners fired at an enemy armored car which was driving back and forth in front of the positions, from north to south, parallel to our line.

"Longer distance!" someone shouted. "Aim higher, aim farther away!"

One of the RPG gunners launched another rocket against the enemy vehicle.

For lack of something else to do, I wasted my time with another cigarette. The bullets from the Russians were still flickering over us, and I tried to orient myself around exactly what was going on in the surroundings. A machine gunner with a scoped PKM had taken cover behind the houses and was struggling, trying to unfuck what appeared to be a jammed gun. He had a cartridge belt wrapped around his helmet, and I did not understand how one would want to make an already heavy helmet even heavier that way. I realized it wasn't the oddest thing I've ever seen, and there was no use in trying to understand it. Why the Russian armored car drove back and forth in front of the position, on the other hand, was something I should be able to figure out. Soon I put one and two together. The Ukrainians shouted that it was driving at a long distance, but still close enough for them to fire AT weapons at it—all of them still missing their intended target, though, of course.

This was by design. It was something the Russians were doing on purpose.

The armored car stayed just outside distance so that the RPG gunners and the rest of the line would lock their focus on it. They were drawing our attention and fire as a diversion maneuver!

The chaos quickly worsened. The Ukrainians began to withdraw from the front line at the same time as a minibus skidded onto the road.

"Help load ammunition and weapons!" shouted one of the Ukrainians. "We are retreating!"

"Retreating?" Långström asked indignantly. "Where the hell are you going!"

We all threw our arms out in protest and loudly questioned the tactics. We had just arrived; the Russian was not even close. Why would the Ukrainians retreat already? We hadn't even gotten into the fight yet. We wanted to stay.

"The Russians have broken through the line in the north!" the Ukrainian replied. "They are coming in with tanks! We have no weapons to beat them here. The line is retreating to the Mayak! Fast! Load up the ammo into the bus!"

North side

Bear had gone back to Mariupol to help with logistics, leaving Chris and Inkasator at the northern part of the village, along with Abdullah and other Ukrainian defenders. Chris had told the Ukrainians the day before that they needed to prepare the entrance road to Shyrokyne with mines, though no one listened. Their intention was to continue forward toward Novoazovsk. Planting mines on the road would slow everything down. The offensive would continue without delay.

Now it was morning. After the smoke from the initial artillery rain settled over the north side, enemy armor appeared on the road leading into the village. The Russian main battle tanks lead the spearhead at breakneck speed, in a column along the road. Behind these followed the armored infantry fighting vehicles, BMPs, with the mechanized infantry riding on top of them.

Chris was in the two-story building closest to the northern entrance to the village and immediately realized that the Russian tanks would hit his location first. They would enter the road he yesterday, in vain, implored the Ukrainians to mine. Now it was too late. He turned to make eye contact with Abdullah but only saw his back as he disappeared out the doorway, down the stairs to the ground floor. Chris quickly assembled his equipment and began to head down to ground level himself, only to be met by the rear of a Russian tank outside the door as he was on his way out. The Russians were already inside the village, cutting him off. He threw aside all the equipment he could spare—if he would make it out of this, he needed to move light and fast.

After ditching his backpack and helmet, he looked at his two rifles—the night vision scope-equipped AK-74, and the sniper rifle. He could only bring one of them. There were many AKs, but the sniper rifle was valuable and unique, so he grabbed it and left. He did not have time to get far before he realized his mistake. There were Russians everywhere, on the streets and between the houses, and he had brought a semi-automatic rifle with a large scope and low magazine capacity—useless for close quarters combat. As he snuck along a house wall, he was surprised by a Russian coming from the front. The Russian was just as surprised, suddenly being only a few meters away from what he perceived as a Ukrainian fighter. Chris threw the powerful sniper rifle to his hip and pulled the trigger. The Russian fell to the ground, and Chris continued his retreat, now even faster than before.

Inkasator had made it a bit further, but he, too, was surprised by a lone Russian. This time, the Russian was faster and put one bullet through his Ukrainian adversary's leg, with a second ricocheting into his cheek, stopping when hitting Inkasator's teeth. The old Chernobyl liquidator handled these injuries like any other of all the rough blows during his life before he spat the bullet out his mouth and ended his adversary. With the Russian dead on the ground, the old man continued onward just like Chris, on his individual retreat west toward the Mayak, albeit a bit slower than before.

The Russian tanks plowed into the village and swung one by one down along the smaller streets as they passed them. Each tank was then followed up by infantry which systematically began clearing the area. Inkasator and Chris were not the only defenders left as the Russians broke through the front line. There were many more Azov fighters now retreating through the rear of the Russian attackers. Chaos ensued as the Ukrainians pushed to withdraw toward the Mayak and the second line of defense. The Russian attack had gone too fast, and it broke down by itself already, only in the initial stage. One of the new Ukrainian BTR-3 fighting vehicles in the village had been hit by artillery before being completely knocked out by one of the enemy's tanks. Another Russian tank tried to continue forward through the village, not knowing that there was another Ukrainian vehicle waiting for it: one of the truck-mounted anti-aircraft guns.

It was an uneven struggle. The Ukrainian truck AA gun had no chance of destroying the enemy tank, but it did hold the moment of surprise to its advantage. The crew waited until the enemy tank was close enough, then the truck reversed viciously into the intersection in front of it with its rear-mounted twin autocannons loaded, unlocked and ready-made. The barrels opened up a hailstorm of fire straight into the front of the tank at close range. The cartridge belts were emptied at breakneck speed as thick 23mm steel cases sprayed out of the guns. The front of the tank was blasted in an instant, all its many glass eyes blinded. Unable to see anything further, the tank reversed slowly in retreat while the Ukrainians continued their withdrawal.

Another Russian tank, probably the command vehicle—a dark green T-72B1 with white propaganda text painted on the sides—didn't turn into Shyrokyne's small roads. It stayed on the main road leading further westwards, north of and past the village itself. As the rest of the Russian armored force behind it had entered into the village, it continued alone along the main road without meeting any resistance. The second Ukrainian line of defense on the heights west of the village and at Mayak was only just now being manned, but no one there really knew what was going on in the village. The tank made it far onto the road before the Ukrainians finally understood what was going on. Armored personnel carriers from the third line opened fire against it with ATGMs but scored no direct hits, which gave the Russian tank time to make a U-turn and pull back.

The T-72's large turret turned backwards as the chassis swung around. It then rumbled back toward the battle, though this time not along the main road, but in through the village. It turned into the same smaller roads the Ukrainians from the first line used to retreat. Ukrainians who saw it initially mistook it for one of their own tanks, because it now came from the west, but were puzzled as to why its turret was pointing backwards. It was only when it met two retreating Ukrainian Geländewagen jeeps that it became clear to everyone what it was actually doing. The first of the jeeps just managed to turn off the road, but the one that followed behind was crushed under the caterpillar tracks of the 48-ton heavy main battle tank. Continuing on through the village after this, the crew's luck had finally run out, as an ATGM from the second line finally caught up with it. The T-72 suffered a hit, causing the ammunition inside the belly under the turret to catch fire. The tank exploded in a huge, violent fireball. Pieces of tracks and support wheels were thrown around to the sides like confetti, and the large, heavy turret was blown up toward the sky.

South side

As the Ukrainians withdrew, we were forced to follow them from the front line, without even firing a shot. We ran back across the field to reconnect with the others from Recon waiting at our house, also in hopes of finding Greek and Tjeck. We ran along the road and in through the gates we had just left. It was the first time I actually got a chance to see the destruction we experienced during the morning from the outside. The steel walls themselves outside were mostly gone, simply littered across the yard. The bathtub that stood outside by the driveway was the next thing I noticed: it had been thrown around and lay upside down far from where it once stood. The nice hammock I had sunbathed in the day before was torn to pieces and lay scattered across. The garage door looked as ruined from the outside as the inside. The worst damage was to the roof over where Långström and I had been sitting. It was almost completely gone. The red panels were distorted by the explosions and lay scattered everywhere. Where they would have sat there, only a broken plank skeleton remained.

We entered the building and looked for our comrades, but our shouts received no answer. Everyone was gone.

"What the fuck? They just left?" Långström wondered aloud, then began yelling, "Hey! Greek! You fucking black asshole, where the hell are you?"

No one answered. He peeked into the garage to see if they were still sitting in the basement, but it, too, was empty.

"Fuck, bro, look at this shit! They even left the car! Absolutely unbelievable!" he said, noticing the pickup still there. "That's a good thing, though. We need a car of our own since Leo crashed the previous one. Here it is. We'll take this one with us. Now quickly, load all personal equipment into the pickup! If there's ammunition or anything else of value, we'll take that, as well! Quickly now, before the Russians get here!"

I hurried into our bedroom to gather my belongings. Above all, I would bring the socks I hung on the back of the chair the night before. I had not forgotten them, but rather had them in the back of my mind ever since we hurried across the field after the artillery stopped. The boots were still not completely dry, and I would bring my expensive socks with me, so I could change as soon as we had a quiet minute. I pulled together the backpack's zipper and turned toward the chair where I had hung my socks. It was gone. Where it stood, there was only loose brick, dust, and broken pieces of wood. Långström came into the room behind me.

"Fuck, man, they really redecorated the interior in here properly," he said, looking around at the destruction.

"I cannot find my fucking socks," I muttered, annoyed.

"Socks?" Långström said in surprise. "Who the fuck cares about your fucking socks? We have to get out of here now, you get that?"

"Dude, those were expensive and unused fucking socks!" I replied angrily, as my commander seemed to have serious problems appreciating their importance to the war effort.

"Fuck your fucking socks! We have to go! Now!"

I cursed the Russian sock marauders while snatching the last candles from the broken kitchen drawers on our way to the garage. It was unclear where we would spend the night now. If it wasn't Valhalla, it would in all probability be in some damn cold and dark place. Each tiny source of heat would count.

We threw everything on the pickup truck, backed out, and began our own unified retreat. I climbed the flatbed and made my RPK ready. We only knew that the Russians were in the village but had no idea where exactly. It seemed quiet enough, but I had a hard time deciding if it was because the fighting had stopped, or if I had lost my hearing during the morning mortar barrage. We took the shortest road, straight west. At the northern edge of the village, I could see a large, black cloud of smoke rising in the sky with something big flying through it. It looked like a tank turret being hurled into the air as a result of a K-kill. The strange thing was I couldn't hear the sound of it, despite it being only about a thousand meters away.

CHAPTER 34: A CHAOTIC RETREAT

The sun was shining outside and its rays were warm and lovely. Fabien drove our new pickup while the rest of us kept an eye out of the windows or from the flatbed. I stood on the platform, constantly focused on every corner of each house we passed, still not knowing anything of how deep into the village our enemies had advanced. The pickup rolled slowly, just so that the wind blew nicely across my face. It was only when we started coming out of the area bombed by the artillery that I realized how violent it must have been. Behind us, we left a thick mist of gray dust mixed with black smoke from burning buildings. The smoke was so compact it was impossible to see through it.

Along the way, we passed by Ukrainian soldiers still retreating on foot—rarely in groups or even in pairs, but more often one by one. We passed a tired soldier jogging westward. He carried a beige canvas backpack full of ready-made RPG rockets. I leaned from the flatbed, stretching out my hand toward him, whereupon he took off his pack and handed it to me. I threw it on the flatbed floor, giving the soldier a little extra power to run faster. Along the way, we continued to pick up more equipment the retreating soldiers were carrying. We even picked up two soldiers who simply could no longer run in their thick, warm winter jackets.

While the pickup was standing still to pick up the laggards, I sensed both the sound of gunfire and rough engines north of us. However, it was difficult to determine the distance to everything. I understood that there were tanks around, but I did not know how close they were to us. They weren't behind us, as we'd already be dead in that case, but did they move through the street north of us? In that case, they were only one hundred fifty or two hundred fifty meters away.

All around us and our pickup, soldiers continued to retreat west. Only a lone figure moved in the opposite direction in front of us. I released my gaze from the direction of the suspicious engine noises and put it forward toward the figure. It was a lone woman, approaching us at a slow pace. She was slim and wearing some kind of short dress. The color was impossible to tell, as she was covered in dirt from head to toe. Her dress, face, and hair were all gray with dust. At the bottom along one leg was a knee sock which had slipped down to her ankle. Her hair stood out in all directions, as if she were a cartoon character who had stuck a fork into an electrical outlet. In one hand, she carried a tin bucket. The woman passed our pickup as my eyes were locked on her. She did not meet them; she never gave us or our pickup any attention. Her gaze lay locked on the horizon, and her face was completely expressionless. Soon she disappeared behind us, entering into the great cloud of smoke we had just left.

We continued on until we came to the school. The largest building in the village in size and appearance, almost identical to the one in Lebedynske. In front of us, we could see the hills and the white Mayak. The road in front of us was divided into a T-junction. It went either straight north or south. The northern road was the fastest out of the village, but then there was a great risk that we would end up right in front of the enemy tanks. The south should be a safer bet, but we lacked a map, and none of us were familiar enough with the village to know where it led. With luck, it would lead us toward the beach, from which we could then move on and regroup at the Mayak. However, it could just as well prove to be a dead end.

Långström decided that we should try south, and the journey continued. The infantry, which had previously been crowded around us, had all already left the village on foot by now. They were not dependent on roads like us in our vehicle. We seemed to be the last ones left in Shyrokyne. We continued straight south until we came to a bridge. In front of us, we could glimpse the Mayak and the eastern border of the prowling landscape with its gradual elevation changes, next to which the village's small Orthodox chapel was located. It was a small but tall white building with a green roof and golden crosses. In front of us, the road ended with a river blocking it. The river was not waterlogged, but the ground was extremely muddy, uneven, and full of trash and debris. The small wooden bridge crossed the river—good enough to run over on foot but far too small and narrow for our pickup.

We resumed discussing how to proceed. The river was far too risky to try to cross. We all had a fresh memory of how badly we got stuck in the fields outside Lebedynske a few nights earlier, and this passage was ten times worse. Långström, however, refused to leave the vehicle behind.

"This is my car now, you understand? We will bring it back with us. I saw us pass a small road west. We'll back up and try there."

We did as Långström decided. We returned north, turned left, and immediately saw the end of the village. We stopped and dismounted. Behind us was a building with a high brick wall around it. Inside the steel gate, we would be able to park the car under cover while we continued to think about what to do. We tore at the gate without success, and the brick wall was solid. Unable to get through it, we were forced to leave the car completely in the open by the road. The road we intended to follow, just like in the previous T-junction, went straight north toward the main road, connecting the Mayak and Shyrokyne. This was far too risky. However, the field in front of us looked good enough to pass. It seemed possible to drive straight across, and at a fairly good speed. If the Russians had positions covering the field, however, it would be suicide. The battle appeared to have subsided, and the village was quiet once more, but we couldn't be sure. Radio contact with the company command at Mayak had broken down, leaving us unable to acquire answers from any of them, as well.

"Richter and I will run up there and talk to them eye to eye," said Långström. "The rest of you stay here and wait. Guard the car. Once we've gotten a hold of the commanders over there, I'll come back down here. Then we'll see how we proceed."

It was almost half a kilometer to Mayak, steep uphill and in completely open terrain. They were in for a heavy and risky run.

One of the Ukrainian laggards we picked up spoke little English and did not like what he heard.

"Come on, we cannot stay here! Let's just leave the car and go, please!"

Långström promptly asked him to shut up and stop being a cowardly little pussy before he and Richter started running across the field, toward the hills of Mayak. The rest of us took positions around and began waiting for their return.

The English-speaking laggard didn't find his courage improved by the long wait.

"Come on, guys, please! We can't just stay here and die!"

He never seemed to cease his complaining, and he completely ignored the fact that no one else wanted to listen to his annoying voice. The fact that engine noises were audible again in the surroundings did not calm him down, either. The shooting had stopped, but the Russians were definitely nearby. His whining dragged on, and I approached the point where I was about to strike him in the mouth with the RPK. I was interrupted by Cuix, who whistled gently and pointed to the field in front of us. There came two figures toward us. Richter's shiny helmet gleamed a bit in the sunshine. They did not run but walked calmly toward us, as if they had all the time in the world.

For my part, I had begun to approach the end of my patience, due to both the engine noise and the nervous Ukrainian next to me. I stepped out on the road, stretching my right arm into the air. It had been a long time since Långström had been in the Finnish army, but I was pretty sure he would remember the hand signals we had both learned there during our late teenage years. I started to lower and raise my fist up and down, signaling the situation required double speed. Långström immediately took the command and began running toward us.

"The Russian has stopped inside the eastern part of the village," he said. "They're keeping an eye on him from the Mayak. The coast is clear. We can drive straight across the field."

I was a little hesitant, but I was also aware that my hearing still hadn't fully recovered from the loud artillery barrages. It was entirely possible that I had simply imagined those engine sounds. Atop the Mayak, they had a great view over the village, and if it looked clear to them, it probably was.

We embarked the pickup again. Still uneasy about the situation, I decided to use my seax knife to cut the sling attached to my RPK. It was a clumsy three-point strap that tended to get tangled, but because it was screwed into the rifle butt, the only way to remove it was to simply slice it off. If we got into trouble going across the field, I didn't want to risk my gun getting stuck on something. I prepared my magazine pockets to be able to reload as quickly as possible, in case we were forced to shoot our way out during the final step of our journey toward the Mayak. Fabien fired up the engine and we started to roll, but after only a few dozen meters the pickup halted again. We were still standing on the road, just in front of the curve from which we were to turn out. Cuix got out of the car and told us to keep an eye out while he investigated something.

"There's some shit over road, maybe IED. I must have look at it," he said with his terrible French accent.

He walked carefully a few meters in front of the vehicle, where he stopped and appeared to carefully investigate something. I tried to focus my gaze but could not see anything at all. Cuix followed a straight line across the road to the left side, then to the right. After only a short while, he returned.

"I'm sorry I see wrong. Just copper wire, from anti-tank missile," he said as he stepped into the back seat again. "I had to see to be sure. Make sure no trip wire and bang! We all dead."

Those anti-tank wire-guided missiles left thin threads—incredibly thin, not even a millimeter thick. I struggled to comprehend how he could see it, let alone from a moving vehicle. The laughing little Frenchman had the eyes of an eagle.

The engine started again, and now we would finally be leaving this godforsaken village. It was a feeling of relief, especially as we had been standing still in the same place for a long time now. Unfortunately, our relief was short-lived, as we couldn't shake the feeling something was seriously wrong. Our concern only grew stronger as another engine noise, not from our pickup, gradually got louder, coming from somewhere around us.

We had come out to the road and were still rolling forward. On the right side, the road continued up past a row of buildings. The other engine noise increased even further—a loud rattling sound. Suddenly, it crashed out from between the buildings onto our right. A Russian BMP-2 rolled out and halted in the middle of the road in front of us. We were only about one hundred meters from the IFV, but at that very moment, it seemed a hell of a lot closer. It stood with its side toward us, with its autocannon aimed high forward, toward Mayak. We had plenty of RPG-7 rockets on the flatbed but no launcher to fire them. The only anti-tank weapon we had, an RPG-26, sat in the back seat with Cuix.

Sound discipline went out the window. As I stood on the flatbed facing forward, I hit the flexing, thin metal roof with my left hand as hard as I could.

"BMP! BMP! Back up! Back up!"

The others saw the enemy vehicle. Fabien threw into reverse gear and slammed the pedal to the floor as the prowling, hilly landscape near Mayak opened up with all the firepower available. The pickup jerked and accelerated quickly backwards. The engine screamed like a stuck pig. The brick wall behind us we never got through before was no match for the car as we crashed straight through it. Bricks flew as the wall collapsed, and the car came to an abrupt halt. We threw ourselves out and took defensive positions as the first anti-tank rockets exploded around the Russian fighting vehicle in front of us.

I checked our rear quickly to see who, if anyone, was covering it. Instantly, I saw the backs of the two Ukrainian soldiers we picked up who were now moving quickly along the street. Seeing those two covering our backs, I hurried up toward the rest of the squad, who had taken cover by some rubble. The BMP soon took a direct hit from an RPG and started reversing back to where it came from.

The other Russians, though, continued to press forward, and the fire from Mayak was deafening. The artillery fire started again, and projectiles from both sides screamed and roared through the air above us. The clatter of small arms fire across the valley never seemed to end. The quiet field we intended to cross was now engulfed in a storm of explosions and high-velocity lead. The Russian tanks pushed through the village once more, and Ukrainian ATGMs screamed through the air as they flew to greet them. After they detonated, their long copper wires slowly fell downwards, glimmering in the sun while being carried by the winds. One of them came down right next to me. It was so petite and thin that I could easily pull it apart with my hands as I grabbed it. I was amazed once again how Cuix had been able to see it before. His sharp gaze probably saved our lives, even though the thread hadn't belong to an IED as he had initially suspected. Had we not stopped to examine it, we would have been out in the middle of the field as the Russian BMP made its entrance and all hell broke loose.

"Well, that's it, no more alternatives left," Långström shouted as he waved us back to the car. "We have to take the road south. We will push pickup over the riverbed!"

Where the BMP almost ambushed us in Shyrokyne, between the buildings to the right.

I turned around to see where the two Ukrainians protecting our rear had gone. It took me a moment before I realized that they never intended to cover us, or interact with us at all. They had deserted as soon as the Russian BMP appeared in front of us. I cursed badly that I hadn't taken the chance to bash that one guy's face with my RPK when I had the chance.

We embarked the car, and Fabien forcefully drove it out of the rubble after crashing into the brick wall. We journeyed to the right and then down to the same place we judged impassable before. The total war continued behind us as we stepped out of the pickup, looking down over the deep mud gruel below us.

"Fuck it, why not just leave the pickup, man? No way we are getting it over this shit, anyway" I told Långström, who hissed right back at me.

"Now, you listen! The car is mine! We will bring it back with us and that's that! Come on now, let's start pushing!"

Unlike the rest of us, Långström did not intend to die on this day. He was going to create more havoc among the Russians for a long time to come, and the pickup was vital to help him do so.

Fabien engaged the four-wheel drive and rolled into the slump. The rest of us followed on foot by the sides. From the Mayak above in front of us, black and gray clouds of smoke rose from where the artillery was striking. Loud thuds followed by bright clouds of smoke rose from where RPGs were launched toward the Russians in the village. The machine guns and carbines sang incessantly. The display was as violent as it was beautiful. Back at our position below it, the view was a lot less grand and spectacular. The bottom of the river was deep, and the mud was wet. The smell of fire, smoke, and burnt gunpowder blended with that of rotten seaweed from the bottom of the deep riverbed. The riverbed was uneven, and the pickup seemed to be about to tip over.

I took the RPK in my left hand and hung off the side of the rear platform. With my arm over the side, supported by my armpit, I tried to throw my body down to create counterweight. We were all screaming and howling through the fire around us, but found the strength to get the car across. We managed to keep the pickup from rolling over on its roof by mere inches. Lebedynske's clay fields were nothing compared to this. Sweat flowed as we pushed on up and forward. Together with adrenaline-pumped muscles and Fabien's excellent performance as a driver, we came back to solid ground after a lot of hard work. My body felt like it was made from rubber as I dragged myself up on the flatbed once more, but we had done it. Now we needed only make the last stretch of the journey, and we still had our wheels.

The valley below the Mayak was still a wild shooting range. We diverted from this and instead passed by the inferno in the direction of the beach. Behind one of the buildings we'd left behind, I noticed that someone had discarded his Kevlar vest, probably too tired to carry it across the field. Considering the artillery on the heights, that had probably been a very unwise decision.

There was a road through the slopes just between the Mayak and the small white church chapel. However, the Ukrainians that had reached it, I thought pessimistically, would at this point in time open fire at anything moving toward them. We were the last ones out of the village by a significant margin. A vehicle that suddenly rolled in would attract attention and presumably fire, as well. Fabien drove at full speed, quickly closing in on the already short distance to our front line. Most vehicles came equipped with some kind of flag to signal their allegiance, and if ever we needed it, it was now—and fast. The flatbed shook as I frantically sifted through the equipment on the floor.

I soon found our flag wedged at the bottom of the soiled and muddy metal floor. It was covered with mud and had torn off in the middle, but it was still in good enough condition that the blue and yellow Ukrainian colors remained, albeit dimly. I quickly got my water bottle out to rinse off the worst of the filth. The pickup rocked as I continually poured water and brushed off the dirt. The blue-yellow stripes slowly begun to surface again. I knelt down, holding the ripped-apart flag above my head as high as I could, letting it flutter wildly in the wind. I held it for everyone to see that we were not Russians, but a Ukrainian vehicle on its way back—the last one out of Shyrokyne.

CHAPTER 35: THE MIGHTY MAYAK

The flag worked. Not a single bullet struck down toward us. We all came out of Shyrokyne unscathed. The pickup continued along the shoreline a good distance past the Mayak. We had only driven a few hundred meters, but the war already seemed like miles away—no destruction, no shooting. The leafless trees around the road swayed calmly and modestly in the winds, smelling fresh of sea air instead of fire and cordite. Everything around the small gravel road was the abode of peace. The pickup stopped at a brick building, secluded in a lightly wooded area where the road ended. We got out and enjoyed a short moment to take a breath. I lit my last cigarette and, checking that my camera still worked, shot a photo of Cuix and Långström as they hung their rifle slings over their shoulders.

"We'll leave the car," said Långström. "It's safe here. We'll go back for it later when we have the chance. Right now, we head back on foot. Back to the Mayak."

Just next to the building where we were parked was a long, white-colored stone staircase leading up to the heights. It connected to a road on top which would lead us straight back. I began heading toward the stairs as Cuix gripped my shoulder firmly.

"Stop! Stop! Look…look at steps. Look at trees around steps!" he hissed.

On both sides of the staircase were short red and white plastic ribbons tied to the branches from the trees around it. The wind blew lightly, and the thin ribbons swayed briskly in it.

"They red and white…is bad sign, I think. It is sure say that mines here. I think it is bad idea to walk here, yes? I would not want walk here, because I like to have feet."

Just like Cuix, I also liked having feet. We all did. The return road instead continued back along the same road we had just used to retreat. After about three

269

hundred meters, we reached a position where we were at the same level as the Mayak, just south of it. Another stone staircase ascended the hill, this one lacking the warning ribbons. No one wished to take unnecessary risks, so we instead moved a bit further east and there began to climb the steep hill. On the average day, it shouldn't really have been a difficult assignment, but we quickly found that the already porous earth had been dug up by the artillery. Like an angled and plowed field, for every two steps we took, we slid one step back down. We were soon down on all fours, crawling upwards. I tried to dig my rifle butt into the ground and pull myself up using the RPK as a cane. I jammed the RPK so hard into the ground that the charging handle hit against the safety lever, but my feet still kept sliding backwards.

Cuix and Långström after the escape from Shyrokyne.

With great effort, we were finally closing in on the top level. The Russians again seemed to have failed in their attempt to push through the village. The Ukrainian defense at the Mayak and on the heights around it had halted them once more, and they now changed their tactics. The enemy's strongest card lay not in his tanks or superior numbers—it lay in his overwhelmingly powerful artillery. If his ground forces could not take the Mayak, his mortars, howitzers, and rocket artillery would instead obliterate it together with everything else on top of the heights. The sounds of large explosions rumbled through the air. I peeked up over the top for a moment to have a look but was met only by a lunar landscape covered in smoke. I quickly covered back down again as large pieces of shrapnel came buzzing past like huge, blood-sucking horseflies. Luckily, no shelling seemed to be aimed at our position. There was nothing of interest at the south side of the hill. All the violence was being focused to destroy the Mayak.

The focus of the inferno remained only about a hundred meters away from us. We lacked all form of actual protection and cover, but luckily the angle of the terrain itself gave us shelter of sorts. The presence of the tall red water tower made from steel towering up above the slope to our left had me a bit unnerved, as a hit there could very well shower shrapnel down toward us. Not much to be done about it, I simply tried to forget about the danger it posed. At the same time, though, for that very reason, it became somehow impossible not to look at it. Just as the deer locks its eyes into the headlights of the car about to hit it, with every scream of an incoming shell about to land, I found my eyes drawn back toward the red steel tower above us.

Outside of the theoretical issue with the water tower, the only real practical problem was the ground shaking so goddamn badly. With every shell creating an individual quake at and around the Mayak, the sheer number of them made the ground shake constantly. The ground moved down the slope, as did we on top of it. While lying down on my belly, I kicked my boots into the ground behind me as hard as I could. Forming my hands into picks, I hacked them into the ground in front of me, cramping my fingers stiff, trying to hold on but to no avail. The only time the ground didn't shake was during the vibrations in between when it did, and we constantly found ourselves slipping down together with the top layer of soft soil. We were forced to crawl back up regularly, but as soon as we had, we simply began sliding back down again.

I turned around and looked down at the beach below us. Nearby, there was a relatively large complex of hostels and hotel buildings, all in varying price ranges. If the Russians were to attack along the beach, they would be able to move freely, as the Ukrainian defense was mainly focused around the village. Well, frankly, all of it was. We were the only unit remotely covering the south side, and we were in a very unfavorable position. There was in practice only two options to choose from: either remain fully exposed to enemy fire and take the fight, or climb the last piece of slope to fight from the hilltop instead, heading straight into the volcanic tornado of flame, smoke, and steel above in doing so.

Oddly enough, the beach and hotel buildings appeared completely deserted. Despite the Russian frontal attack having failed twice, they did not seem to think of simply flanking us. I had a hard time understanding them, but most of all I kept thinking about a simple cigarette. The weather was still nice, and the sun was warming. The view was pleasant over the calm, clear, blue water below. We couldn't really do anything else, anyway, and all in all it was the perfect setting for a relaxing smoke break. To be fair, I wasn't entirely out of smokes. In the back pocket of my helmet cover, I still retained some cigarillos: my two last Al Capones. These, however, I wanted to save for something a little bit calmer. A stage where we could relax completely. Relaxing to that degree right now remained difficult, as our large white brick HQ was being smashed to dust by the heavy Russian artillery just a hundred meters away.

And it was evident that this was indeed heavy artillery. As the heavy mortars screeched when dropping from the sky, the heavy howitzer rounds were something completely different. They sang much louder, and it wasn't so much that high-pitched screeching as it was more like a dark, toned roar. The ground still shook, and we kept sliding down and crawling back up the slope.

After some time, it began to quiet down again. The Russian artillery needed to cool down their (at this point) likely red-hot, glowing barrels. Långström stood up and waved to the rest of us.

"Right, people, let's go! Quickly, before they start again! Toward the Mayak!"

The thick dust and smoke had already began to settle as we began moving across the broken-up landscape. The ground looked like it had been attacked by hordes of hungry wild boars—craters everywhere, all of it turned upside down. The trees that remained on the hill were all badly hurt. While the trunks usually remained standing, the bark had been ripped apart, leaving large, bright wounds everywhere on their bodies. The ground crackled where we ran on top of all the twigs and branches the shrapnel and shockwaves had ripped off and thrown around. All sorts of rubbish lay scattered around. A badly broken-apart Kevlar helmet without an owner. Random car parts missing from their associated vehicles. Large pieces of heavy mortar tail fins sticking up from the ground, still smoking-hot, pouring out fumes. A steel hurricane had pulled over the hill and destroyed everything in its path. Surely, I thought, nothing could have survived here. Then we saw it: the mighty Mayak emerging from the dust, still standing there in front of us, a powerful white lighthouse, a beacon for us to gather. Its thick, strong walls had taken countless hits—everything from main battle tank cannons to heavy howitzers. But it still stood there just as steadfast as before.

We approached and entered through the same entrance we had used the previous morning. Outside next to it stood one of our pickups. The body had been torn apart by the artillery as it withdrew from the village, but it had escaped. Its seats were stained with fresh red blood just like the white paving stone outside the building. Kirt, the young company commander in charge of the defense, stood outside briefing his men during the respite we'd just been given. His eyes were bloodred, but his body language was confidence-inspiring. It was obvious he was leading his men well, a good military commander.

We tried to inform ourselves about the general situation and reconnect with our lost friends. Soon enough, a familiar profile emerged from the darkness inside the Mayak. It was the Greek, soon followed by Tjeck.

"Where the fuck have you losers been?" Långström asked indignantly.

The Greek's head slid back on his shoulders as he struck his arms out forward of him.

"What the fuck, bro! You just disappeared!" Långström answered his body language with taunting laughter.

"Admit it, you just got lost!" The Greek tried defending himself. "Everyone else pulled the fuck out. They said the Russians had broken through and that we were all fucked! What the hell should we have done? Stay there, two fucking guys, waiting for you to maybe come back? For all we knew, you were all dead, anyway. Obviously, we went with the others."

It sounded fair enough, but Långström wasn't having any of it.

"You got lost, you bum. That's the only thing that happened!"

The morning had led to heavy losses, the heaviest Azov ever had in such a short time. Bonn had been badly hit by shrapnel from some kind of shell, possibly from an enemy tank. Another soldier from Recon with a helmet cam recording watched on as he received a bandage around his bleeding throat.

Coughing blood out of his mouth, Bonn looked at him and said, "Post this to Facebook!"

Becks, our new drone operator-turned-infantryman, had also been seriously injured by artillery as he and another young soldier tried to retreat. While he was still alive, his comrade had not been as lucky, losing both his legs—and then his life. Kozak, the young Ukrainian with the fancy mustache, was also dead. I found it especially strange to hear about his death. I had previously held the strange feeling that he would somehow become an important person in this story. As it turned out, I wasn't entirely wrong about that. It was true that his name would never be forgotten, because after today it would be carved in stone.

The total number of casualties was unclear. The situation was still critical, and nobody was keeping proper records. Someone said that the company had already taken losses upwards of sixty men, out of which at least five to ten were already dead.

"Inkasator got shot in the leg, too," said the Greek.

"He got injured? Again?" I asked. "But he made it, right?"

"Yeah, man! You should have seen it. When he came back, there was already a fucking camera team in place. Fucking journalist rats! Fuck! Anyway, they just shoved a camera in his face and asked him what was going on. Inkasator even gave them a short interview, then he just said, 'Please excuse me, I need to go to that ambulance over there. I've been shot!' And then he just walked off! Like a fucking boss!"

The Greek laughed heartily and clasped his hands tightly in a simple but honest applause for the old man.

"Oh! Oh! And do you remember that fat guy we saw yesterday? That guy wearing two fucking plate carriers on his chest, who I told you got hit here last year?"

I remembered seeing such a man outside the Mayak before we were transported over to our now mortared-down villa inside the village.

"Yeah, I recall. What about him?"

"Fuck, man! He got hit again!" the Greek yelled out loudly, looking as if we would fall over from laughter.

We stood upstairs and looked out through a window opening toward the village. Dark smoke from fires rose from many burning buildings. The artillery had calmed down, but small arms fire continued, machine guns singing constantly. A lone Ukrainian T-64 tank which had come from the rear lines to support us rolled slowly toward the village center to probe the enemy, but was soon forced to back off again.

"What about Bear and Chris then? Have you seen them?" I asked, and the Greek's smile sank slightly.

"Well, Bear went back to Mariupol to help with some logistics bullshit before the Russian attacked, so I'm sure he's fine. As far as Chris is concerned, nobody knows. No one has seen him since the Russians attacked. For all we know, he's missing in action."

"Tell it like it is, you moron," Långström interrupted. "He's obviously dead. Fuck him now—he doesn't matter anymore."

Långström went down the stairs toward ground level. As harsh as it sounded, he was probably right. A lot of people were dead, and it was highly likely that Chris was simply one of them. We had far more pressing issues to think about than that. Despite that, though, the Greek and I looked at each other, still doubtful about it.

"I think he's alive," I said, and the Greek nodded in agreement. "If any us could survive that, it would be him."

Those of us around regrouped outside at the two short stone steps leading up to the narrow asphalt road. It ran parallel outside the Mayak, down toward the easternmost stone staircase by the beach we had previously avoided. Unlike inside the belly of the Mayak, the outside was warm, and we sat down to rest for a moment in the sun. It was almost quiet again—no shooting, no artillery. Everything was calm. I took off my helmet and got the thin plastic box of cigarillos out of its back pocket. Opening it, I grasped one of the sticks and held it up toward the others.

"I'm almost out of smokes now. This is my second last. But let's share it, now that we finally have a moment of peace to do so."

I lit the cigarillo with my old Finnish Jaeger-engraved Zippo lighter and began passing it around for those who wanted a puff. Behind us came a sudden croaking sound from the direction of the beach. An old Soviet UAZ minibus that had been converted to an ambulance was standing still on the narrow road. The driver got out and shouted something in our direction in Ukrainian—a call for attention we completely ignored. We didn't know what he wanted, or if it was even us he was calling. Above all, we didn't really care. We were on a smoke break.

The driver switched to English: "Come on guys! Please! Help!" he shouted in a strong yet resigned voice while throwing his arms downwards. He looked like a man inches away from giving up on everything.

Our smoke break came to an end before it had barely begun. Ignoring such a desperate cry for help was simply bad manners. Besides, it was impossible to relax with an adult man crying toward us down the street, anyway, so we hurried over to help him.

Right next to the narrow road ran a thick gas pipeline on the ground. The shelling had torn it apart. The twisted steel remains, still bolted firmly to the ground, had now torn into the thin metal side of the ambulance. After ripping a long hole straight into the side belly of the vehicle like a saw through a canned mackerel, the pipe had then wedged itself into the chassis. Thick blood ran down through the long cut down to the ground. The ambulance had come to a halt, stuck to the metal pipe.

"There are many people still hurt," cried the driver. "I have to go help them! Please help me! There are lots of them. I must help them!"

I grabbed the thick, stiff metal pipe as the others pushed the ambulance, trying to wiggle and force it free while the driver hit the gas in reverse. The wheels skidded and rubber burned. We thrusted like animals, roaring and screaming at the top of our lungs, trying to grant our muscles that last extra bit of power needed.

"Perkele!" I cursed loudly in Finnish as I reached for the last ounce of strength in my body.

We finally got the ambulance unstuck through sheer force and violence. The driver skidded off northward across the bumpy and bombed-out asphalt.

"Well then," I said while catching my breath. "About that smoke break then..."

Just as I finished the sentence, I noticed my voice being overpowered by an increasingly loud, sparking noise above us. It was the Grads—the Russian rocket artillery—coming in fast.

"Grad! Take cover!"

We threw ourselves down right where we were standing as the heavy rockets passed over us. They overshot the Mayak and landed just west our position. Between the narrow trees next to us shone large, golden-yellow explosions followed by black and gray clouds of smoke. The branches still standing fluttered as the pressure created bright white waves on top of the ground, heading toward and hitting us with a powerful and loud force. The ground shook violently. The Russian artillery barrage resumed once more.

"Were heading back to the southern side to make sure they don't flank us!" Långström shouted. "There's trenches for cover there. Follow me!"

The artillery increased in strength. We hurried back toward the slope where the trenches would be, only to find nothing of the sorts. Near the edge were some shallow pits, not even sufficiently deep to lie down in. Either the trenches and foxholes had never been finished, or all the barrages had filled them back up again. We were simply forced to hit the dirt. I cursed myself for not carrying a shovel like I had done for years during peacetime training. I had gotten rid of it after seeing how no one who ever did carry one ever used it. For a moment, I considered taking off my helmet to use it as a makeshift entrenching tool, but the whistling shrapnel around made me reconsider it, as I preferred a non-punctured skull. I instead began digging with my bare hands, but for each shell that landed, the small hole I made collapsed and filled itself back in.

"Gah, fuck this shit! We can't stay here," Långström stated wisely, an insight to which no one else objected.

We crawled down on the other side of the edge again, exactly the same point as before—just as safe from the artillery but still equally vulnerable to an attack from the beach. Långström lay on his back, looking down at the hotel buildings below us.

"There is no point in lying here like a bunch of retards," he said. "Let's go down toward the beach and capture those hotels instead. Come on now, follow me!"

CHAPTER 36: A HOTEL WITH A SEA VIEW

We hurried down the hill where the porous soil made our advance extra quick. Our boots sank into the sloping ground, and we quickly descended several meters for each step, sliding down toward the beach. The hotel complex was rectangular in shape, about three hundred fifty meters long along the beach and one hundred meters wide. A long, straight street ran through it from west to east, surrounded by the densely packed hotel buildings.

Although the area looked quiet, we couldn't be sure that the Russians weren't already inside themselves. We therefore moved carefully through the streets, looking for a suitable building to provide defensive positions. The large street which led through the entire area seemed clear, but moving through it would leave us exposed from all sides. It also offered basically no place to take cover from small arms fire, should we end up in an engagement. Långström instead led the group further toward the beach around the complex. From there, we could advance with good visibility toward the enemy side with our right flank secured by the sea.

Most buildings were one or two stories high, with the exception of a large hotel boasting a third. The building was white and did not look very stable, but its location was good enough. Located in roughly the middle of all the hotels, its west-facing entrance was protected from the enemy side. Decently sized balconies facing the beach gave a decent line of sight toward the east.

Långström grabbed my shoulder and pointed at the corner of a high steel fence next to the beach.

"Take position here. Cover the beach while we move up to the hotel."

I unfolded the bipod and laid down while the group continued onto the smaller street left of me, toward the closed gated entrance. The Greek was the last in line. He kneeled down next to me where I had my eyes and RPK focused in the enemy direction.

"Needless to say, if someone comes from that direction, you'll just drop them immediately without asking, right?" he asked, to which I nodded without releasing my eastern-facing gaze.

"Yes. Obviously."

The entrance door was locked, but a snappy blow from Långström's muzzle brake against the window next to it opened up an alternative way inside. The floor inside was clean, lacking dirt marks from muddy boots or even gravel. No Russians had been in here. I rejoined the group as we searched the floors. Most of the doors to the hotel rooms were locked, but we found some keys at the reception. The keys we found didn't open every door in the hotel, but among the keys we did find, luckily enough, were the ones to the most important rooms—the ones with balconies on the top floor.

"This is a hotel," I said aloud to myself. The rest of the group glared at me.

"Yes?" asked the Greek. "Is there anything further obvious you feel the need to inform us about? The color of the wallpaper? The size of your cock? Or maybe even the general failure of liberal Western democracy? I, for one, a man interested in the world around me, am here, ever eager to learn!"

"It's a pretty big hotel," I continued without responding to his Hellenic nonsense. "A hotel like this must have had many guests in the summer. In a hotel like this, there must be a bar somewhere."

The group had slowed down a bit after the stressful morning but now picked up the pace again. Fabien, in particular, lit up at the thought of an open liquor cabinet. Långström, Richter, and I looked for enemies while the rest of the group searched high and low for the promise of adult beverages.

The balcony consisted only of a rounded concrete floor with a thinly spread-out steel fence surrounding it. There was nothing around that could stop bullets. We were at least able to create some concealment on the eastern side with some furniture we'd thrown white curtains over, though. Everything in front of us was calm, the only movement being the large fires from Shyrokyne, which loomed over and behind the gentle hills in front of us. Narrow streaks of black smoke rose upward, where they spread like a wide blanket at low altitude above the village. North of us, the heavy artillery fire still rained down onto the main Ukrainian positions.

Tjeck was in a nervous state, which Långström took notice of.

"Hey, Tjeck. We need to get some food, ammo, and water down here. Why don't you go and get some?"

"Where from?" Tjeck asked with a stammering voice.

"The Mayak, of course. They've got lots of supplies there."

"What! I go back to Mayak alone?"

"Yes, of course. What's the problem?"

Cuix showed up next them and turned over to Tjeck: "It's okay, friend, I'll come with you."

Both then ran back out again, making their way back up toward the Mayak while us others stayed in our positions down by the hotel.

"Wow! Are they wasting the Mayak or what?" Långström laughed behind me as I kept watch to the east from the balcony. Suddenly his gaze was drawn to the sea.

"Fuck! Did you see that?"

I turned my gaze to the south as I was struck by a powerful shockwave.

"Something just exploded out there," Långström said excitedly. "Fuck, man, it was fucking huge!"

The shoreline was covered by thick pieces of broken-up ice which had floated ashore, but the open sea began again just a short distance past. A little further out, I could also glimpse what looked like a cloud of fog slowly sinking down to the water's surface. Soon a new explosion came, and it was as Långström described. I had never seen anything like it in real life. A huge tower of water rose toward the sky, as if a naval mine had detonated. The pressure wave quickly traveled above the water surface and soon hit us again with a huge bang, making our very skin vibrate.

Cuix on the hillside next to the Mayak with the hotels in the distance, icy sea behind them.

There were most certainly naval mines in the water outside, but that they suddenly began detonating one by one seemed suspicious. The explosions came one after the other, at long but seemingly regular intervals. I started counting the seconds between explosions and soon timed it to exactly one minute each. I only knew of two kinds of weapons which would be capable of creating such an impressive amount of violence in a single shot. One was the "Pion," a tracked 203mm howitzer, which was disqualified here by its two-minute reload time. The even thicker 240mm mortar called "Tyulpan," or Tulip, on the other hand could be reloaded in approximately one minute. Ukraine lacked these weapons, but it was known that the Russian Federation operated a rare few of them, less than a dozen. I had also seen the OSCE reporting that one of these self-propelled siege monsters had recently been sighted outside of Donetsk. Judging from the explosive display in front of our eyes, the Tyulpan had since moved south since it was first spotted, now visiting its powerful ways of destruction upon us. If there was anything that the seemingly indestructible Mayak could not survive, it would be a direct hit from a Tyulpan. Luckily for us, though, its aim seemed to be completely off. Shell after shell came in, but they all landed in the water outside.

"Ha! Fuck me, look at that shit!" Långström laughed and lowered his binoculars from his eyes. "Look at the seagulls! They're really having a fucking blast right now!"

I redirected the RPK from the enemy direction to look out toward the sea through my scope. The pressure from the huge 240mm shells must have killed indescribable amounts of fish, all coming to the surface as easy pickings. Huge swarms of hungry seabirds now appeared over the set table, diving into it in a display of wild gluttony. It was a dangerous game, however, as the shells still kept coming. If each new impact tore a hundred seagulls to shreds, though, there were always a hundred new ones to take their place as soon as the soiled and dirty gray-black water fell back down again. The gruesome imagery and foul stench from a sea of exploded fish guts outweighed any self-preservation these rats of the skies might have had.

Up at the Mayak, Cuix and Tjeck were collecting the items they were to bring back to us. Tjeck, not being a smoker himself, also took the opportunity to look for more cigarettes. On his way down to the basement, he met a Ukrainian friend.

"Hey, man, do you know if there are any cigarettes anywhere? Our guys are out down by the hotel buildings."

The Ukrainian immediately went back down into the dimly lit large basement where most of the other defenders were covering from the artillery.

"Hey! Guys! We need to collect cigarettes for the foreigner guys holding the right flank!" he called out.

The Ukrainian soldiers immediately began picking up their packs, surrendering one or two smokes each into the collection.

With everything collected, they were ready to return back down to us. Cuix exited the safety of the Mayak first with Tjeck, keeping a short distance behind. As he left the building, though, a large shell landed just outside. Tjeck jumped back inside, seeking cover behind the thick walls. As the dust settled, he cautiously peaked back outside, seeing Cuix standing there waving to him.

"It's okay, friend, come on now! Let's go!"

They then began making their journey back down again, covering between shell impacts and small arms fire, and somehow they made it. The food was Ukrainian army rations: plastic-wrapped packages of the usual stuff, namely canned buckwheat and crackers that tasted like paper. But that wasn't all. Out from the balcony, I heard Cuix whistling from inside the room. Even though all the hard liquor seemed to have been looted (probably by the hotel staff themselves before they closed up), his fine-tuned French nose had successfully sniffed out wine. In his hand he held a single bottle of white wine he managed to scavenge from downstairs. The grape-based drink had never been a passion of mine, and as a Frenchman, Cuix seemed hesitant about the bottle's Georgian origins. Alas, the entire group was thirsty enough that we quickly downed the bottle between the lot of us. Fabien probably had the biggest sips, and his face was soon adorned by a red, drunken nose.

We took turns guarding the balcony.

"Hey, shouldn't we take the opportunity for a group photo?" I asked.

Långström, being on guard at the moment, strongly refused: "Take pictures all you want. I'm keeping an eye on the beach."

"Come on, man, there's nothing happening over there," I said while setting up the camera on top of a bench, pointing it toward the other side of the room. "I've got a timer on the camera, then I'll get a picture of all of us. It will only be a few seconds."

"Just take your fucking photo already. I'm not going to be in that shit, I told you."

Carolus (left) and Cuix (right) hold up the bottle of Georgian wine.

It was obvious Långström was refusing not because of military strategic reasons, but simply out of the principle of not having his face pictured in general. There was no way to argue with him about it. After I took the group photo without Långström in it, it was my turn to keep a lookout. Unlike the Mayak boys, we had ourselves been spared most of the artillery, but soon it came creeping up on our porch, as well. Outside on the beach and the sand below, shells from light mortars began impacting. Initially it wasn't so bad; they were still at a safe distance. Compared to all the other shelling we had received recently, they frankly didn't seem like much as they popped around in the sand below. They were, however, gradually coming closer. As the impacts passed one hundred meters, their presence was beginning to create a bit of discomfort.

"Hey, Långström!" I turned my head around, shouting indoors. "They're starting a run with light mortars outside. Maybe I can come inside or something? We still have a fair view from the windows. I can keep a lookout from them."

Långström stuck his head out the doorway and observed the explosions down below, now about seventy meters away and still closing in.

"Bah," he snorted. "You won't see shit from inside here, and that's not so bad. They're still far off. If they start shelling the roof or something, you can come in, but stay here for now."

Group photo at the hotel. From left to right: Richter, Fabien, Greek, Cuix, Tjeck, and Carolus.

I looked over at the impacts which were still landing one by one with short intervals between them. He was probably right. The soft, sandy ground wouldn't even produce any secondary shrapnel. The speed with which they were approaching had also slowed down, and most impacts now seemed to land in about the same spot, the small craters they left behind piling up mostly in one place. The impacts soon turned from being a nuisance to something entertaining to observe. The Russians had still not tried to flank us, and the explosions were far more entertaining than watching the empty beach to the east. The great Tyulpan mortar had stopped firing a long time ago, and even the artillery crashing down on top of the Mayak slowly began to wither. The sun lay low over the water outside. It seemed that the Russians had exhausted themselves. The war was over for the day.

Långström, having come to the same conclusion, brought another chair out on the balcony and joined me.

281

"It seems they're done for today."

The sun moved rapidly downwards, soon coloring both the horizon and the great sea in a bright, golden red light.

"Did you by any chance have any smokes left?" Långström asked.

I picked up the cigarillo case and clicked open the plastic lid, picking up my last Al Capone.

"Just this one, then I'm all out."

"Do we share it then?"

"Sure. We share it."

CHAPTER 37: BACK TO SHYROKYNE

"I wonder where that fucking pussy who was complaining all the time is," I said.

"Who? Which one?" the Greek asked, after which I told him about the two Ukrainians we had picked up in the village, who later deserted us as we ran into the BMP.

"Fuck me, I would have liked to bash my RPK into that coward's face."

"You won't get the chance. Didn't you hear?" Långström said. "After he lost his wits and ran off, trying desperately to save his own life by heading to the Mayak, he got promptly killed by the artillery."

It became evident that the hotel would be where we would also spend the night. The Ukrainians defending Shyrokyne had suffered heavy losses, and the defense was already thin without anyone else to defend the flank. Just like our old barracks in Urzuf, this was a building built around summer tourists, not a winter war. The walls were thin, and the temperature dropped quickly indoors. We searched the hotel for blankets and mattresses, which we threw into a pile in a smaller room on the top floor, next to the one with the balcony which would serve as a guard post during the night.

We had not heard any news about Chris, and the question remained if we would do anything about it. The enemy had clearly taken a firm grip on the village, and he was still in there somewhere. We were far too weak to make a counterattack; our numbers weren't even bolstered by the other Ukrainians. It was out of the question. The enemy was far too many, and their tanks and artillery made them all the more superior. Our only option now was to try to sneak back in and search for him secretly, under the cover of darkness. The group discussed how to proceed, to Långström's great annoyance.

"Are you still going on about that shit? Can't you just fit into your thick skulls that he's dead? There is nothing we can do. Just fuck it. He had a great run, now forget about him already!"

Långström's conclusion was cold, but he was right. No one had received any signs of life from Chris. With all the bullets and shrapnel ripping through the air that morning, he was far from unlikely to have fallen. From a purely tactical point of view, it was unreasonable to go back into the village to look for someone who in all likelihood was already dead.

"We have to try something, at least," I said. "We have to try."

Långström pointed a firm finger at me.

"Fuck, man! Do you think he would have gone back into Shyrokyne to help you, had you gotten lost out there?"

The answer to that question was easy: Chris would have done it for any of us, no matter how bad the situation looked. That was probably also why all the rest of us were anxious to at least try to get back in. Not even Richter, who otherwise didn't care about any living thing, objected to it. He was not a driving force in the discussion, but neither did he actively oppose going back in to possibly face the Russians. Tjeck's nerves were on the verge of rupture after the morning artillery, but even he wanted to do something. However, everyone was tired after the long day, not to mention that we'd been scouting and fighting constantly for at least a solid week straight. Above all, we were scared. Going back alone against the enemy who had beaten the entire company so brutally could quickly turn into a suicide mission. I wanted to go, but I was far too scared to be the deciding voice in what we would do. That burden fell on the Greek, who resented how we bickered instead of coming to a decision when a comrade's life was at stake. He raised his voice even more than usual and pointed toward Långström with a great fury in his eyes.

"You shut up! I will let you know a single, very simple thing right now! I am not leaving Shyrokyne alive without Chris!"

"Sure, sure!" Långström replied resignedly and annoyed. "We will do as you please, retards. Fuck it. We'll go back in and look for him. Just keep in mind, you fucks, that when we all die in the middle of that fucking piece of shit village, it wasn't my decision. You don't fucking blame me for any of it! This was *all* your decision! You did it to yourselves."

Chris wasn't the only one who was missing: the Recon & Sabotage platoon had also lost a soldier during the retreat. Unlike Chris, however, this one had provided signs of life via radio and revealed that he was hiding inside the village. Besides asserting he was alive, he didn't offer much for his potential rescuers to go on, though. During the morning chaos, he had sought shelter in a basement, but he seemed to have no real idea where in the village this was located. He also couldn't find the courage to go outside and look around. A Russian armored car and infantryman had passed right by him, and he was convinced that the enemy was still nearby. Recon were going in after him, looking for a needle in a haystack. We foreigners would accompany them, looking for a needle we weren't even sure still existed.

We left the hotel and headed up toward the Mayak to combine our numbers. A squad from Recon went first and we followed behind. The column moved to the edge of the heights, five meters between the warriors. It was the middle of the night and pitch-black outside. Behind the horizon, the dark sky shifted more into dark blue as the sun's presence was glimpsed from the underworld below. In Shyrokyne, many of the buildings were still burning in a bright orange. The largest source of light was the village school, in which the long brick building had also caught fire. Dark smoke flowed out through the many windows upstairs while being lit up in red by the fires inside. To the right of the field of view rose the dark side of the church chapel, its golden Orthodox cross in the spire gleaming in the light from the flames from below. The black profiles from the warriors moved forward against the background, illuminated by the flames. The view meeting my eyes was as beautiful as it was strong and significant.

We took the narrow wooden bridge across the riverbed back, next to where we had previously pushed our pickup across. The dark mud below us seemed even more devilish in the black night. Recon passed first, then us following behind. Upon entering the village, the squads split up—Recon covering the right side, with us on the left. The surroundings were all quiet and eerily peaceful, us moving silently, slowly, and carefully through them. No one knew for sure how far into the village the Russian infantry had gotten. Their tanks surely remained in the rear position, our ATGMs blocking their way. It wasn't, however, far-fetched to assume their infantry would try to close the distance under the cover of darkness, just as we did right now.

Our Ukrainian colleagues repeatedly attempted to contact their comrade through the radio, but no signs of life were given back. We couldn't know if his batteries had run out, or if the Russians had found him. They were now, just like us, searching blind. Seeing as we had already made it to the village, however, we might as well continue. After all, we had come this far, and both of our objectives still remained. Pushing deeper inside the village and moving between the buildings, a large stone staircase soon appeared in front of us. It was down this route that many Ukrainian infantrymen had retreated that morning, and it would lead us toward the school we had passed with our pickup shortly after.

There were many steps, many of them damaged by artillery fire, but soon enough we came to the top and reached the main road. In front of us, the long school building towered above. The fires were still raging inside, the otherwise dark smoke reflecting the brightness from the flames down toward the road below and in front of it. We proceeded east, again along the road with us foreigners on the left side and Recon on the right, still slowly and carefully. The night was quiet, the fires around us crackling peacefully, just like our war stove we'd used as a fireplace had done our first night in Shyrokyne. We advanced forward one hundred meters, two hundred meters, three hundred meters. The houses were deserted on both sides of the road. The night lay quietly around. Only the muffled crackle of fire followed us.

The nice silence was suddenly broken, just like it had been during our raid in Kominternove, by barking dogs. Just like before, it was soon followed by another, and then another. In contrast to Kominternove, however, the Russians weren't as sleepy and were way more attentive to this sign of coming trouble. They quickly woke up and responded to it.

To our left, on the other side of the houses, there was an open field between the houses on our side of the street and the one north of us. The uncut lawns of tall grass and gardens lay empty, with the enemy covering them instead of the roads. Immediately a cascade of tracer fire opened up over the fields next to us, a fiery response causing us to instinctively take cover. An intense amount bright green and red lines of tracers shot past us, flickering in between the buildings and bushes next to us.

It took a moment for us to realize it, but none of the Russian fire seemed to be directed at us. All of it, however fierce it was, went over the field to our left. Not so much as a bullet crossed the street toward us. We looked at each other, confused looks on our faces, flickering in both red and green from the violent lightshow next to us. Each man, wise from experience, soon realized what was going on. The Russians were always afraid of the night, and when they were made aware of our presence by the wild dogs, they answered in turn by shooting at the sounds that had startled them—the dogs, not us.

Discombobulated as the Russians' violent behavior was, there was no telling how long their initial confusion would last. If they hadn't been previously, they were at full swing right now, and it was likely only moments before they would get their shit back together. It seemed wise to take advantage of their short moment of panic to quickly pull ourselves out. None of us had found what we came looking for, but at least we now knew how far into the village the enemy infantry had crept forward.

The Shyrokyne school ablaze at night.

On our way back, we once more passed the burning school where we regrouped. I took the opportunity to bring out my camera to try and take a picture

286

of the burning building next to us. To avoid the bright flash from giving me away, though, I had to press the camera against my chest. After all, if there was some Russian nearby, I was reluctant to let the white glow light up my face by placing it against my eye like how one would normally shoot a photo.

It was a little trickier than I had initially thought, but once I got my picture, I turned around to rejoin my group. What I saw around me was just a deserted road. As I had tuned out my surroundings to focus on shutter speeds and angles, the others had already continued and I was left alone. To say the least, an uneasy feeling arose, and I hoped that the picture of the burning school would be worth the now very possible consequences. To my luck, however, I soon found the others again just down by the end of the staircase below.

Soon we got back to the Mayak and paused for a while outside.

"Fuck me, I could use a fucking cigarette right now," I said.

It wasn't the first time I had moaned and bitched about it, but this time I was approached by Tjeck, who tapped me on my shoulder. In his hand, he was holding a pack of some unknown Ukrainian brand of smokes. The lid was open and inside were something like fifteen cigarettes, all apparently mixed of different colors and brands.

"What the hell, man! You're my fucking hero, you know that! Where did you find these?"

"I collect them earlier today, when we did our UPS run to get water and ammo," Tjeck said, noticeably proud, and rightly so.

"Fuck, man, you mean you've had cigarettes all this fucking time? And you didn't say anything!"

"Yes. I didn't say, because you would have just smoked all of them at once, and then you would be complaining again, so I wait a while."

Långström remained at the Mayak for a briefing as the rest of us returned down to the hotel. We managed to create a cozy feel in our little retreat full of mattresses and blankets. Bear, who had returned from Mariupol, helped me set up the candles I had brought from our now wrecked previous home. They came to serve as our major source of heat, and the small room drawer became a pedestal for the living little fires. All the candles almost made it look like a somewhat awkward sacristy. We were warm and comfortable, needing nothing more.

Långström returned and opened the door. Cold air drew in from ground level while a warm bulge erupted above him, quickly escaping outside.

"Alright, now you'll finally get what you wished for!" he said happily and hurriedly with a big grin on his face. "You wanted to go back into Shyrokyne and save Chris? Well, guys, now you're about to get what you wished for! Ha! Tomorrow at dawn, we return to Shyrokyne. We're going to counterattack! We're going to strike back at the fucking Russians!"

"Have we received reinforcements to do so?" I asked Långström expectantly. It would have been nice to assault the enemy just as he stormed us, and take the village back through superior numbers.

"No," he said and laughed. "What the fuck do you expect? We will attack them with what we have."

"What?" the Greek asked from his reclining position on the mattress next to me. "But we have fucking nothing left, no?"

"Exactly, but the Ukrainians will scrape together whatever they can find and attack with that."

"With what's left? The Russians still outnumber us, even if we brought every single man available. Fuck, man, this just isn't possible."

"No, exactly! We'll see how it goes tomorrow. There's no fucking way this will work, but tomorrow we will find out for sure!"

There was no point in asking questions about the upcoming task, and there were not many hours left to rest. We let the candles burn and crawled down among the blankets we laid out. We kept two men on watch shifts with the last ones to wake us up at 4:30 in the morning. Outside, the sound of a tank engine could be heard, something Tjeck took notice of.

"I think it's a tank outside," he whispered nervously.

"Yes, I hear it, too," I replied.

"Is it ours?"

"Don't know. I'm sure it's fine. Forget about it and go back to sleep," I replied, after which we all slept like babies.

We were soon woken up by the last guard shift. We got up, gathered our equipment, and left the hotel again—first up the long white stone stairs, then the last bit toward the Mayak where the soldiers who were to carry out the counterattack slowly gathered around. As I stepped through the Mayak's gate, I immediately smelled a familiar scent. It wasn't uncomfortable on the nose like the fires of Shyrokyne, and it smelled better than even burned gunpowder and cordite. It was the smell of hot food. I let my senses guide me through the dark rooms to one of the corners, where I saw a man standing with a ladle next to a large insulated metal pot shining bright like silver. Next to him were some paper plates on the table. I immediately grabbed one and placed myself in front of him. He said something in Ukrainian to which I couldn't answer. I instead simply and eagerly held up the cardboard plate in front of him. He smiled and emptied his large ladle into it, so much so that the thin paper plate bulged inward and spilled some of the contents.

It was some kind of soup; more accurately, it was just broth with a few small pieces of potato floating around in it. It didn't look like much, but the soup was still hot, much unlike the cold February night outside. I had not seen hot food for days, let alone smelled it, and after drinking the wonderful broth, I quickly tried to herd the rest of the group in there to experience the wonders for themselves. Some of them took what was available; others had a poorer appetite.

The force that would carry out the counterattack was gathered outside the Mayak. It was small and motley, much smaller than we had hoped for.

"Apparently, some say they're sick and can't come. What the fuck is that all about?" the Greek mumbled in a loud and annoyed tone of voice. "Well, fuck me then, but isn't this a fucking great time to get a sudden stomach ache?"

Tjeck was so scared that he had a difficult time even speaking.

"How are you doing?" Fabien asked in his thick French accent. "Are you okay? You can do this, right? It's okay. You come with us only if you know you can do this."

Tjeck smiled nervously and struggled to utter his words. "I think artillery is the hardest for me. That sound it makes—it is difficult for me. Enemy bullets will be easier, I think." He stammered as he spoke, looking away. His facial muscles twitched involuntarily as he tried to squeeze out a confident smile.

I knew inwardly that this wouldn't work, but I couldn't help but be impressed by his courage. He was so scared to the point he struggled to create actual sentences, but he was still determined to move forward with the rest of us. He was, in a sense, perhaps the bravest of all of us. I tried to calm him down by pointing toward Shyrokyne and the valley below us, showing how some thick and dense fog had slipped in from the sea to cover the village.

"You see the fog over there? That's a really fucking good thing. As long as the fog is there, the Russians will have a really hard time lining up their artillery against us. We still have a chance."

Tjeck laughed nervously and again said how he really hadn't been enjoying getting shelled by artillery: "Bullets I think will be better."

There seemed to be no more than four groups in total gathering in preparation. A loosely assembled crowd of men from different units formed an ad-hoc platoon-sized unit. Outside the driveway stood one of our BTR-3 fighting vehicles. When we had laid eyes on them for the first time just days before, they were all shiny and new. Now, though, this one was as worn as any of the archaic Soviet vehicles, with its large road wheels punctured and its chassis torn by shrapnel. It would, like us, however worn and dirty it was, take part in the counterattack. Bear grabbed the large steel fences surrounding the vehicle and began climbing on to it.

"Hey, man, what the fuck do you think you're doing?" asked the Greek.

"The BTR is damaged," Bear replied as he sat down on the front of the roof next to the gun turret. "Its vision is fucked. The driver sees nothing. I need to sit here and tell him where to drive."

"W-w-what the fuck are you saying!" the Greek shouted while waving his arms. "You can't fucking sit there!"

I also joined in the shouting: "If we get shot at, you'll fucking die! You'll get stuck behind those steel slats. No way you'll get out of there alive!"

Even Långström broke in, firmly ordering Bear to step down from the vehicle: "You won't stand a fucking chance! You are dead as soon as we get contact! Don't you fucking get it?"

Bear, the young college student, stood up holding the metal cage around him and pointed his other hand toward us. It mattered nothing what we said or how loudly we shouted. He was determined to continue as it was.

"I fucking know I will die, yes. It doesn't matter! Someone must tell the driver where to go, and I will be the one who does it! You do your shit, I'm doing this!"

The engine started and the many wheels began to roll as Bear grabbed the slat armor which locked him into the cage. The BTR-3 took the lead on the road down toward the village as us other warriors followed like a long snake behind it. We followed the road until we came to a small depression at the northern part of the village. There, a number of other vehicles had already taken up positions: a

Ukrainian T-64BV main battle tank and two BMP infantry fighting vehicles—one a BMP-1 and the other a BMP-2.

Fuck, this is nice, I thought. If we actually had a tank with us, maybe this could still work. Sure, the Russians had several more of them, but even still, it just might. We might just survive this.

Assemblage of men and vehicles in position before the Shyrokyne counterattack.

CHAPTER 38: THE ROAD TO GLORY

Långström tapped me on the back of my helmet, and I turned around to see what he wanted.

"We are going to die! Do you understand that?" he shouted over the engine noise.

"Yes," I replied bitterly, "that's pretty evident."

Långström had finally returned to his normal combat ecstasy. He was sharpened and back on the chopping block again.

"We're all going to die! It's so fucking great!"

The coarse diesel engine of the fighting vehicle roared away, spewing thick white smoke just like the exhaled air from our own heated lungs. The steel tracks started squeaking, the chassis jerked backward, and our BMP began rolling forward. Forward toward the village, toward the enemy, with a squad on foot on either side. All toward a certain death. All determined to face it as men rather than retreat as cowards.

There were three streets leading into Shyrokyne from where we began, and our ad-hoc assault group split up to cover two of them. The BTR-3 and half the infantry would move up on the main one south of us, while we would take the middle street. The northern road was left empty, as we were already way too few in numbers. The T-64BV and the BMP-1 stayed in reserve.

Every single muscle in my body was dead tired. My boots felt heavier than usual. Even though the worn and dirty BMP did not drive faster than a brisk walking pace, my legs were close to cramping just trying to keep up. The only joy of it was that I forgot how sore my cold feet were. It was as if all the energy my body had left to give was focused on keeping my eyes functioning at maximum performance. As we passed the buildings, I could clearly see every single window, every single corner. Every single nook and cranny was sharper and clearer than I had ever seen anything before in my life, all to search for the slightest movement where an enemy might suddenly appear.

The fog still covered our advance quite well, but we remained quite vulnerable. Our large fighting vehicle took up considerable space on the narrow road with us next to it. We were all excellent targets for the enemy waiting ahead. My arms were tired, and I began to think it was a mistake to cut the sling during yesterday's retreat. The RPK felt heavier than usual, and being unable to relieve the weight on my back and neck began to take its toll. I couldn't find the strength to carry the weapon against my shoulder in a ready fashion, but I contented myself with knowing I was just as quick and accurate from the hip and carried on. If any Russian with an RPG appeared in front of us, I was certain I would be able to gun him down at least before his rocket hit us. That was some kind of justice, at least.

I was so focused on the windows and corners in front of us that I failed to see closer objects. The BMP-2 was driving far to the right side of the road, leaving us little maneuvering space next to it. Being forced further and further to the side, I soon walked face-first straight into a sprawling bush. I found myself blinded for a moment and was forced to back away after its sharp branches had almost picked my eyes out. I let my gloved fingers travel over my eyes—they seemed to both be there. Slowly opening them proved I could evidently still see, and the first thing I saw was that there was no blood on my fingertips. After wiping away some tears, I focused extra power to my legs, hurriedly moving forward the few meters of ground I had lost, now ready for further obstacles. I ran crouched under low-hanging tree branches and tore myself through thick brush.

Långström wanted me further forward, closer to the front end of the BMP-2. He made this evident by physically pushing me forward. It was, however, impossible, as the engine exhaust was located just in front of me, blowing hot fumes straight into my face. As pleasant as the hot exhaust was this cold winter morning, the burned Eastern European diesel fumes made it impossible to see anything. I backed up again, something which Långström probably took to indicate I was afraid, and he pushed me back forward again. Hot gases stung and teared up my eyes and burned like fire in my already struggling lungs. I screamed at him to stop pushing me, but my voice was completely drowned out by the loud engine and squeaking steel tracks. He pushed me forward again, forcing me to turn around and yell in his face to fuck off. As I roared through a diesel-powered respiratory system, spitting snot while excessively pointing at the exhaust, he finally got the memo. We returned to normal, continuing the advance with me at a comfortable distance behind the hot fumes.

We soon reached the end of the three hundred-meter-long road. The BTR-3 and infantry were still moving on the road south of us, closing in on a small bridge which they would then be forced to cross. Even when crossing the bridge, their approach was still pretty well-protected by the buildings around them. We, on the other hand, had no such protection. In front of us lay a valley we would now be forced to cross—two hundred meters of open terrain, the deep lowland making us all the more vulnerable. We'd first have to run downhill, then uphill again before once more approaching built-up areas offering cover. While in the valley, we would find ourselves exposed in every direction. Some fog still remained in the valley, but the brisk winds from the sea was pushing more and more of it inland.

The sun also began to rise. It was only a matter of minutes before our only advantage and cover would dissipate completely.

The BMP halted, and each squad took up firing position to cover their respective field of view around it—Metro on the left and the Foreigner Group on the right. I got down on the ground and carefully examined every window while searching the surroundings in front of us through the scope. The Greek came and sat down next to me, tapping lightly on my helmet to get my attention.

"Two o'clock," he whispered lightly while pointing with his hand. "Enemy BMP."

I lifted the gun and bipod over in the direction and began looking through the scope. As the fog dissipated more and more I also soon saw the contours of the enemy fighting vehicle. It was only one hundred fifty meters away. I had been so busy looking for movement and human figures that I had completely missed the stationary, potentially most dangerous thing in front of us.

"It seems abandoned, though, right?" the Greek whispered.

As the fog blew further out of the valley, I could also see that the turret hatches were open. Same thing with the rear crew compartment doors. The vehicle—maybe the same BMP-2 we almost collided with yesterday—was knocked out, and I drew a sigh of relief with my humiliation somewhat subsided.

Other than the knocked-out BMP, everything looked clear. No movements or other armored vehicles were visible, but going down into the valley still remained a bold and risky move. The minutes ticked, and the fog cleared out faster and faster. If we were to go further, we would have to do it now.

"Greek! Bear!" Långström hissed and waved to the two of them. "You two head down the valley as scouts! Hurry up, move!"

The Greek nodded and advanced first, stopping at a lone tree, waiting for Bear to follow and pass him. They then moved cautiously but quickly one by one down the valley to look around. No Russian opened fire. As they had reached the deepest point, they waited a moment and then waved to us. Everything seemed clear enough. I folded my bipod, got up on heavy legs, and readied myself. This run would be the riskiest moment so far—not only during our counterattack, but throughout my entire time in Ukraine.

The engine from our BMP-2—not the broken-down one in front of us— roared as thick, white smoke sprayed from the exhaust straight up into the air. The drive wheels and the shiny, gravel-polished tracks squeaked and screamed. The pointed chassis jerked violently as the 14-ton steel body was pulled forward by the rocking start of its caterpillars. Rapidly increasing to full speed, the tracks ripped beams of torn ground and soil several meters high into the air behind it. The fighting vehicle threw itself down the valley, us warriors following it on foot. Where all my body's energy had previously been focused to power my visual receptors, it now switched over to drive my legs instead—my thigh muscles tightening convulsively, painfully, as if there were fire inside of them. Thrusting forward with long strides through the frost-covered grass, we began assaulting down the seeming infinitely long, open valley in front of us.

Just as we neared the lowest point of the valley, I heard the first bullet flicker through the air among us. I did not let myself be distracted by it, hoping it was just one of us who had let off a round by mistake. I tried to focus all my energy on keeping my legs working, but as more and more bullets began to whirl and strike through the air around us, I realized we hadn't been so lucky. We had played a reckless game by hoping the Russians would be further into the village, which would have allowed us to cross the valley unhindered. The gamble had been unsuccessful, and the cards were already on the table. We were, to use a suitable expression, fucked.

There was no other way to move but forward as the air around us begun to storm with enemy fire. Bullets scurried around more and more as the ground began to slope uphill. Our BMP had already reached the top, but for us legging it, we found ourselves struggling to keep up speed. Like in a nightmare, no matter how hard we tried to run, we seemed to move only slower and slower. Gasping for air to power my tired flesh, I saw a figure quickly run out in front of us all. At first, I thought it was Långström, but the man was taller in stature. It was one of the Ukrainians who had advanced on our left flank, his face previously unknown to me. He seemed to have thrown off his combat   vest and equipment and moved much faster than the rest of us.

He sprinted about twenty or thirty meters in front of us, until he suddenly stopped and turned around to face us. Contrasting with the sky behind him, he stood still for a brief moment, showing a dark full beard over his worn and tired face. With his AK in one hand   , waving his other   toward us and yelled, pushing us to follow him up the slope like a shepherd leading us out of the valley of death. The Ukrainians, characteristic of the Slavs, roared a loud and death-defying "oorah," overpowering the violent snapping sound of enemy bullets. Tired legs and strained muscles gained momentum once more. Speed increased again. We kept moving up the slope—still slow, but faster than before.

By a miracle, we made it up to level ground, ready to finally face the enemy, but we couldn't see any of them. Everything was quiet and still in front of us, but still the bullets continued to strike past us. I found Långström's gaze, hoping it would lead me in the right direction, but I found it just as confused as mine was. Unlike in the valley below, there was finally cover available around us, but as long as we couldn't figure out where the enemy was, we could hardly make any use of it. How would we know which wall to take cover behind when it still seemed as if bullets were flying at us from every direction?

We simply took kneeling positions on the road where we stood. I tried to listen for the origin of the actual gunfire around us, but I couldn't hear anything beside that sharp snapping noise bullets make as they whiz past one's ears. I threw my eyes around, looking for some kind of indication where the enemy was firing from but found nothing. The sun was rising. Tracers and muzzle flashes which had been visible in the dark night were now all drowned out by the natural light. If any smoke from burned rifle powder was around, it married into and disappeared with the fog. There was no indication as to where the enemy was, and only the loud cracks of hypersonic bullets could be heard.

I listened closely. I faced back toward where we had come from and realized the bullets came from somewhere to our right side. The northern road above us— the one we had lacked manpower to advance on earlier—continued a bit further out than ours south of it. At the end of that road, only one hundred meters from our position, lay three large two-story villas. The direction of the sound indicated that those had to be the enemy positions. The Russians had been much further into the village than we had anticipated. Our BMP had turned in on a road, standing still with its rear facing us. It seemed the crew hadn't perceived the fact we were taking fire from behind, so we couldn't rely on any help from its 30mm autocannon.

I looked at Långström, who nodded at me, having come to the same conclusions I did. I unfolded the bipod, and we quickly threw ourselves down next to the road. Though we still lacked any real cover, at least we were finally firing back at them. The broken windows were dark inside, and I could not see any enemies, but it had already been way too long. We had to respond now. We began laying down suppressive fire on all the windows and other suspected areas, and as soon as we did, more and more of our friendlies started doing the same.

Creeping up from behind us, a Ukrainian emerged from the smoke and fog—a tall man, gnarled and crooked with a rough, worn, and dirty face framed by a tousled brown beard. The helmet on his head was adorned with a large Odal rune, and in his hands he carried a PKM. Wrapped around his neck hung a long cartridge belt, its copper-colored 7.62x54 cartridges shining in the dim light. He was called "Kelt," and he approached Långström with slow and relaxed steps, calmly sitting down in a kneeling position next to him.

"Ah, the AK," he said, looking down at Långström, who was firing rapidly at the enemy position. "AK is good…PKM is better."

Kelt lifted his PKM and pointed the barrel in the enemy direction. At the same time as he gave off a loud, diabolical laughter showing some teeth visibly missing, he held down the trigger. The machine gun thundered as large puffs of smoke ejected from the gas ports to its sides, from the flash hider in front, and from the ejection port also spewing spent cases from the receiver side. He never stopped laughing, loudly and heartily, dirty teeth shining from his mouth, as he slowly mowed the gun from side to side, back and forth around the enemy positions. It was obvious he was finding great pleasure in his profession.

The enemy's fire began to subside and soon ceased altogether. Whether we hit any of them or just forced them into cover was impossible to say, but our response had worked. The calm was, however, short-lived, as soon other bullets began to land around us, now coming from different angles instead. We were still under fire, though now from a longer distance. More Russians had been alerted to our presence, but luckily for us, the enemy fire seemed completely disoriented and was very ineffective. The enemy was probably as surprised by our attack as we were. They didn't expect us to go forward and hit them right back, especially not as we emerged from the valley. Their response to our attack was strong but completely chaotic, and thus much less effective than it should have been.

This time, we returned fire much more quickly. I still couldn't see any Russians, but just listening to their bullets and firing back in that direction had worked before, and it could work again. I threw myself into cover behind a pile of gravel and let the RPK sing against all suspected enemy positions. We returned such heavy fire that even these Russians soon also backed away. Enemy fire slowed down more and more; it began to seem as if we might just get out of our fucked situation in one piece. Maybe.

I was now lying prone with my back to our BMP, which still held the position on the small road eastwards. Just a few meters in front of me was Bear, also firing toward the Russians on our right side. My gun was close to running dry. As I began reloading, I saw Bear turn his face in my direction. His gaze fixed on something behind me, whereupon his face turned into a chalk-white color. His body cramped up as he gasped for air in order to let out a loud call.

"Tank!"

The word alone made the blood in my vessels freeze.

Behind us, less than a hundred meters away, an enemy tank broke out between the bushes and buildings to the left of our BMP. The tank's engine roared as it swung toward us, approaching at a rapid pace. Sixty meters, fifty meters, forty meters—within seconds, it was only thirty meters away, and it did not seem to stop.

"Take cover! Fast!"

Two small buildings with thin brick walls were the only ones available between us and the tank. We hurried there, purely on instinct now. Just as we threw ourselves behind the corner, an enemy RPG missed its mark just next to us, but the surge in stress levels caused by the tank was so huge the explosion was barely noticeable. We took cover on top of a pile of broken asbestos roof panels, likely blown off the building by the artillery the day before.

We had made it through the valley and then, even without any cover, had managed to take fire superiority against the enemy, not just once, but twice. An enemy main battle tank, on the other hand, was something completely different. We suffered a severe shortage of anti-tank weapons. Långström carried the only one: a light single-shot RPG-26. It might have worked against a BTR or BMP, but against an actual tank? We needed more than that, and we did not have it.

The tank engine could be heard on the other side of the building, growling and hissing loudly. The last we saw of our BMP, it was still standing there on the road. The two vehicles couldn't be far apart at this point, only a few meters. I lay down, pressing myself hard against the ground next to the wall, waiting for our fighting vehicle to save the situation. Its 30mm autocannon had no chance of piercing the tank's thick frontal armor, but perhaps it could scare it away? Or it could open a long burst of automatic fire straight at it, just as our self-propelled anti-aircraft gun had done the day before. Even better, if the BMP-2 could flank the tank and pierce its thin side armor, that would actually destroy it. I waited and waited, eager to hear the autocannon open fire. At first, I waited nervously, but my unease soon turned into anger and frustration. Seconds passed, seeming like an eternity.

What's the fucking hold up? I thought. *Open fire, goddammit, before it's too late!*

An extremely loud, razor-sharp sound cut through the air. It sliced right through my ears, despite my wearing 3M hearing protection—electronic earmuffs which muffled high-decibel noises and amplified low-decibel ones. They dulled explosions well, but the muzzle energy of a Soviet 125mm tank gun at such a short distance was way outside their performance levels. My ears ached and rang loudly.

You fucking idiots, I thought of the BMP and its crew. *You should have opened fire but you never did. Now you're all fucking dead, just like the rest of us.*

We were now completely trapped. The only way back to our own lines lay across the valley below. Running across it under small arms fire as we had done earlier was dangerous enough, but now there were even more Russians in the vicinity. With a tank at their disposal, we didn't stand a chance to retreat that way. We had nowhere to go.

It finally hit me. It was here on top of a pile of broken roof panels made from carcinogenic asbestos, in a shattered and dirty little piece of shit village in eastern Ukraine that the adventure would come to an end. Where everything would end. It was right here, right now where we were all going to die. The hourglass of life counted down its last few trembling seconds. For the third time in my life, I saw everything pass in review in front of my eyes. It seemed like there was a lot I had missed out on. I wished I had seen pictures of a family of my own shimmer over my eyes, but I never had one. I felt a bit disappointed over things like this, but I was never bitter. These days had been the best of my entire life. I had gained the opportunity to experience things others would only dream of, and my friends I was lying next to were the best men I had ever gotten to know. All in all, I remained very pleased with my life as I calmly closed my eyes to wait for the coming explosion.

CHAPTER 39: BACK THROUGH THE VALLEY

The ground shook, and the sound cut in my ears just like before, but as I opened my eyes again, we were all still there. The poisonous dust from the roof panels began to slowly settle again. We were still alive, at least for the time being. Nothing had really changed; however, our situation was still exactly the same. We remained stuck with nowhere to go. Everything just repeated. We waited for the third shell to hit us.

As for the Azov fighters on the southern road below us, their situation was almost as miserable despite their better starting position. They did not have a long valley to pass like we did, only a small bridge. There was also plenty of cover in the form of buildings and the like around them. On their approach, however, they met an enemy who had been at least somewhat more prepared. The Russians immediately met their attack with focused fire, both from the front and from the sides. The infantry had suffered losses, and the bridge had collapsed under the belly of the blinded BTR-3, forcing it to call for help from the Ukrainian T-64BV which had remained as reserve. The attempt to tow away the stuck BTR using the tank in the middle of a firefight was bold and brave, but instead simply ended with both vehicles getting stuck. Their situation critical, the blind BTR fired off its smoke launchers, creating a thick, white cloud of phosphorus smoke covering the terrain around them.

The tank next to us fired its third shell, the second one to impact near us. Strangely, though, just as before, nothing happened. As the ground stopped shaking and the dust settled, we were all still there, still alive. I looked at Långström, who simply seemed just as confused as I was.

"Is he going to hit us at some fucking point then, or what the hell is his problem?"

Loud explosions went off around us, but they didn't originate from the Russian tank—they came from the other two remaining Ukrainian T-64BV tanks giving us fire support from the top of the Mayak heights. On each side of the strong white building, they drove shooting and scooting back and forth to land fire on the enemy around us inside the village. The 700-horsepower engines roared as they propelled the steel beasts toward the edges. There they halted to swing their large gun turrets around before unloading their deadly 125mm projectiles down toward the Russians next to us. As their cannons roared, the tanks then backed away slowly, hissing as new shells were brought into their breaches by the autoloaders inside. Rolling back and forth, the Ukrainian main battle tanks fired so quickly and violently that the thermal shields around their gun barrels simply broke off. Their heavy shells howled as they passed above our heads slamming down among the enemy positions around us. The Russian tank slowly began to back away, returning to where it had come from.

All the explosives created large clouds of dust and smoke, which got even thicker as the heat and burning cordite bound particles of water to them. As these clouds were then quickly cooled down by the freezing winter air around them, they began to, like streams of water, slide down into the valley below us. Down there, these clouds soon married the phosphorus smoke from the BTR-3's smokescreen, now being pushed inland by the same sea breezes which had pushed our previous misty covers away from us not long before. The fog which Mother Nature created had left us be, but the war itself had given birth to a new one which soon lay like an even thicker blanket over the valley.

T-64BV on the hills near the Mayak, thermal sleeves blown off the barrel.

The Russians' vision was obscured. This was our chance and we would grab it. We carefully moved out of our hiding place to make sure the coast was clear. A loud engine noise was heard from where the Russian tank had previously stood, but it did not sound like a tank. As the vehicle came out from behind the house wall, the already loud noise increased significantly—a crackling engine and squeaking steel tracks. It was our BMP that came rolling, full speed in reverse tearing through the smoke. Its speed was so high that the fighting vehicle left the ground for a brief moment where the road ended and the valley began. It landed with a thump so hard that improperly attached equipment was thrown off of it. The dented side plates which hung loosely down covering the upper part of the tracks wobbled up and down. The chassis rocked around violently with its tracks tearing through the soft ground below it, spraying rivers of dirt everywhere.

We foot soldiers followed our inexplicably resurrected fighting vehicle's example, running down through the smoke back to the road we had come from on the other side of the valley. As I came back up where we started from, I turned around to see that everyone followed. One by one, the soldiers emerged from the cloud of smoke below, familiar faces, as well as unknown ones. Out of the corner of my eye I saw a handful of men moving along one of the buildings on the other side of the valley. I had a hard time figuring out why anyone would still be remaining behind. In order to see better, I lifted my RPK and scope to my shoulder like many times before.

This time, though, the lens wasn't helping. I couldn't see anything through it. The inside of the scope was full of dust, and there seemed to be some kind of pressure damage in the middle of it. The Russian tank had finally rendered my old Weaver scope obsolete to the point of uselessness. I lowered my gun and instead tried to focus my tired eyes alone. The men did not seem to move away from the Russian side, but rather looked as if they were heading toward the corner where we had been covering. The distance was close enough that even without a scope, I could see that they lacked yellow armbands.

Still standing, I quickly lifted the RPK back up while pushing the selector down to automatic fire, lining the barrel up with the men in front. Armor-piercing and tracer rounds began feeding from the magazine into the gun as I let them strike back and forth over the house wall and among the Russians next to it. While firing, my index finger began aching indescribably. I glanced quickly at my right hand while still firing and saw how the trigger struck my index finger forward into the trigger guard for each shot. The RPK fired as it should, but something was wrong with the trigger group. I moved my hand up on the pistol grip and kept the fire up using my middle finger instead. The blows to the hard metal trigger guard went so fast that the individual hits could not be discerned. My cold fingers felt as if they were being pressed into a vise. Before even the first 45-round magazine finally clicked empty, I had once again been forced to switch, firing the last bursts using my ring finger. The loud cacophony of the enemy's fluttering bullets from their other positions erupted all around again, and I pulled back into cover to reload.

The last soldiers began to rise from the valley, and we seemed to have made it without heavy losses. We were now preparing for the Russian counterattack and began to spread out across the edge of the valley. Our group would hold the right side. To get to the buildings on the side, we had to cross an ordinary-looking fence. It was not especially high and looked like it shouldn't have been a problem to climb over. Or so it seemed. My body was tired and the equipment heavy. Cuix seemed to think I was taking a bit too long to pass the obstacle. As I was about halfway over, the small Frenchman took charge and tackled me from below the last bit over the fence. I thumped onto the ground, landing on my back on the other side. It knocked the air out of me, but I nevertheless still appreciated the helping hand. We took positions and waited for a while, but the Russians did not seem to follow us. The rest of the Ukrainians began to withdraw further back along the same street we came from, and we did the same.

The BMP gave covering fire with its machine gun while reversing as we followed it. Occasional cracks of small arms fire pursued us during our withdrawal, but nothing much heavier. Fabien fired wildly around until he suddenly fell next to me. For a moment, I was sure he'd just been hit, dead before hitting the ground. Thankfully, I soon noticed he was still moving, agonized but not in the way a dying man would be. I realized that he had simply stumbled over a piece of thick steel wire. We arrived at the end of the road, back at the western edge of the village. Bear sat down by the roadside, exhaling heavily.

"I just can't stop thinking about that tank," he gasped.

I was still unsure exactly what had happened.

"Which tank?" I asked, not sure if there had been more than just the one.

Bear looked at me like I was an idiot. "*The* tank, of course!"

"Oh, you mean that one?" I replied, still a bit dizzy.

"I just can't stop thinking about it."

Tjeck was in extremely poor condition. His last nerve had broken and his eyes flickered. I could see Långström arguing with a man that had to be Metro about what we should do next. It dawned on me that it was him, Metro, who had run in front of us during the assault, and he waved to us to follow him. He was still wearing all his equipment, but as his combat vest was all black, it hadn't been visible in contrast with the dawning light behind in the valley before. One of the Ukrainian soldiers next to us moved around in a strange manner, limping with one of his legs. At some point, it seemed that he understood it was more than just a stretched muscle. He unbuttoned his trousers, where blood ran from a tiny hole in his thigh. A 5.45 bullet had gone straight in and then out again, without making more than a small, needle-like puncture. Bear was quick to put a bandage around the injured man's leg. After being patched up, the warrior buttoned up his pants again and resumed his duties, though still limping somewhat.

Although the thick morning fog masked our initial approach to Shyrokyne through the valley, the fog had since lifted, and only man-made smoke from artillery, tank cannons, and the BTR's smoke dispensers covered our retreat back through it—and that artificial fog was rapidly dissipating. The enemy's artillery could start pounding again at any moment, and we needed to find cover quickly. Bear grabbed a sledgehammer from the BMP and headed toward the house on our right to see if he could get inside. The house seemed abandoned, but a large, ferocious, and—as luck had it—chained-up dog outside had no desire to let a stranger into the yard. The rest of us looked for other opportunities. Most of the buildings nearby seemed weak. They would not provide much protection from artillery, with one exception. A bit further up the street on the left side, a proper two-story brick building towered high. The walls looked solid, and its upper floor would give a good overview of the area.

"We'll take that one," Långström decided as we moved forward again along the street.

Next to the roadside, the building we wanted was enclosed by a tall, red brick wall. We could, of course, have tried to climb over it, but it would be best if we could get through it completely. Pushing it, the top bricks seemed loose. Perhaps the same was true for the entire wall? Bear had given up on his own futile attempt at breaking and entering and joined back up with the greater gang of home intruders. He and I now took charge from across the street and threw ourselves against the brick wall. The wall flexed somewhat, and we managed to knock the upper part of it down, but the cement which locked the brick in place was stronger at the bottom half. Pushing it over wouldn't work, at least not with mere manpower. We needed to find another way in.

"Where is Tjeck?" Cuix asked. "Has anyone seen him?"

I offered to go search and soon found him huddled against a shed at the back end along the road.

"Hey, man," I said to him. "What's going on? How are you doing? Everything okay, man?"

He looked up at me, the muscles in his face cramped as he tried to squeeze a smile. He responded in a stammering voice.

"Heh, yes, I thought bullets would go better for me, but I think I was wrong. And then came that tank and all that other shit happen, too."

I sat down next to him and patted him firmly on the shoulder.

"It's okay, man, you're with the A-team now. You know that, right? Just stay close to us and everything will be fine. We'll fix all this. Just stay close to us and you'll be safe."

Tjeck laughed, still nervously, but he seemed to have calmed down a bit. I pointed to the house we headed into.

"The rest of the group is over there by that house, trying to get in. We are going over there, too, okay? You just follow me now. Let's go. Stay close to me."

Tjeck nodded gently. I started moving forward as he followed me, and suddenly a burst of bullets flickered in the air past us. I threw myself on the ground and waited a moment. The air turned quiet again. It seemed to have been just a random shower of lead, nothing more. I continued a bit forward before halting to see how Tjeck was doing. He was nowhere to be found. I turned back to search and found him in exactly the same place from where I had just picked him up.

"Hey, Tjeck! What are you doing?" I shouted and tapped him on the helmet. "You're supposed to follow me, remember? You need to stay close to me, you hear that?"

Tjeck cautiously apologized and promised to come along, and thus we began our journey anew.

We managed to advance a bit further this time before there were again disturbing sounds from the surroundings. This time it wasn't flickering and passing bullets, but something exploding in the tall, white, frost-covered grass to our right. The grass was so high that the detonations were barely visible. They were small but many and came at a very fast pace. After experiencing all sorts of heavy artillery, my synapses no longer connected trying to figure out what was creating these small impacts next to me. I simply got lost attempting to place them. I just stood there staring at how the grass fluttered as thick puffs of black smoke rose up from below it. Had I understood that it was an automatic grenade launcher, a 30mm AGS-17, I would, of course, have sought cover immediately, but it simply hadn't dawned on me. I just stood there like an enchanted idiot, watching how the frost covering the tall grass melted from the heat while the long straws were cut down and thrown around by the maze of sharp shrapnel.

Once I came to my senses, I turned around to look for Tjeck again. Not to my surprise, the situation had repeated itself, and I returned for the third time to bring him forward with me.

"Okay, man, what the hell?" I said, feeling like an annoyed parent who had lost all patience. I grabbed him by the collar and dragged him forward all the way with me, this time ignoring any explosions and snapping bullets around us.

I saw Cuix and Fabien sitting in some cover further ahead and dragged our shaken friend over there.

"Hey, Cuix, keep an eye on Tjeck, will you?" I said. "He's a bit jumpy. You'll have to hold on to him, because he wants to run away every time something happens. I'll go and see if we can find another way in toward that building. Just hold on to him."

Cuix smiled and nodded, grabbing Tjeck by his arm pulling him down to sit next to him.

The crew of the BMP which had covered us from the street was eager to go back home. Bear interpreted what the crew commander was saying.

"He say that the BMP was damaged in the battle. The gun is broken, and the gearbox has locked in reverse, so it can only drive backwards."

The Greek laughed loudly and scornfully.

"Ha! How timid that when you come across a tank, the only gear working is the reverse! That's just a great fucking coincidence, isn't it?"

He continued to mock the crew by calling them "moral casualties"—men who were badly injured, but not physically, rather only in spirit and soul. Cowards, as a man of lesser vocabulary might have said.

As if by a miracle, we hadn't suffered any dead during the chaos, only a few wounded, which the BMP now evacuated from the battlegrounds. It was unnerving to watch it leave us behind. Broken cannon or not, from now on we had no more armor support.

We advanced through the yard of a smaller neighboring house to get around the brick wall outside our goal building. The field between the houses was open with a short steel fence in the middle. Långström and a few others were on their way over it. However quickly they climbed it, it would be dangerous to be stuck on top of the fence out in the open, even if only for a few seconds. There was some debris in the small yard, including a large transport pallet. To make it easier to cross the fence in the future, I left my RPK by the house wall and told the others to cover me. I ran out, picked up the large pallet, carried it forward, and threw it over the fence. Lying on top of it now, the pallet functioned as a bridge of sorts, making crossings a lot faster. I returned back to my position by the corner wall, where I covered with my RPK. The rest of the soldiers who were going over could now pass in a single leap without having to stop and climb over anything.

The street above us was the same from which the enemy further east had previously fired upon us. I took the opportunity to enjoy a tasteless Ukrainian cigarette while I kept a lookout to the north. Långström was looking for ways into the house in front of us. The Greek was waiting behind me, and soon Fabien, Cuix, and Tjeck also joined in. Tjeck had calmed down much better under Cuix's direction than mine, the happy little Frenchman having not shied away from using physical force to restore order, common sense, and sanity.

We could still hear the enemy tank in the distance in front of us, and it sounded as if it was getting closer. Bear came walking behind us, mumbling something.

"What?" asked the Greek.

"We must destroy tank," Bear mumbled a bit louder.

The Greek repeated the question again: "What? Is that the order?" But he received no clear answer.

Bear stopped just behind me and shouted past me: "Långström! We must destroy tank!"

Långström paused his attempted break-in and turned around, questioning what he had just heard.

"And how the hell do they expect me to destroy a tank?"

We still only had one single-shot RPG in the whole group. It hadn't been enough during our earlier engagement, and it remained just as insufficient now.

"The only thing we can do is try to knock out its tracks!" I shouted, feeling rather numb about the entire ordeal. Should we attack again, it was at least no crazier than the attack we had already performed and failed at. Surely, we could attack the Russians once more if that was called for, even if the result would be the same.

Bear started making loud laser noises: "Pew, pew, pew!" On the street below, Metro shouted something about someone being a fucking cunt.

"P-p-positive thinking, maybe?" asked the Greek.

An anti-tank unit soon arrived to give us support. These handful of men were properly equipped with thick, black snowmobile goggles covering their eyes and loaded RPG-7 launchers in their hands.

Lucky for us, though, the tank was only driving around in the valley in front of us, and we decided to leave it there. Långström told Bear to translate to Metro, telling him that the tank crew was scared as well, as they didn't know we were "dolbajobs (idiots) without anti-tank weapons." The most important thing now was to find cover, quickly, as artillery was going off again nearby. We hurried across the field and over my makeshift wooden bridge toward the house. I again took position looking north, facing the houses and street above us. I did not have to lie there for long until I heard voices coming from the enemy side. The distance was short, less than one hundred meters. Through the tall grass, I could glimpse figures moving around. I waved to Långström and warned that we had unidentified people moving around by the houses above us, something which he took into account.

"Stay here for now. If they try something, you just kill them all."

With the scope busted, I couldn't clearly make them out. The figures were visible in front, but I could not identify them beyond reasonable doubt. I did not want to shoot a civilian in the head by mistake and therefore avoided opening fire, although I should have understood that there were no old women walking outside to talk about the weather right now. The few civilians in the village all lingered in their basements. The men on the other side of the short field were, of course, Russians.

Chris had previously prepared a number of door charges made from detonating cords—long plastic tubes filled with explosive material. Långström now took the last one of these and applied it to the thick wooden door.

"Fuck trying to get in quietly! This door is fucking getting it now! Fire in the hole!"

Everyone took cover and the door buster went off in a large explosion. The air turned hot as pieces of wood chips sprayed from the gate out over the rough-hewn concrete stairs in front of it.

The wooden door turned out to be much thinner than it initially seemed, but behind the thin facade, a much thicker one made from thick, red-colored steel revealed itself instead. The charge was powerful and had completely obliterated the outer layer, but it hadn't managed more than to merely dent the actual steel door behind it. After a short moment of laughing surprise, irritation set in to replace it.

Our improvised plan to break into the building during the Shyrokyne counterattack.

"It's fucking great how people make their fucking houses intruder-proof these days, isn't it?" muttered Långström while the sound of heavy mortar impacts echoed through the village.

Where the door had previously been locked, it was now even more wedged shut in its new and currently dented format. As such, we put into action a new, improvised plan: a ladder was requisitioned from the neighboring house, with which Cuix and Tjeck were sent in through one of the high-positioned windows. We collected our hand grenades and gave them to Cuix and Tjeck, with which they would try to blow the door open from the inside instead.

The rest of us waited while it thundered from inside the house. The many loud explosions did not seem to do any good, and the door remained fixed in place. The plan almost seemed like it was about to fail, but then the thick steel door finally rocked open. Through the smoke, Cuix looked out with a big smile shining across his thick black beard.

"That was the last grenade I had! I did not think it would work, but look! We are open! Welcome inside! Come, come!"

We entered the house proper and were met with an unfinished interior. It must have started construction shortly before the war and never been completed. There was no furniture except for a kitchen island in the dining room. The walls were unpainted in gray concrete. The only thing adorning them was a large portrait of Joseph Stalin hanging by the entrance.

"These people were rich!" a raging voice was heard saying behind us.

It was Kelt, the PKM-toting madman, who came stepping in through the still dusty doorway. He looked around for a moment before contentedly continuing, "Not anymore!" followed by a loud and scornful laugh.

The kitchen was like the rest of the building: mostly empty and unfinished, with the exception of some cutlery and the like. Kelt's eyes fell directly on a proper meat cleaver which lay on the sink. He held it up against the dim light from the kitchen window. It was shiny and chrome and fit his hand very well. Making the meat cleaver his own personal sidearm, he laughed with further content.

I took a stroll to look around and examine the inside, as well. On the large kitchen island stood a solitary white enamel mug with a black-colored top. Its small size combined with the cartoon characters painted on its sides made it clear it was a children's cup. It would have been neat to bring some memory from that day back home, I thought, and I had a cousin with a son who was about the right age to use such an item. I dropped it in my vest's big dump bag and thought that the boy could get it for his next birthday—if we survived the rest of the day, that was. It was still only morning and the day still had many more hours left.

"I have good news," said Richter, who was listening to the radio as the rest of the group gathered around him. "Chris is alive. He came back by himself a while ago."

We were all surprised, Långström probably most of all.

"I knew it!" the Greek shouted triumphantly. "I told you! If there was anyone who would survive that, it would be Chris!"

"Where is he now?" I wondered, whereupon Richter informed us that he was resting at the Mayak.

"There is more news, as well," he continued. "A new cease-fire agreement has been signed in Minsk. It will take effect today at 12 o'clock."

The dull thuds of mortar shells in the village could be heard outside. I looked at the clock. It was still a couple of hours left until 12.

CHAPTER 40: THE CEASEFIRE

Tjeck was looking for the safest space in the building, worried that the artillery could start again at any time. He found an appropriate little hideout, but to his great disappointment, it turned out to already be occupied by a man. Space wasn't really the issue, and Tjeck could easily have squeezed in himself. Sharing a room with a man manically jerking his cock, seemingly completely out of touch with reality, however, was something he'd rather avoid. The search continued.

I kept an eye from the upper floor while we waited for the hour to reach midday, for the war to end. It was an uncomfortable position and wait. I had made some unsuccessful attempts at cleaning my scope lens but got nowhere. Too much dust and grime had leaked into it, and the damaged center of the lens covered the crosshairs. The sight was so filthy it was unfit to merely observe through. I couldn't shake the uncomfortable thought that one of the Russians could be sneaking around outside with his own scoped rifle. I backed further into the building, hoping that the darkness would shield me from whoever might be watching from outside.

Greek passed the time holding provocative monologues.

"The Taliban will win. It's that simple. What does the West even want? Women to go to school and learn to read? How do you win a war with that? Ha! The West can stay there for another five or ten years, whatever. It won't matter in the slightest. The Taliban want to win the war, and they will do anything to make it happen. That's the only thing that matters, and I admire them for it."

In addition to the Greek's prophecies on the US war in Afghanistan, the second major topic of conversation was whether the cease-fire was even real. Information about it was contradictory. The first report said it would take effect at noon, but then came another which said it was at midnight. A third one came in claiming that the cease-fire had already been in place since the last day, then a fourth saying it wouldn't come until Sunday. No one knew for sure.

Mortars fired sporadically over the village, but the heavier artillery seemed to be absent. After a while, the mortars also ceased. Calm began to settle over Shyrokyne. The clock kept ticking and soon it hit midday. We left our safe building, cautiously venturing outside. The low cloud cover which had covered the sky and gave the village a completely gray hue had begun to crack. Color had returned to the landscape. Everything was completely quiet.

We hurried toward the western edge of the village along the road. The nearest cover outside it lay two hundred fifty meters away in the form of a small number of buildings next to the road leading to Mayak. The uncertainty whether the cease-fire was actually in place made the short but open crossing seem obnoxious at best. Everyone had our previous experience of moving in open terrain near the Russians in fresh memory. For all we knew, the Russians could still launch a counterattack at any time. There wasn't much else to do, however. We needed to run it.

The long rest while waiting for the clock to strike twelve had given renewed strength to my legs, and I managed to move forward with real speed. I even outran all the others in the group, something which otherwise rarely happened. However, my newfound strength soon turned out to be like putting empty batteries in an electrical appliance. The little juice that was there gave energy as usual, but only for the briefest moment. The fact that my feet and legs got tangled in the many copper wires the anti-tank missiles had left scattered across the area did not make things easier. A single thread was easy enough to tear, but they came in great numbers. Already halfway over the open ground, I felt my legs begin to give up again. Not even the thought that the Russians could be right behind us could motivate me to go on. I needed to stop for a rest, if only for a second.

Retreat from Shyrokyne.

As useless as the situation was, I thought I could make some good out of it. To be constructive, I grabbed my camera as I began to slow down. As the others passed by me, I managed to snap a neat photo of our flight across the field. The few seconds of rest also turned out to be the only thing I needed. I could keep running again.

When we finally reached the small houses west of the village, we went straight down into the ditch next to the road. Fabien looked really worn out from the run. Bear seemed unusually merry, even for him. A man came walking past us between the houses. He carried a PKM on his shoulder and the long cartridge belt hung around his neck. It was Kelt. As he was passing by us, he saw the camera pointing at him, met it with his gaze, and gave off an abysmal roar while raising his right arm in a Roman salute.

We continued further up toward the Mayak on the road, passing by a smashed Geländewagen jeep. Perforated by shrapnel all over its body, it still didn't look as if it was the artillery which had knocked it out. Some dried blood was visible next to the passenger seat. The entire chassis was compressed somehow, especially in the middle. It looked almost as if a tank had run over and crushed it, but I reminded myself that stuff like that only happens in movies.

Kelt.

310

Fabien examines the crushed Geländewagen.

A bit further up, we passed an abandoned Ukrainian BMP-1. It was surrounded by the same kind of steel cages as BTR-3s, but these had provided little protection. The thick metal slats were winding and bending as if they had been made out of weak tin. The gunner's hatch on top of the turret was open, and on it someone had painted a white Autobots insignia, the good guys from the *Transformers* series of cartoons. It was a great picture on its own; all that was needed was the group gathered in front of it. I waved to everyone and pointed toward the knocked-out fighting vehicle.

"Get in a line in front of the BMP! I'm going to take a picture!"

To my great surprise, even Långström decided to get in line. With a broader smile than ever before, he took position to the left, leaning his arm against Bear, who was holding up a shot-through AK magazine. Tjeck hung forward, completely exhausted in both body and mind, but his smile was genuine. Likewise, the Greek smiled with satisfaction at still being alive. Cuix lit a cigarette and blew a thick cloud of smoke through the corner of his mouth as I snapped the picture. Fabien framed the picture on the right with his AK in the air and a tired look on his face. Richter knelt down in front of the others with, like usual, a completely bland expression on his face. The picture was perfect.

We continued up the road. To our left side, we could clearly see much of the village. Everything was white from frost and calm. It seemed very much like a peacetime exercise had just ended, each side having packed up their belongings and returned back to their camps. We soon reached the heights at the Mayak. The destruction around was complete and impressive. The ground had been torn apart by both artillery and the caterpillars of the tanks. Crushed and burned-out vehicles stood between trees with their bark peeled off by the artillery. The ground was littered with debris. Pieces of Grad rockets and mortar shells protruded from the soil to such an extent that care had to be taken not to trip on them.

Left to right: Långström, Bear, Greek, Tjeck, Richter, Cuix, Fabien.

Back at the Mayak, I asked for Chris and was shown down to the basement. Below ground in the damp, dark room lay the command center itself, and in a sleeping bag rolled out over some ammunition boxes was Chris. Having had even less rest than us, he was sleeping heavily. I wanted to embrace our friend returned from the dead, but it was good enough to just see him, to know he was actually back. I returned out and back to the surface, leaving him to remain in the underworld for a little while longer.

"Donbas is on their way to replace us," someone said—the other of the two most prominent Ukrainian volunteer battalions. Finally getting relieved was as welcome news as any. Kirt's company, which would become known as the "Iron Hundred" for their defense of Shyrokyne, were all tired and badly in need of rest. The same was true of us.

"Is that them, then?" asked the Greek, listening in the direction of the village where the silence abruptly ended in a sudden, intensive firefight.

Greek was right. It was. They were on their way to Mayak in trucks, only for one of them to miss the exit and drive straight into the Russians instead. Their exact number of casualties from this single mishap never really became clear.

The ones who didn't make the wrong turn soon began rolling up next to us on the driveway outside the Mayak. Donbas soldiers disembarked and began to get acquainted with the surroundings. We stood outside the entrance talking, joking, still excited about just being alive, relieved to be able to get out of here soon. I longed for a shower and dry socks more than ever. The atmosphere was great and the weather very pleasant.

A tall, older, and kind of fat man who seemed to be some kind of commanding person in Donbas approached us and asked us something. We couldn't understand him and thus had no answer, other than to explain that he needed to speak English. As he didn't seem bilingual, he instead continued in either Ukrainian or Russian. There was something about his body language and tone that seemed very imposing, or at least he appeared to try and make it so. Richter had remained some distance away but was now approaching to explain what the other battalion commander was on about.

"He says you should hand over your guns to him," Richter explained.

We just laughed.

"What? We should give him our guns? What is he saying really?" asked the Greek.

"He says they have a shortage of weapons and ammunition, and that you foreigners are not fighting, anyway. You are just here to take photos, a bunch of posers he says. Therefore, you should hand over your guns to him."

We asked again if he was joking, but Richter calmly explained that he translated everything literally. The powerful commander looked at us firmly, waiting for us posers to hand over our guns to him.

The mood turned really sour, and we made it clear that the fat, unshaven potato man would not be getting anything from us. Besides, we had more important matters at hand, anyway. We were going back into Shyrokyne again, third time being the charm. Chris was back safe and sound, but the man from the Recon platoon still remained somewhere inside the village. His nerves had calmed down a bit, and he had turned his radio back on again. Having panicked during our first rescue attempt, he had turned it off for fear that the Russians would find him, and he now asked his platoon comrades, and us, for help one more time.

The sun was up, and there was no good way to sneak into the village like during the darkness of the night. None of us could really have cared less. We took the same route as we had done during the night, completely open, and headed back across the fields into the village. We came back to the small footbridge, crossed it, and then took up position on both sides of the road ahead. This time the Foreigner Group went first with Recon following behind us. Cuix and I took the lead. We looked at each other, nodded, and began moving forward in a quick, synchronized fashion—houses, corners, walls, intersections, everything covered quickly and professionally. We advanced aggressively forward, and if any of the Russians were in our way, they would be the ones in trouble. We were out for blood.

Suddenly, we both got an awkward feeling that something was wrong and began to slow down. We couldn't place it. It didn't seem as if there were enemies nearby, though. It was something else. We stopped, looked around, then turned our heads backward to see if there was something happening behind us. There, far behind, stood the Ukrainians waving at us. Out of one of the houses moved a lone figure joining up with them, whereupon they turned and began moving back. We all looked at each other, surprised. We had only moved a few hundred meters into the village. We were still on the same street where we had been driving back and forth in our pickup during the general retreat.

It slowly began to dawn on us as we, too, moved back. He had not been caught in the middle of the village at all. He had been right on the edge of it, but without the sense to be able to move the last few meters out by himself. And those Russians with the BTR he claimed to have heard right outside—who were they, anyway? There had been no Russians on this street. We had been right here at the same time—we would have known. Was it us he had mistaken for Russians? And our pickup for a BTR? Had we risked our lives, not just once but twice now, just because he had lost his mind?

It boiled inside us during our withdrawal, even though the mission for once had been successful. As we reached the heights, where the Russians could no longer shoot us in the back, the Greek burned off. He pointed at the man in an anger I had never before seen on him and roared furiously.

"He should go back to that house, *alone*! And then come back, *alone*!"

We returned to the hotel afterward. The Russians were apparently exhausted, and the cease-fire actually seemed legit.

"Fuck, man. We really fucking stink, don't we?" said Långström, and I agreed.

"Hell, yes, I could use a hot shower back at Yurivka right now, let me tell you."

"Hot shower?" Långström asked in a way that clearly meant he was up to no good again. "No way, bro, we don't need to wait that long. Don't you remember? This is a hotel! We have a goddamn indoor pool on the ground floor!"

I laughed in hopes that he wasn't serious, but regretfully soon saw that he was. The rest also protested—Greek the loudest—but this only made Långström even more rigid. The only one who didn't say anything was Richter, who, clever as he was, used the commotion as cover and immediately vanished into thin air.

"Everyone takes a bath! It will be good and refreshing. Everyone will enjoy it. That's an order!"

The indoor pool was large and would certainly have been nice in the summer, but now it was mid-February, and there was no heat in the water. The surface wasn't quite frozen, but it wasn't far from it.

"Come on, now, you have that Coastal Jaeger tattoo on your arm. You should enjoy this. Or are you afraid of water or something?"

Långström laughed and jumped down into the cold water, my shame forcing me to follow. The cold water radiated through my entire body. Lungs and muscles cramped. Only moments later, though, it turned out to indeed be very refreshing, just as Långström ordered. Fabien even took a little swim around the pool.

"Where the hell is the Greek?" Långström asked. "Hey! Greek! You're going to swim, too, you fucking faggot, or I'll kill you!"

Långström's threats echoing through the ground levels did little, though, as the Greek had effectively hidden himself, not to show up again until after the washing session was well over.

"I went to stay guard," the Greek said, trying to save face afterwards, which didn't stop Långström from calling him a pussy.

After getting dried up, some of the guys went over to get the pickup while the others collected our gear. It was getting dark, and we began getting ready to leave. Chris also soon rejoined the group, coming down from the Mayak.

"Good to see you, mate," he said in a very tired tone of voice as we wrapped our arms around each other.

"It's fucking good to see you, too, man," I said. "I'm glad you're alive. We tried to go look for you, you know?"

"Yeah, I know, man, I know."

Darkness fell and we left Shyrokyne. We stopped in Mariupol for dinner in the base mess hall: overcooked macaroni and some kind of broth. As simple as a simple meal could get, but still amazing. We all sat down together around one of the oblong tables on the side. We still had a lot of bottles of Pepsi brought back with us, and this caught the attention of some Ukrainians. One man approached our table, walking slowly and very strangely. He looked like an old man with Parkinson's disease, struggling to put his one foot in front of the other. Coming over to us, he uttered some kind of sounds which could hardly be considered words. Not knowing how to respond to the frail man, his friend soon moved up and translated.

"He asks if you have a bottle of Pepsi to give him."

He was not the only man to be shattered by the artillery; there were several who had problems with their motor skills. I hadn't seen anything like it other than in old black and white movie reels showing shellshock victims from the First World War. His body was fine, but it was as if his brain had been damaged by the force and pressure of the artillery. Cuix handed the man a bottle of Pepsi.

We gradually began to try and piece together everything that had happened the past few days. We laughed heartily at all the insanity and the madness that we all, in some strange way, still seemed to have survived.

The laughter stopped as a Ukrainian appeared at the table.

"Everyone will return to positions west of the front to rest, but the foreigners will go back to Shyrokyne."

Långström smiled and looked up at the man behind him.

"Ha! What are you talking about? Are you joking?"

"No, it's an order. You must get ready now, to go back to Shyrokyne."

Cuix asked, still with food in his mouth: "What? Everyone else go home and sleep sweet in their beds jerking their cocks, but we go back to fight alone?"

Tjeck broke down completely.

"Who in the hell gave that order?" Långström asked indignantly.

"I do not know," replied the Ukrainian. "I was just told to tell you guys need go."

"You there!" I said and got up. "I will go back to Shyrokyne. I will attack Novoazovsk! I will fucking go attack the Urals for you if that's what you want! But fuck all of you! I'm not doing any of it, not before I've gone to Yurivka first, taken a shower and a shit, and changed my goddamn fucking socks!"

Långström also got up.

"Come on, let's find the fucking cunt that gave this order," he said, and we grabbed our guns, heading outside to hunt.

We never got a clear answer as to who it was that had ordered us back.

Around us, rumors probably began to circulate how the idea had not sit well with us, and it was soon withdrawn. After having calmed our nerves, our journey continued, back to our warm rooms in Yurivka. The pickup was so filled up with gear that we could barely fit on the flatbed. Without the sun, it was cold outside, and we wrapped ourselves in everything we had so as not to freeze to death—jackets, blankets, sleeping bags, everything. Sitting on top of a mountain of stuff was difficult, and we held on for dear life not to fall off, fingers cramping up while trying to not fall asleep and let go.

After about an hour's drive, we were finally back. We stepped through the front door of our building for the first time in days which had felt like months. Chris picked up the whiteboard that had been used for educational purposes next to the entrance and began writing something on it. He turned it over and knelt beside it with the sniper rifle in his other hand.

"Can you take a picture of me here? I want to remember this moment," he said.

On the board was a short summary of what he had experienced during the battle of Shyrokyne, caught behind enemy lines.

"You know, I sat for several hours in just one room, hiding, All alone, with the Russians outside. The only thing I couldn't decide on was if I would empty my Makarov at them when they came in and then kill myself with the F-1, or if I should throw the grenade at them and then shoot myself."

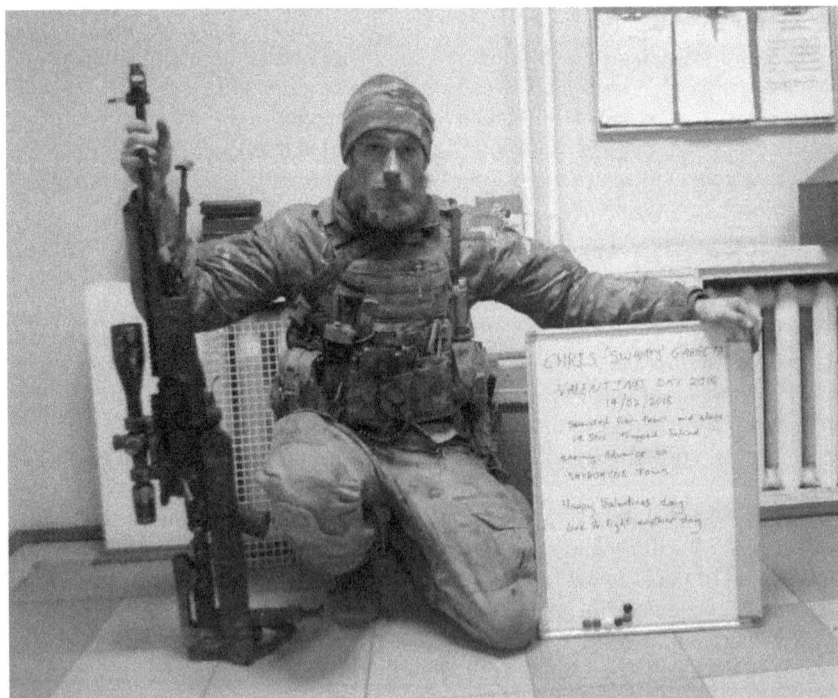

Chris after making it back from being trapped behind enemy lines at Shyrokyne.

"You should have thrown the grenade at them and then shot yourself," Långström broke in. "Less chance of something going wrong. If you had saved the grenade for yourself and it turned out to be a dud, you would have been fucked."

The door to the left of our room had a large hole through it, full of soot burn marks all around it.

"I think the guy who had the only key was wounded or killed," said Richter. "When his roommates came here, they couldn't get in, and apparently one guy was tired of shit and opened it this way instead."

"What the hell did he do then? Did he throw a grenade at it, or what?" I asked.

"No, it's not that big. I think he just put the barrel of his AK against the door and emptied the entire magazine."

With the exception of Inkasator—who, as far as we understood, was hanging in there in spite of the bullet hole through his leg—nobody in our group was neither dead nor seriously injured, strangely enough. We picked up our keys, and the doors to the dwellings of peace opened. When only cold water came from the shower at first, I suspected for a moment that the man who had fired his AK through the door next to us had damaged something. Luckily, though, it turned out that a simple turn on a lever on the hot water mixer was enough for it to start. I stepped into the bathroom and enjoyed the really, really long and really, really hot shower I had dreamed of for so long. When I came out, I put on a pair of dry socks and went out to have a normal Marlboro-brand cigarette. It felt good to be alive. Life was perfect.

EPILOGUE

We slept over breakfast and did not rise until lunch. Being able to sleep in without being woken up by new orders or artillery was a godsend luxury. Being indescribably glad to still be alive, the situation called for a feast to celebrate. Considering everything we had done, we could not see it any other way than that this was fitting despite the alcohol ban. We had earned a feast. Långström grabbed the Recon platoon commander, a man called "Messer," to ask permission for us to get really fucking shit-faced.

"We want to head down to the store and buy a lot of beer. Both you and I know that we have worked hard these last few days. The guys really deserve this. They've earned a drink. What do you say?"

Messer frowned a bit and pondered for a moment before rejecting the request.

"No. I'm sorry, but I cannot allow you to drink beer here at base."

Långström shrugged in exactly the same way as he had when the commanders said that he wasn't allowed to blow up the ammunition transport at Kominternove. He stepped back into the room, closed the door behind him, and lifted his backpack.

"Come on, we're heading to the store to get vodka and whiskey. He said nothing about us not being allowed to get pissed on booze."

So it was, and after coming back, we gathered in the other room, the bigger of the two, and began the process of getting piss drunk. The booze took quickly and brought back the memories from previous days which we continued trying to piece together.

"Wait a minute," Cuix asked. "It sounds like when we counter-attack, everyone know we go on impossible mission. I not know that. What the fuck! You are all terrible friends. Thanks."

"Well, if someone would have shot the tank with their RPG," Richter answered, "we would have won that engagement. Who carried the RPG? You?

318

Thank you, Cuix."

"Me? No, that was 100 percent Långström," Cuix defended himself.

There was stiff knocking on the door, and we thought for a moment that we had been caught and prepared for battle. Långström cautiously opened the door slightly, and it turned out he was the only one being sought after. Apparently, the commanders upstairs were doing exactly the same thing as us—not boozing their brains out, that was, but trying to piece together previous events. Långström had been asked to come upstairs and participate.

"Hey, man, you can't go like this. You're fucking drunk," we all said, but Långström just ignored our worries.

"Shut up, faggots, it's nothing. I can act sober! It will be fine! Besides, I have one or two things to tell those people, as well!"

Långström swayed out with a strong whiskey breath and slammed the door behind him. The rest of us left behind pondered about how to proceed. Should we try to hide the alcohol and pretend like nothing happened? Or should we just keep drinking while there was still time and then simply take our punishment afterward?

Långström soon wavered back in with a big smile on his face, pleased to have pointed out a number of errors and shortcomings to the higher-ups, which he also spelled out for us.

"Didn't they say anything about you reeking of booze?" we asked.

"Huh? Well, yeah. Or, hmm, no. Fuck do I know? Who cares, anyway? They were talking a bunch of bullshit how they wanted the car back, but I made it clear that it's ours now. Those losers left it behind for the Russians; it was us who brought it back. I told them it was mine now, and they could have a look in their assholes to find some other answer from me!"

It was quite obvious that the other Ukrainians very well understood that Långström was drunk, but it was also easy to assume that no one wanted to be the one who dared to point it out in a problematic way. Had his insanity previously only been a simple flying rumor, he had demonstrated the fact a lot in the past couple of days. It was likely they were simply afraid of the sinewy man with the ugly beard and hobo-esque haircut.

The feast continued. Bottles of Pepsi remaining in large numbers from the front were mixed with Red Label Scotch. It had been exactly one year since my friend from conscription had tragically died: the event which perhaps most of all had created the circumstances leading me to where I was at this very moment.

Helvete, I thought. *You should have been with us, man. Fuck, man, you would have loved this shit.*

The days went by. The armistice agreement, Minsk 2, had been implemented in much the same way as Minsk 1 in September the year before. The fighting had slowed down, albeit with clear exceptions. Just as the Russians had still continued attacking around the airport in 2014, they still kept hammering on the encircled Ukrainian forces around Debaltseve. The Russians pressed hard, and within just a few days, the Ukrainian pocket would fall with similar, though not quite as disastrous, results as at Ilovaisk. In Shyrokyne, both sides maintained largely the same positions. The battles were fought between the Mayak and the village, without one side having the will or force to push the other away. In time, the front line seemed to begin quieting down again.

We continued with our shooting drills and similar, simpler, and relaxing tasks. There did not seem to be any more action worth the name coming, and I started thinking about going back home again. The war seemed to be coming to an end. Since nothing of special value was happening, anyway, I suggested that the others in the group head with me to Kyiv. If nothing else, just for a couple of bar runs before my flight left, an offer everyone took up. All except Långström.

"You know, I think I'll stay," he said. "This war won't be over for a long time, I'm sure of it. There's still some action left."

We shook hands outside our little hotel building and said goodbye and good luck. I climbed into the car that would drive me and the others to the train station in Berdyansk. Långström remained for a moment, alone outside the gate, and watched us as we left before he turned back inside. Once at the station, we visited a nearby store and bought plenty of beer so as not to risk thirsting for the long journey back west. We had two cabins with four beds in each at our disposal. The train started to roll slowly, and we corked our first two-liter bottles.

Once in Kyiv the next morning, we met the late Leo's ex-girlfriend, who showed us to the central apartment she had arranged for us. Then we began anew, with bars, fine food, women, and cheap booze. While we were indulging the worst, and as some Azov guy they called "Jesus" puked on the bouncer outside the bar, Långström was leading an attack together with Ukrainian tanks to retake the hotel complex by Shyrokyne. He had been right. The war was not about to end, but would last a long time, albeit in a much stiffer form. The initial mobile stage of the war was over; the trench war had only just begun. Shyrokyne would for a long time come be one of the most active scenes for this new conflict. If the village had been destroyed in just the first days of fighting, the coming months would grind everything left to the ground.

For my part, I boarded my flight as I had planned. I expected the war to end and had neither the desire nor the opportunity to sit on a base behind the front and wait like I had done for most of 2014. When I landed and got off in Helsinki, a man coming from the same plane approached me at the airport.

"Regiment Azov?" he asked, whereupon I asked if it was so obvious. He replied that it was, probably from my still dirty brown smock, and asked if I had been to Shyrokyne.

After meeting some friends in the capital, I went on to visit relatives near the brigade where I had done my military service many years prior. I stayed with my uncle out by the countryside. Even if I was standing on solid ground, I assumed I hadn't really quite landed, something that was apparently noticeable. In such a case, most people would probably have suggested some kind of psychologist, pills, or some other trendy and retarded solution. As it was, though, my uncle knew better and knew exactly what I needed. Showing me a seat by the kitchen table, he went down to the cellar and picked up two cans of cold, high-quality Finnish beer. He placed one of them in front of me and sat down on the other side front of me.

"Tell me about it. What was it like?"

And I started telling my stories—the same things I have now told you, though probably in a messier and much less organized manner. He didn't say much but mostly just sat there, listening. As soon as my beer was empty, he picked up the empty can. He didn't ask me if I'd like a new one, he just went downstairs and got one. That was all. That was all I needed.

When I then returned to Sweden, I had to be met by the Swedish Security Service in Stockholm. They came in the form of two young men, one playing "good cop" asking the questions based on a form, with the "bad cop" reading my body language and taking notes. The questions from the form were clearly written to deal with Islamic terrorists—or rather some naive and comedic view of what they were like.

"Do you want or plan to carry out religiously motivated terrorist acts in Sweden?"

I had a hard time discerning if they were joking, but Bad Cop's harsh supervision indicated they were actually serious. When asked if I could handle explosives, I answered yes, something that made them jerk back hard. I was a soldier, and a good one with both broad education and experience. I had served in the military forces of three nation states. I had just come back from an actual war—why wouldn't I be able to handle explosives? It slowly began to dawn on me that I was living in a clown world.

After that, I went home to my apartment, where I waited every day for the war to end. I waited as long for it as I longed to return. I had lost many of my friends, few from the war itself, but most from my platoon in the Home Guard. Either they looked down on me contemptuously as some kind of dirty, less worthy person, or they didn't dare to be seen talking to me at all. If I by chance ran into someone at the bar I knew and asked how their life was going, they would anxiously begin to look around for fear of being seen with me. Just my sitting at another table was enough for former close friends to vacate the entire restaurant. I missed not only the war, but most of all my friends from it. At least I got to see the new *Mad Max. Fury Road*, it was called. Really cool movie.

The war turned into a stalemate of trench fighting, and after a year of fighting around Shyrokyne, Azov was finally withdrawn from the frontline. The Foreigner Group soon dissipated afterward.

The two Frenchmen, Cuix and Fabien, stayed and participated in many missions in Shyrokyne. Cuix would even, despite his keen eye for mines, step into a trip wire during a firefight and find himself injured by shrapnel from an OZM bounding mine. He would, however, heal up and return to fight. He would do so with an extra marker for the men wounded in action on his arm, carrying it with pride. Cuix would go back and forth between fighting and civilian life, and he also took part in battles in 2022 after the Russian invasion of Ukraine.

Because of the widespread Russian propaganda, Chris would suffer all sorts of harassment from authorities back home in Britain. Even though they certainly knew better, they still chose to treat him like a terrorist by, among other things, confiscating his belongings and closing his bank accounts. It went so far that he essentially could not return home and was forced to stay in Ukraine because of it. When fighting was not going on in Shyrokyne, he spent a lot of time disarming and clearing minefields around the front line. After a while, he had a falling out with the leadership of Azov Regiment and finally left Ukraine. In 2022, though, as the Russians invaded, he returned once more. Initially, he stayed in the west of the country, working to distribute aid and incoming equipment. Soon enough, though, he would be back doing what he did best: helping the Ukrainians deal with the massive amounts of unexploded ordinance left behind after the Russians retreated from Kyiv.

After some time, the injured old driver Inkasator again asked the group to kidnap him from the hospital to allow him to return to the front. He remained in the Foreigner Group and was soon injured a third time. As a light mortar shell landed just next to him during a firefight, he was hit all over his body, but was luckily saved by his helmet and Kevlar vest. After that, he once again demanded to be kidnapped from the hospital in order to return to the front, which he did.

The two young Ukrainians, Richter and Bear, had contracts with the National Guard and stayed for a long time as translators for the Foreigners. For Richter, he had no other interest besides the war. He would stay as long as he could. He would also later become the squad leader of the Foreigner Group. Bear would leave the war in 2016 to continue his studies at the university, and later found well-paid and high-prestige employment outside the country. In 2022, he initially aided the Ukrainian artillery from abroad, locating Russian invaders and transmitting their positions to Ukrainian guns. He would, however, soon leave his work and return to his native country to once more partake in combat.

The Greek complained a lot how the war had slowed down and become boring, but he would still stay and try to squeeze out as much fighting from it as possible.

Think about it," he said. "This still isn't a bad way to live. You get a salary—not much, sure—but you get free housing and free food, so you don't really have any expenses. You also get good military experience over time, as long as you're invested in it. If you're lucky, you can even do some actual fighting. No matter how you twist and turn it, it's better than being middle class and gay. It's a boring war, sure, but it's way better than no war at all."

After going back and forth between civilian life and fighting at the front, he would remain in Kyiv and was there during the Russian invasion. As the Russians were at the gates of the city during the first days of the invasion, he and some others were frantically trying to acquire weapons to fight the enemy. When no guns were to be found, the Greek left his comrades and began walking alone toward the outskirts and the enemy outside Kyiv, quoting Hector and vowing to go down with the city.

We all thought that Tjeck, who had such problems with his nerves, would never be able to come back to the front, but he would surprise us all. Although he was constantly terrified and on the verge of shattering, he somehow kept coming back to follow the others. The first time, he stayed constantly in the basements, ever fearful of the mortars outside. Despite his almost paralyzing fear, however, his courage never ceased. He got better and better, and after a while, he, too, had turned into a very fine and capable warrior.

Mikola, having gone back to Sweden to cut his ties there, would return to stay in Ukraine permanently, although he soon transferred away from the Foreigner Group to another unit. Still in active service by 2022, he held the longest service record of any foreign fighter in Azov. He would remain in Mariupol and become besieged in the factory works of Azovstal.

Tor and the Finn both returned for a while after the winter had passed, and both took part in fighting in and around Shyrokyne. The Finn even took command of the Foreigner Group for a while after Långström had left. Tor would find and bring back another dog in the area, a small puppy seeming to be a mix between a Basenji and some kind of mid-sized Sighthound. A veterinarian named the dog "Kozak," and after Tor had brought the pup back with him after his tour ended, I adopted the dog as my own. Tor would later enroll in the Ukrainian Foreign Legion and partake in the fighting in the Kherson region in 2022.

"Becks," the young new member of our group who was badly wounded during the defense of Shyrokyne would survive, but his foot was pierced by shrapnel and would never heal up completely. He would still compete in athletics for wounded veterans and, however not returning to Shyrokyne, once again pick up a rifle on the 24th of February, 2022

"Bulbash Malinki," as we called him, the smaller of the two Bulbash boys, became the first Belarusian citizen to die in combat during the first days of the invasion in 2022. He fell in fierce fighting in the suburbs of Kyiv. His real name was Ilya.

"Metro," the commander who guided us through the valley of death in Shyrokyne, would remain in Azov as a Lieutenant. He would fight in Mariupol in 2022 and fall trying to protect the city. His real name was Stanislav.

"Kelt," the mad machine gunner with whom we fought during the battle of Shyrokyne, was one of the many Azov veterans who demanded to be flown into the Azovstal steelworks in Mariupol after the city became surrounded by the Russians. He, like many others, must have known the battle was hopeless, but nevertheless, he willingly and eagerly went into the fight, anyway. His real name was Oleg.

Långström continued as commander for a long time and was constantly annoyed after his crazy plans and ideas were shot down by the higher-ups. Someone told me that at some point he swam naked across a river to the enemy lines to plant mines, simply because nobody else dared to come with him. As the fighting in Shyrokyne began to die down, though, he moved on, unclear where or to what. He returned to fight in 2022.

I would myself would return to Ukraine for my last fighting in 2017. At that time, Azov was being deployed to the area west of Donetsk, and the Foreigner Group—some old, some new—would reassemble for another chance to fight.

THE END

ADDITIONAL PICTURES

Yuri.

During exercises. Tall Serbian in the middle.

Tor.

Mikola (left) with French Special Forces veteran (right).

Greek.

Bear.

Metro.

Chris, Tjeck, Greek and Långström after Pavlopil capture

Bulbash boys.

Inkasator, Fabien, Cuix, Chris and Tjeck after Pavlopil.

Damaged car outside Mayak.

Bulbash malinki.

Richter.

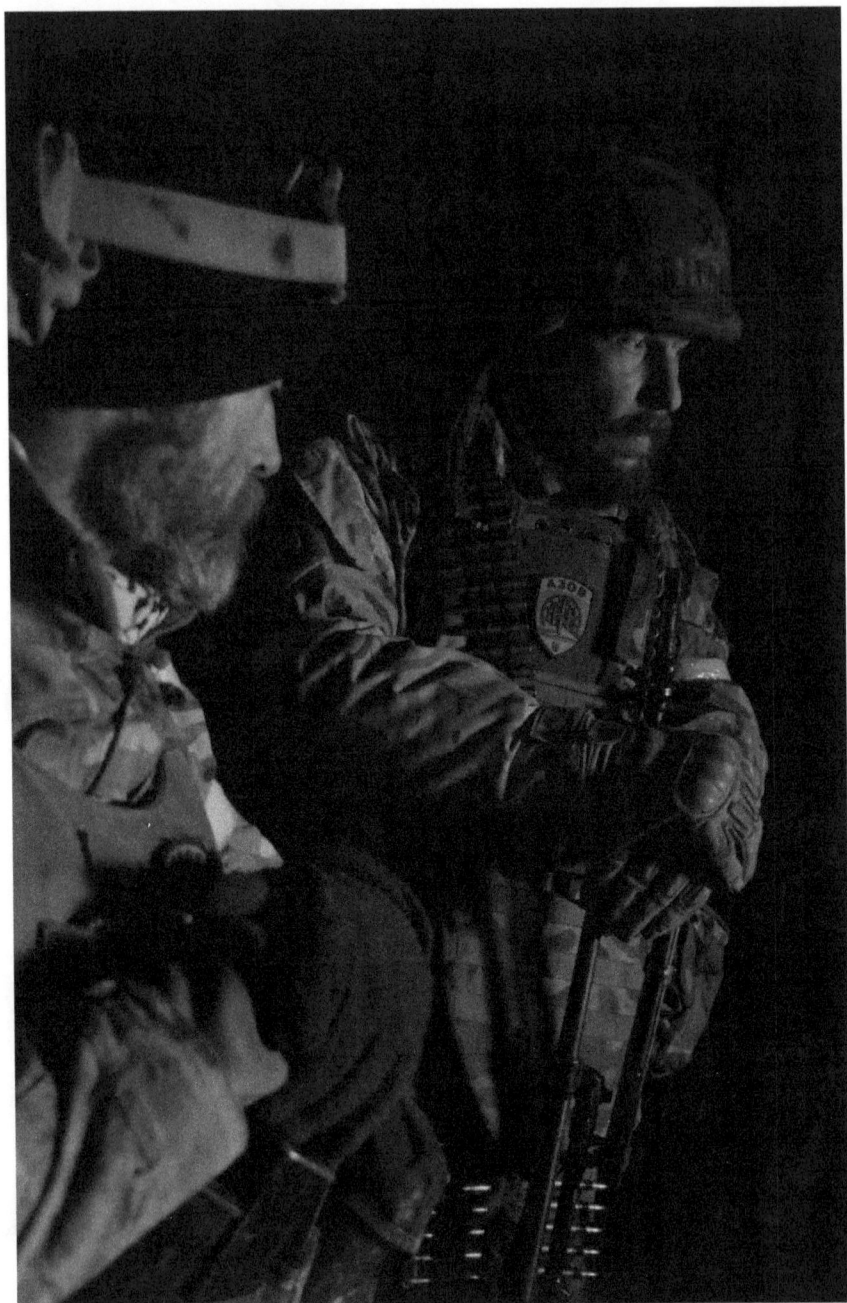

Fabien and Kelt.